Nondestructive Evaluation of Ceramics

Related titles published by the American Ceramic Society:

Hybrid Microelectronic Materials (Ceramics Transactions Volume 68)
Edited by K.M. Nair and V.N. Shukla
© 1996, ISBN 1-57498-013-0

Magnetic Ceramics (Ceramic Transactions Volume 47)
Edited by Bhaskar B. Ghate and John J. Simmins
© 1995, ISBN 0-944904-81-5

For information on ordering titles published by the American Ceramic Society, or to request a publications catalog, please contact our Customer Service Department at 614-794-5890 (phone), 614-794-5892 (fax), <customersrvc@acers.org> (e-mail), or write to Customer Service Department, 735 Ceramic Place, Westerville, OH 43081. Visit our on-line book catalog and web site at <www.acers.org>.

Ceramic Transactions
Volume 89

Nondestructive Evaluation of Ceramics

Edited by

Christopher H. Schilling
Iowa State University

Joseph N. Gray
Iowa State University

Published by
The American Ceramic Society
735 Ceramic Place
Westerville, Ohio 43081

Proceedings of the Symposium on Nondestructive Evaluation of Ceramics held at the 99th Annual Meeting of The American Ceramic Society in Cincinnati, Ohio, May 4–7, 1997.

For information on ordering titles published by the American Ceramic Society, or to request a publications catalog, please call 614-794-5890.

Printed in the United States of America.

1 2 3 4–01 00 99 98

ISSN 1042-1122
ISBN 1-57498-037-8

Dedication

Honoring Otto Buck

Dr. Otto Buck
(May 14, 1933–November 20, 1997)

Born on May 14, 1933, in Stuttgart, Germany, Otto Buck received his bachelor's and master's degrees in Physics from the University of Stuttgart in 1956 and 1959. He then received a doctorate in metal physics from the Max-Planck-Institut for Metallforschung in 1961. After receiving his Ph.D., he was a Research Assistant at the University of Stuttgart for three years and a Member of the Technical Staff at Siemens A.G., in Munich, Germany for two years. He then embarked on a successful, twelve-year career at the Rockwell International Science Center, where he advanced to the position of Group Manager and Principal Scientist. In 1980, he became a Professor of Materials Science and Engineering at Iowa State University and a Senior Scientist at the Ames Laboratory of the U.S. Department of Energy.

Otto Buck made pioneering scientific contributions in the areas of (a) ultrasonic measurements of internal stress and crystallographic texture, and (b) ultrasonic scattering and harmonic generation by fatigue cracks to provide information on crack tip shielding and its influence on crack growth rate and detectability. He published several, special topic journals and books on nondestructive evaluation and fracture mechanics of materials. While at Iowa State University, he became one of the founding fathers of the NSF Center for Nondestructive Evaluation. He was a member of several scientific and engineering societies, a fellow of the American Society of Metals, and a member of the editorial board of the Journal of Nondestructive Evaluation. On November 20, 1997, he died after an extended illness.

While his technical contributions are generally well recognized, perhaps less known is his impact in advancing the careers of the many people who worked under him, particularly the editors of this volume. We'll ever be grateful for his encouragement to pursue research in nondestructive evaluation of ceramic powder systems. Thank you, Otto.

C. H. Schilling
J. N. Gray

Contents

Surfaces and Films

Advanced Characterization Techniques

Preface

Facing stiff competition, ceramic manufacturers are currently seeking affordable, on-line methods of detecting and regulating production defects such as cracks, warpage, and spatial variations of porosity and composition. Breakthroughs in defect sensing technology are needed to increase the efficiency of existing production operations and also to open up new markets for ceramic products that are impossible to fabricate using traditional methods for quality control.

The objective of this volume is to present a collection of papers on recent advances in the nondestructive evaluation (NDE) of ceramics from participants at the Symposium on Nondestructive Evaluation of Ceramics held at the 99th Annual Meeting of The American Ceramic Society on May 5–7, 1997, in Cincinnati, Ohio. The goal of the symposium was to bring together researchers and industrial colleagues from the NDE and ceramic manufacturing communities to discuss needs and opportunities for NDE in the production of ceramics. This proceedings is third in a series of American Ceramic Society symposia proceedings on NDE of ceramics; these earlier symposia were held on August 25–27, 1987, in Boston, Massachusetts, and on July 9–12, 1990, in Columbus, Ohio.

As part of the effort to identify needs and opportunities for NDE in high volume manufacturing of ceramics, a workshop on this topic was held during the symposium. Panels composed of research and industrial leaders provided overviews in the workshop followed by lively discussions among the audience and panel members. Several of the main points from the workshop are summarized in the overview paper by Schilling and Gray in this proceedings.

The symposium organizers would like to express our appreciation to the participants of the panel discussion by recognizing each of the members: Bob R. Powell of the U.S. Automotive Materials Partnership and General Motors R&D Center, John Pollinger of Allied Signal Ceramic Components Incorporated, Edward Kraft of Kyocera Industrial Ceramics Corporation, Dennis Tracey of Saint Gobain-Norton Industrial Ceramics Incorporated, Said Jahanmir and Gabrielle Long of the National Institute of Standards and Technology, Muhammad Alim of the Ohio Brass Company, Christopher Schilling, Joseph Gray, and Bruce Thompson from the Center for Nondestructive Evaluation at Iowa State University, William Ellingson of the Argonne National Laboratory, Stewart Stock and Rosario Gerhardt of the Georgia Institute of Technology, and Cameron Hubbard of the Oak Ridge National Laboratory. We would also like to thank the workshop moderator, Arvid Pasto, of the Oak Ridge National Laboratory.

This Ceramic Transactions volume is organized by topic with collections of papers relevant to each of the following aspects of ceramic NDE: (i) an overview of needs and opportunities for NDE of ceramics, (ii) suspension and green-body characterization, (iii) analysis of sintered ceramics and composites, (iv) characterization of surfaces and films, and (v) advanced characterization techniques.

Any undertaking of this nature requires the combined effort of many individuals and organizations. The editors are grateful for the sponsorship of the symposium by the Basic Science, Engineering Ceramics, Whitewares, and Electronic Ceramic divisions of the American Ceramic Society. We are also grateful to the Metallurgy and Ceramics Division of the Ames Laboratory of the U.S. Department of Energy, the Department of Materials Science and Engineering at Iowa State University, the Department of Mechanical Engineering at Iowa State University, and the Center for Nondestructive Evaluation at Iowa State University for financial support in organizing the symposium. We are also grateful for the many individuals who contributed their time and expertise as reviewers of the manuscripts, and would like to recognize the authors of each manuscript for the excellent quality of each submitted document. Finally, we would like to thank Sarah Godby, Jackie Davis, Mary Cassells, and other members of the publications department of The American Ceramic Society for their patience and support during the review and editing process. We also wish to thank the other symposium organizers for their assistance: Thomas Watkins of the Oak Ridge National Laboratory and Rosario Gerhardt of the School of Materials Science and Engineering at the Georgia Institute of Technology.

It is the sincere hope of the editors that the readers of this volume will find it informative and useful in their educational, research, and manufacturing endeavors.

Christopher H. Schilling
Joseph N. Gray

Overview

NEEDS AND OPPORTUNITIES FOR NDE IN CERAMIC PROCESSING

C. H. Schilling[1,2] and J. N. Gray [2,3]

[1] Ames Laboratory[+] and Department of Materials Science and Engineering
Iowa State University, Ames, Iowa 50010

[2] Center for Nondestructive Evaluation
Iowa State University, Ames, Iowa 50010

[3] Department of Mechanical Engineering
Iowa State University, Ames, Iowa 50010

Abstract

Post-fire defects in ceramics (e.g., cracks and warpage) can usually be traced to microstructure heterogeneities in the green state, e.g., isolated pores, packing-density gradients, and so on. Breakthroughs are needed in the way such heterogeneities are measured and understood during manufacturing, breakthroughs that could improve production efficiency and open up new markets for ceramic products. Traditional methods of quality-control are difficult to use when attempting to optimize the many different processing variables that ordinarily affect microstructure-defect evolution in ceramic production. These methods usually involve slow and costly trial-and-error procedures that are off-line in nature. They generally do not provide enough information on the spatial- and temporal-evolution of green-microstructure defects from one unit operation to the next. In this paper, we discuss needs and opportunities for process monitoring sensors in ceramic manufacturing. We emphasize the need for a new generation of process development tools to provide a better, fundamental understanding of defect evolution during processing, tools that involve the combined use of manufacturing process simulations and NDE.

[+] Ames Laboratory is operated by Iowa State University under the contract number W-7405-eng-82 with the U.S. Department of Energy.

Green-Body Defects: A Review

Green-microstructure heterogeneities include agglomerates, impurities, spatial variations of porosity and composition, delamination, segregation of different sizes of powders and granules, and preferred orientation (texturing) of asymmetric particulates (1-6). It is well known that such heterogeneities produce nonuniform shrinkage stresses during drying, debinding, and sintering, resulting in cracks and warpage.

Microstructure defects in green-bodies can be introduced during powder synthesis, or they can form and grow during subsequent unit operations including granulation, shape-forming, drying, and debinding. Most of the published literature on the evolution and elimination of green-body defects concerns dry-powder pressing, the most common, high-volume method of nonclay-ceramic production. There are several, excellent review publications in this area (3, 4, 7 - 10). By comparison, there is less active research on green microstructure evolution in the less prevalent, slurry-based forming operations of slip casting, pressure filtration, centrifugal casting, injection molding, tape casting, and gelcasting (1, 2, 5, 6, 11 - 32).

Restrictions with Conventional Quality-Control

Traditional quality-control methods suffer from practical drawbacks that prevent new breakthroughs in understanding the evolution of green-microstructure defects during ceramic manufacturing. Such methods are commonly slow, costly, and provide little information regarding the time and spatial evolution of green-microstructure heterogeneities during a given shape-forming process. Such methods usually involve trial-and-error correlations of measurements obtained well before and well after shape-forming. For example, in slip casting, measurements of the particle-size distribution, the specific surface-area, the zeta potential as a function of pH, and the viscosity as a function of shear rate are typically correlated with the green density, green strength, measurements of pore-size distribution, measurements of post-fire properties such as the density, strength, microstructure analysis, and so on. Very little information is provided regarding the effects of processing variables on the evolution of microscopic features such as particle-size segregation, packing-density gradients, the existence of flow channels in filter cakes, and so on. We should mention that fractography is a common, quality-control approach: after eliminating the biggest, strength-limiting defect by trial-and-error, the green-body strength is controlled by a smaller defect, which can then be identified by additional fractography and eliminated by more trial and error.

Another concern with traditional, quality-control methods is that they provide volume-averaged measurements of small (subcentimeter) specimen volumes, measurements that may or may not represent all other regions within a single production part or for that matter all parts on a production line. This is a serious problem for an engineer whose task is to determine, for example, if cracks in production parts should be attributed to nonuniform microstructures (e.g., density gradients occurring over several centimeters). To draw such conclusions, the engineer would be faced with a horrendous job of cutting a production part into many small pieces and then performing microscopy, density measurements, and so on, with every piece. The engineer would also face the problem of obtaining "representative" green microstructures for microscopy analysis. Because of the fragile nature of green bodies, specimen cutting usually requires the prior application of microstructure preservation techniques, such as (*i*) use of chemical fixation, or (*ii*) drying, followed by partial sintering, cutting, and polishing. Consequently, there are usually questions as to whether or not such "preservation" techniques introduce artifacts (e.g., human operator bias, polishing pull-out, changes in particle packing associated with impregnation of fixation chemicals, and so on).

Complicating the matter is the sheer number of processing variables that typically control defect evolution in ceramic greenware: (*i*) powder synthesis variables such as the particle-size and -shape distribution, powder composition and crystal structure, and changes in surface chemistry; (*ii*) granule- and slurry-processing variables such as non-uniform mixing of powder suspensions, rheological properties, pH, polymer additive concentration and composition, gravity-induced segregation; and (iii) shape-forming variables such as the properties of a slurry (e.g., composition, rheological properties, surface chemistry variables) and forming-equipment variables (e.g., the level and rate of stress application, the temperature, mold geometry). The above problems with conventional quality-control techniques make it difficult to sort through the many processing variables and determine which ones are in fact critical in a given production operation. Again, costly, trial-and-error methods are normally used for this purpose. We believe that a clearer approach for defect elimination will be possible if a better understanding of defect evolution can be established.

Market Opportunities

Building this understanding will likely provide significant opportunities to establish new markets, especially for nonclay ceramics. Consider the unprecedented variety of high-purity, nonclay ceramic powders that are just recently commercially available (33 - 38). Such powders include various borides, carbides, nitrides, and oxides for applications such as piezoelectrics, microelectronics, automotive parts, and so on. For the most part, clays generally do not meet the demanding electrical- or mechanical-properties requirements for these kinds of applications.

Despite the recent proliferation of new types of nonclay ceramic powders, there is a critical factor preventing market expansion for these raw materials: experience indicates that, compared to clays, the nonclays require much more stringent control of manufacturing process variables to remove green-microstructure heterogeneities and in turn avoid cracking during drying and debinding (5, 6). Factors that typically require more stringent control include the powder purity, particle size, slurry chemistry, mold design, and so on. Despite many years of empiricism behind the development of quality-control strategies for clay-based ceramics, these strategies often fail for nonclay ceramics. It is just recently understood that differences in interparticle friction are responsible for the remarkable difference in the fracture sensitivity of clay- versus nonclay greenware. Clays have an enhanced ability to dissipate drying and debinding stresses by particle rearrangement, a process that is inherently controlled by the stacked-platelet morphology of clays, the nature of the surface charge, and the adsorption of structured water molecules on clay surfaces (39 - 41). Nonclay ceramic powders are usually equiaxed in shape, and they do not possess the same surface chemistry necessary to reduce interparticle friction.

Consider the following trend, which is common to both clays and nonclay ceramics alike: the bigger the green body, the larger the absolute shrinkage during drying, debinding, and sintering, and the more crack-prone the body becomes during these operations. Experience indicates that, in order to prevent such cracking, larger shapes require much more stringent control of green-microstructure heterogeneities than smaller shapes. A consequence of the increased crack-susceptibility of nonclay green-bodies is that it is difficult to produce dense, engineering shapes as large as those routinely made from clay ceramics. Many years of empiricism led to the development of quality-control guidelines for clay-based ceramics, particularly meter-scale objects such as sanitaryware. However, such guidelines are not well developed for nonclay ceramics, as is evident from the scarcity of crack-free, meter-scale, dense parts that are commercially sintered from these powders (e.g., a ceramic engine block).

In order to overcome this problem, manufacturers of nonclay ceramics are faced with two alternatives: (*i*) design new chemical additives to produce more clay-like, crack-resistant greenware, or (*ii*) develop a better understanding of what the critical defects are in a given production part, how they evolve during unit operations prior to sintering, and how they can be prevented or eliminated during ceramic production. Working towards the former goal, researchers are currently developing a science base for the design of chemical additives that mediate interparticle friction and in turn produce clay-like rheological properties in suspensions of nonclay ceramic powder (14 - 16, 29, 42 - 52). However, much more research is needed before generic guidelines are perfected for the commercial development of green-strengthening additives that will allow mass production of dense, crack-free, meter-scale parts from nonclay ceramic powders.

NDE Needs & Opportunities

Let us explore option (*ii*) above: develop a better understanding of what the critical defects are in a given production part, how they evolve during unit operations prior to sintering, and how they can be prevented or eliminated during ceramic production. We believe that a combination of improved, on-line sensors integrated with a better understanding of defect evolution throughout different unit operations will help achieve this goal.

Major advances in nondestructive evaluation (NDE) equipment have been made in the past decade (see, for example, papers in this proceedings and reference 53). For example, better methods of fabricating piezoelectric sensors, semiconductor x-ray detectors, higher-performance computers, and so on, have resulted in significant improvements in measurement speed and sensitivity. In many cases, major reductions in equipment costs have also resulted. Despite these advances, on-line NDE sensors are rarely encountered in high-volume, ceramic manufacturing applications at the present time. In many cases, NDE instrumentation is too delicate and too expensive for routine use on the factory floor. If these barriers can be overcome, opportunities exist to improve manufacturing efficiency by reducing down time, energy costs, waste of ceramic raw material, and loss of value added in defective parts.

Several articles have been published on NDE of ceramics, however, most of the techniques reported are used in an off-line manner for laboratory research (see, for example, papers in this proceedings and references 5, 6, 9, 12 - 14, 17 - 19, 21 - 32, 54 - 82). We believe that significant opportunities exist to improve ceramic manufacturing efficiency if many of these NDE technologies are advanced to the point of allowing economical usage either as on-line sensors or as off-line process development tools. A critical review is needed to assess the engineering and economical barriers preventing the implementation of each NDE technology in these two respects.

We believe that the “ideal,” NDE technology (or set of technologies) for on-line measurements should meet the following general requirements:

(i) It should be capable of differentiating between different types of defects, e.g., packing-density gradients, isolated pores, surface and subsurface cracks, spatial variations of composition, and so on.

(ii) It should also be capable of monitoring the time- and spatial evolution of defects throughout a succession of unit operations prior to sintering.

(iii) On-line, NDE equipment should be reliable, easy to use, and provide rapid, economical measurements without slowing production.

Specific requirements for on-line, process sensors in primary unit operations of ceramic manufacturing are summarized in Table I. Beginning with powder synthesis, traditional characterization techniques are typically off-line in nature and entail such methods as thermogravimetric analysis, pycnometry, x-ray diffraction, various types of spectroscopy, and so on. Unit operations of powder synthesis and particle-size control (e.g., sieving, milling) can benefit from the introduction of nondestructive, on-line sensors that are capable of detecting subtle variations in particle size and shape, packing-density, moisture concentration, and chemical composition. For example, improved, on-line sensors are needed to detect sporadic contaminants in powder feedstocks (e.g, dust). Another example is the need for a robust, on-line technique for morphology analysis, a method that can quickly yield numerical indices of morphic features (e.g., elongation, roughness) of powders and agglomerates. A rigorous mathematical basis for this type of analysis was recently developed (83, 84). A commercial, laboratory instrument was recently introduced that applies these mathematics to optical microscopy images, a technique that has significant potential for on-line characterization and process control (85).

On-line sensors are also needed to detect changes in the chemistry, density, and rheological properties of feedstock suspensions that are used to spray- or freeze-dry granules. For example, one can envision a system to detect changes in slurry density and solids concentration using ultrasonic sensors or even a simple, gravimetric sampling device (pycnometry). On-line sensors for colloidal chemistry parameters such as the zeta potential, electrical conductivity, chemical composition, and so on would also be useful. In addition, on-line sensors of rheological properties are needed; methods involving oscillatory tube flow (85) and torsional waveguide analysis (87) appear to have significant potential in this regard.

Improved, on-line techniques for characterizing granule properties are also needed, properties such as the moisture concentration, mass density, porosity, granule size-distribution, and morphological features such as the granule shape,

Table I. Summary of NDE Needs in Ceramic Manufacturing.

UNIT OPERATION	ON-LINE SENSORS NEEDED
• Powder Synthesis & Sizing	• Particle Shape, Size, Chemistry • Bulk Density • Sporadic Contamination of Powder
• Granulation	• Slurry Chemistry & Rheology • Granule Morphology, Size • Size Segregation of Granules • Segregation of Powders within Granules • Moisture Concentration • Flow Properties of Granules • Bulk Density
• Powder & Granule Pressing	• Microstructure Evolution & Uniformity • Porosity Mapping • Powder Composition Mapping • Moisture Concentration Mapping • Binder Concentration Mapping • Bulk Density
• Slurry Casting and Plastic Molding	• Slurry Chemistry & Rheology • Microstructure Evolution & Uniformity • Porosity Mapping • Composition Mapping • Warpage & Cracks • Bulk Density
• Drying, Debinding, and Sintering	• Microstructure Evolution & Uniformity • Moisture Mapping • Porosity Mapping • Composition Mapping • Warpage & Cracks
• Machining	• Surface & Subsurface Crack Detection • Grain Dislodging

the presence of hollow granules, and the segregation of granule constituents including binders, different types of powders or different-sized powders within a granule. Three methods of moisture and carbon-analysis are recently introduced that appear to have significant potential for on-line measurements in ceramic manufacturing: near infra-red reflectance (88, 89), nuclear magnetic resonance spectroscopy (90), and microwave spectroscopy (91).

Maintaining consistency of granule compaction and flow behavior is a key problem for many manufacturers. Traditional methods of flow-behavior characterization are typically off-line in nature (e.g., angle-of-repose, direct shear tester, compaction measurements, and so on) (10, 92, 93). On-line methods are needed which can directly measure compaction and flow properties of powders and granules. Often it is poorly understood which processing variables are critical in this respect. The ceramic industry would benefit from having a better, fundamental understanding of how granule compaction and flow behavior are controlled by critical factors such as the granule shape, microstructure, moisture concentration, and so on. It is also important to develop a better understanding of the effects of critical processing parameters on the evolution of granule shape and microstructure during spray- and freeze-drying. To achieve this goal, technology is needed that couples on-line, NDE measurements with computerized models of microstructure evolution during granule synthesis.

On-line sensors are needed that provide mapping / diagnostics of the following defects during dry pressing: packing-density gradients, local variations in the structures of pores and agglomerates, composition gradients, binder- and moisture distribution, preferred orientation of assymetric particles, and so on. We can make the same general statements regarding the need for on-line, nondestructive sensing / diagnostics of microstructure evolution in (*i*) the slurry-based forming operations of slip casting, gelcasting, tape casting, injection molding, and so on; and (*ii*) subsequent unit operations of drying, debinding, sintering, and machining. It is likely that different sensor technologies will be optimum for different unit operations. Integrating sensor measurements from each unit operation to the next will be needed in order to monitor / diagnose defect evolution between unit operations. In turn, this can help in the development of better measurement standards, better processing tolerances, and a better database of statistics on a given unit operation.

Concluding Statements

We have made the point several times that there are serious difficulties using traditional methods of quality control to sort through and optimize the multitudes of processing parameters affecting the evolution of green-microstructure defects in ceramic production. We stressed the need for better sensors capable of tracking defect evolution from one unit operation to the next. We then summarized specific detection needs for each of the primary unit operations in ceramic manufacturing.

Of course, there are two ways we can address the use of NDE in high-volume ceramic manufacturing: (*i*) use of NDE for on-line process monitoring alone, and (*ii*) use of NDE and other appropriate theoretical and experimental tools to build a better understanding of defect evolution *off-line*, and then develop some simpler on-line monitor. With respect to case (*i*), we mentioned earlier that NDE instrumentation is often too delicate and too expensive for routine use on the factory floor. The fragility (and hopefully the cost) of on-line equipment should be less of an issue for case (*ii*) above.

Addressing case (*ii*), we believe the key issue is a determination of what constitutes a critical defect in the first place. Frequently, not enough is known in this regard. For example, consider a hypothetical situation where a critical defect in one production part may not be critical for another part having a slightly different shape, composition, porosity, and so on. Another example is where a non-critical defect suddenly becomes critical after some aspect of a unit operation changes (e.g., temperature, humidity, throughput). For some production operations, not knowing what constitutes a critical defect may lead to serious increases in inspection costs. In particular, safety-critical parts or high-value-added parts are usually "over-inspected" to eliminate as many defects as possible, regardless of whether they are critical or not.

As an alternative, let us consider a design paradigm which is recently being developed for safety-critical metal castings (94). In this case, the engineer first establishes what constitutes a critical defect in the finished casting using a combination of mechanical properties experiments, NDE, microstructure analysis, and so on. Specific aspects of the critical flaw are ascertained in detail; e.g., it may be a solidification shrinkage pore of a certain shape and orientation within the part, it may be surface roughness of a particular characteristic, and so on. The engineer then incorporates this information into a computer simulation to analyze how the critical defect arose in the first place and how it affects the evolution of the mechanical properties and the evolution of the microstructure during subsequent stages of processing. Of course, the validity of the simulation is rigorously checked against laboratory standards involving NDE, microstructure analysis, and mechanical properties tests of specimens seeded with defects of known geometry. Once the validity of the simulation is established, the engineer is in a position to analyze design trade-offs involved with optimizing manufacturing process variables. As a final step, the engineer can decide how much inspection is needed and which inspection methodology is affordable.

Let us consider a similar paradigm for the manufacturing process design of high-performance ceramics. The first issue would be a definition of what constitutes a critical defect in a production part. For example, questions such as, "How small of a defect is critical," and "Which orientations of the defect are important," would typically need to be answered. The next step would involve building a strong, theoretical and experimental understanding of microstructure evolution and mechanical-properties evolution throughout different unit operations in ceramic production. This is obviously not a small undertaking. For example, this step might include measuring the fracture behavior of green specimens seeded with defects, experiments that would reveal the likelihood a certain defect would lead to green-body failure. Having a strong experimental and theoretical foundation, manufacturing process simulations could then be created and validated. In turn, reasonably informed methods of process optimization and reasonably informed accept / reject criteria can result.

Significant progress has already been made towards achieving this design paradigm for powder compaction operations. Several excellent review articles have recently been published summarizing progress in the process control of powder pressing operations by the application new methods of powder characterization in combination with compaction simulations (3, 4, 7 - 9). New developments in powder compaction science are summarized in these articles, which are written from the perspectives of a broad range of experts in ceramic engineering, powder metallurgy, and pharmaceutical processing. We hope that researchers will eventually build upon this combined experimental / theoretical approach to develop simple, on-line sensors for powder pressing operations.

We have the same hope for slurry-based forming operations such as slip casting, injection molding, extrusion, and so on. There are several models of slurry consolidation and plastic flow which appear to have significant potential in this regard, models that involve a range of both continuum and discreet-element approaches (7, 11, 20, 95 - 111, and several articles in reference 112). A critical review is needed to assess the potential of using these models in conjunction with NDE and other experimental methods to establish a better understanding of the evolution of green-microstructure defects during ceramic manufacturing. We hope such models will form the basis of a new set of process development tools, tools that would eventually enable the creation of simple and effective on-line sensors for process control.

Acknowledgements

This research is supported by the Office of Basic Energy Sciences at the U.S. Department of Energy. The authors wish to thank M.F. Berard and S. P. Huss for assistance.

References

1. F. F. Lange, "Powder Processing Science and Technology for Increased Reliability," *J. Am. Ceram. Soc.* **72** (1) 3-15 (1989).
2. M. J. Hoffman, A. Nagel, P. Greil, and G. Petzow, "Slip Casting of SiC-Whisker-Reinforced Si_3N_4," *J. Am. Ceram. Soc.* **72** [5] 765-69 (1989).
3. K. G. Ewsuk, "Compaction Science and Technology," *M.R.S. Bull.* **22** (12) 14-18 (1997).
4. S. J. Glass and K. G. Ewsuk, "Ceramic Powder Compaction," *M.R.S. Bull.* **22** (12) 24-28 (1997).
5. V. K. Pujari and D. M. Tracey, "Processing Methods for High Reliability Silicon Nitride Heat Engine Components," *ASME Trans, Proc. Int. Gas Turbine and Aeroengine Congr. And Exposition*, ASME 93-GT-319, Cincinnati, Ohio, 1993.
6. V. K. Pujari, K. E. Amin, and P. H. Tewari, "Development of Improved Processing and Evaluation of Silicon Nitride," in *ASME Trans., Proc. Int. Gas Turbine andAeroengine Congr. and Exposition,* ASME 91-GT-317, Orlando, Florida 1991.
7. I. Aydin, B. J. Briscoe, and N. Ozkan, "Modeling of Powder Compaction: A Review," *M.R.S. Bull.* **22** (12) 45-51 (1997).
8. W. R. Blumenthal, H. Sheinberg, and S. A. Bingert, "Compaction Issues in Powder Metallurgy," *M.R.S. Bull.* **22** (12) 29-33 (1997).
9. J. J. Lannutti, "Characterization and Control of Compact Microstructure," *M.R.S. Bull.* **22** (12) 38-44 (1997).
10. N. G. Stanley-Wood, editor, *Enlargement and Compaction of Particulate Solids* (Butterworths, London, 1983).
11. I. A. Aksay and C. H. Schilling, "Colloidal Filtration Route to Uniform Microstructures," in *Ultrastructure Processing of Ceramics, Glasses, and Composites,* edited by L. L. Hench and D. R. Ulrich (John Wiley and Sons, New York, 1984), pp. 439-447.
12. F. M. Auzerais, R. Jackson, and W. B. Russel, "The Resolution of Shocks and the Effects of Compressible Sediments in Transient Settling," *J. Fluid Mech.* **195** 437-82 (1988).

13. K. Been and G. C. Sills, "Self-Weight Consolidation of Soft Soils: An Experimental and Theoretical Study," *Geotechnique* **31** (4) 519-35 (1981).
14. L. Bergström, C. H. Schilling, and I. Aksay, "Consolidation Behavior of Flocculated Alumina Suspensions," *J. Am. Ceram. Soc.* **75** (12) 3305-14 (1992).
15. J. C. Chang, B. Velamakanni, F. Lange, and D. Pearson, "Centrifugal Consolidation of Al_2O_3 and Al_2O_3/ZrO_2 Composite Slurries vs Interparticle Potentials: Particle Packing and Mass Segregation," *J. Am. Ceram. Soc.* **74** (9) 2201-04 (1991).
16. J. C. Chang, F. F. Lange, D. S. Pearson, and J. P. Pollinger, "Pressure Sensitivity for Particle Packing of Aqueous Alumina Slurries vs. Interparticle Potential," *J. Am. Ceram. Soc.* **77** (5) 1357-60 (1994).
17. C. F. Cooper and S. F. A. Miskin, "The Slip Casting Kinetics of Alumina by Gamma-Ray Absorption," *Proc. Brit. Ceram. Soc.* No. 3 (Fabrication Science), pp. 135-54 (1965).
18. R. F. Deacon and S. F. A. Miskin, "Gamma-Ray Absorption for the Continuous Observation of Slip Casting Kinetics," *Trans. Brit. Ceram. Soc.*, **63**, 473-86 (1964).
19. K. E. Davis, W. B. Russel, and W. J. Glantschnig, "Disorder-to-Order Transition in Settling Suspensions of Colloidal Silica: X-Ray Measurements," *Science* **245**, 507-10 (1989).
20. K. E. Davis and W. B. Russel, "A Model of Crystal Growth in the Sedimentation and Ultrafiltration of Colloidal Hard Spheres," in *Advances in Ceramics, Vol. 21: Ceramic Powder Science,* edited by G. L. Messing, K. S. Mazdiyasni, J. W. McCauley, and R. A. Haber (American Ceramic Society, Westerville, Ohio, 1987), pp. 573-85.
21. A. G. Haerle and R. A. Haber, "Ultrasonic Real-Time Monitoring of Cake Structure during Slip Casting" *J. Am. Ceram. Soc.*, **78** (3) 819-822 (1995).
22. K. Hayashi, K. Kawashima, K. Kose, and T. Inouye, "NMR Imaging of Advanced Ceramics during the Slip Casting Process," *J. Phys. D, Appl. Phys.* **21**, 1037-39 (1988).
23. C. W. Maranville, J. N. Gray, and C. H. Schilling, "Novel X-ray Imaging Method for Evaluating Defect Evolution in Ceramic Tapes," in *Review of Progres in Quantitative Nondestructive Evaluation, Vol. 15*, edited by D. O. Thompson and D. E. Chimenti (Plenum Press, New York, 1995).
24. A. S. Michaels and J. C. Bolger, "Settling Rates and Sediment Volumes of Flocculated Kaolin Suspensions," *Ind. Eng. Chem. Fund.*, **1**, 24-33 (1962).
25. A. A. Parker, G. H. Armstrong, and D. P. Hedric, "NMR and Sedimentation Studies of a Polymeric Steric Stabilizer for Alumina," *J. Appl. Poly. Sci.* **47**, 1999-2003 (1993).

26. C. H. Schilling, W. D. Samuels, and I. A. Aksay, "Analysis of Packed-Particle Rearrangement During Colloidal Ceramic Processing: Gamma-Ray Densitometry and a Review of Alternative Methods," in *Ceramic Transactions, Volume 5, Advanced Characterization Techniques for Ceramics,*, edited by W. S. Young, G. L. McVay, and G. E. Pike (American Ceramic Society, Westerville, Ohio, 1988), pp. 181-98.
27. C. H. Schilling, W.-H. Shih, and I. A. Aksay, "Advances in the Drained Shaping of Ceramics," in *Ceramic Powder Science IV, Volume 22* , edited by S. Hirano, G. L. Messing, and J. Hausner (American Ceramic Society, Westerville, Ohio, 1991), pp. 307-21.
28. C. H. Schilling, W. -H. Shih, J. J. Lannuti, and I. A. Aksay, "Stress Density Variations in Alumina Sediments: Effects of Polymer Chemistry," in *Mechanical Properties of Porous and Cellular Materials, MRS Symp. Proc.,* edited by D. J. Green, L. J. Gibson, and K. Sieradski (Materials Research Society, Pittsburgh, PA, 1991), pp. 151-6.
29. C. H. Schilling, "Plastic Shaping of Colloidal Ceramics," Ph.D. thesis, University of Washington, Seattle (1992).
30. C. H. Schilling, S. P. Huss, and J. N. Gray, "Regulating Density Gradients in Slip Consolidation: A New Methodology," in *Science of Whitewares*, edited by V. E. Henkes, G. Y. Onoda, and W. M. Carty (American Ceramic Society, Westerville, Ohio, 1996 a), pp. 117-31.
31. C. H. Schilling, R. M. Smith, and R. A. Roberts, "Ultrasonic Analysis of Alumina Powder Compacts," in *Nondestructive Evaluation and Material Properties III,* edited by P. K. Liaw, O. Buck, R. J. Arsenault, and R. E. Green, Jr. (The Minerals, Metals, and Materials, Society, Warrendale, Pennsylvania, 1996), pp. 17-26.
32. C. H. Schilling, V. J. Garcia, R. M. Smith, and R. A. Roberts, "Ultrasonic- and Mechanical Behavior of Green- and Partially-Sintered Alumina: Effects of Slurry Consolidation Chemistry," *J. Am. Ceram. Soc.* 1998, in press.
33. T. Abraham, "Advanced Ceramic Powder and Nano-Sized Ceramic Powder: An Industry and Market Overview," in *Ceramic Transactions, Volume 62, Science, Technology, and Commercialization of Powder Synthesis and Shape Forming Processes*, edited by J. J. Kingsley, C. H. Schilling, and J. H. Adair (American Ceramic Society, Westerville, Ohio 1996), pp. 3-13.
34. G. Geiger, "Commercialization of Powder Synthesis Processes," in *Ceramic Transactions, Volume 62, Science, Technology, and Commercialization of Powder Synthesis and Shape Forming Processes*, edited by J. J. Kingsley, C. H. Schilling, and J. H. Adair (American Ceramic Society, Westerville, Ohio 1996), pp. 15-20.

35. S. Somiya, *Hydrothermal Reactions for Materials Science and Engineering: An Overview of Research in Japan* (Elsevier Science Publishing Co., New York, 1989).
36. M. Schwartz, *Handbook of Structural Ceramics* (McGraw-Hill, New York, 1992).
37. D. Segal, "Chemical Preparation of Powders," in *Materials Science in Technology, Volume 17A, Processing of Ceramics, Part 1*, edited by R. J. Brook, R. W. Cahn, P. Haasen, and I. J. Kramer (VCH Verlagsgesellschaft mbH, Weinheim, Federal Republic of Germany, 1996), pp. 70-98.
38. D. Segal, *Chemical Synthesis of Advanced Ceramic Materials* (Cambridge University Press, Cambridge 1989).
39. H. van Olphen, *An Introduction to Clay Colloid Chemistry, Second Edition* (John Wiley and Sons, New York, 1977).
40. R. M. Pashley and J. N. Israelachvili, "DLVO and Hydration Forces Between Mica Surfaces in Mg^{+2}, Ca^{+2}, Sr^{+2}, and Ba^{+2} Chloride Solutions," *J. Colloid Interface Sci.*, **97** 446-55 (1984).
41. W. G. Lawrence, "The Structure of Water and its Role in the Clay-Water System," Chapter 16 in *Ceramic Processing Before Firing,* edited by G. Y. Onoda, Jr. and L. L. Hench (John Wiley and Sons, New York, 1978).
42. J. C. Chang, F. F. Lange, and D. S. Pearson, "Viscosity and Yield Stress of Alumina Slurries Containing Large Concentrations of Electrolyte," *J. Am. Ceram. Soc.* **77** (1) 19-26 (1994).
43. G. V. Franks and F. F. Lange, "Plastic-to Brittle Transition of Saturated, Alumina Powder Compacts," *J. Am. Ceram. Soc.* **79** [12] 3161-68 (1996).
44. H. Goel, C. H. Schilling, S. B. Biner, J. A. Moore, and B. K. Lograsso, "Plastic Shaping of Aqueous Alumina Suspensions with Saccharides and Dicarboxylic Acids," in *Science, Technology, and Commercialization of Powder Synthesis and Shape Forming Processes*, edited by J. J. Kingsley, C. H. Schilling, and J. H. Adair (American Ceramic Society, Westerville, Ohio 1996), pp. 241-54.
45. T. Kramer and F. F. Lange, "Rheology and Particle Packing of Chem- and Phys-Adsorbed Alkylated Silicon Nitride Powders," *J. Am. Ceram. Soc.* **77** [4] 922-928 (1994).
46. Y. K. Leong, P. J. Scales, T. W. Healy, D. V. Boger, and R. Ruscall, "Rheological Evidence of Adsorbate-mediated Short-Range Steric Forces in Concentrated Dispersions," *J. Chem. Soc., Faraday Trans.* **89**, 2473-78 (1993).
47. E. P. Luther, T. M. Kramer, F. F. Lange, and D. S. Pearson, "Development of Short-Range Repulsive Potentials in Aqueous, Silicon Nitride Slurries," *J. Am. Ceram. Soc.* **77** [4] 1047-51 (1994).

48. E. P. Luther, J. A. Yanez, G. V. Franks, F. F. Lange, and D. S. Pearson, "Effect of Ammonium Citrate on the Rheology and Particle Packing of Alumina Slurries," *J. Am. Ceram. Soc.* **78** [6] 1495-5000 (1995).
49. C. H. Schilling, S. B. Biner, H. Goel, and J.-L. Jane, "Plastic Shaping of Aqueous Alumina Suspensions with Sucrose and Maltodextrin Additives," *J. Environmental Polymer Degradation* **3** [3] 153-60 (1995).
50. B. V. Velamakanni, J. C. Chang, F. F. Lange, and D. S. Pearson, "New Method for Efficient Colloidal Particle Packing via Modulation of Repuslive Lubricating Hydrating Forces," *Langmuir* **6**, 1323-25 (1990).
51. B. V. Velamakanni, F. F. Lange, F. W. Zok, and D. S. Pearson, "Influence of Interparticle Forces on the Rheological Behavior of Pressure-Consolidated Alumina Particle Slurries," *J. Am. Ceram. Soc.* **77** (1) 216-20 (1994).
52. T. K. Yin, I. A. Aksay, and B. E. Eichinger, "Lubricating Polymers for Powder Compaction," in *Ceramic Transactions Volume 1, Ceramic Powder Science II,* edited by G. L. Messing, E. R. Fuller, Jr., and H. Hausner (American Ceramic Society, Westerville, Ohio, 1988), pp. 654-662.
53. D. O. Thompson and D. E. Chimenti, editors, *Review of Progress in Quantitative Nondestrutive Evaluation, Volumes 1-16* (Plenum Press, New York, 1981-1997.
54. G. Busee, M. L. Lyamshev, and J. Stanullo, "Remote Photoacoustic Testing of Ceramics in the Course of Sintering," *Acoustical Physics* **42** (1) 35-39 (1996).
55. R. W. Davidge, "Defects in Ceramics- The Targets for NDE," *Br. Ceram. Trans. J.* **88**, 113-6 (1988).
56. W. A. Ellingson, P. S. Wong, S. L. Dieckman, J. L. Ackerman, and L. Garrido, "Magnetic Resonance Imaging: A New Characterization Technique for Advanced Ceramics," *Bull. Am. Ceram. Soc* **68** (6) 1180-84 (1989).
57. A. G. Evans, B. R. Tittmann, L. Ahlberg, B. T. Khuri-Yakub, and G. S. Kino, "Ultrasonic Attenuation in Ceramics," J. Appl. Phys. 49 [5] 2669-79 (1978).
58. A. M. Gaudin and M. C. Fuerstenau, "The Transviewer - X-rays to Measure Suspended Solids Concentration," *Eng. & Mining J.* **159** (9) 110-12 (1958).
59. K. Goebbels, "Nondestructive Evaluation of Advanced Ceramics," in *Proc. 2nd European Symposium on Engineering Ceramics*, edited by F. L. Riley (Elsevier, London, 1989), pp. 99-149.
60. M. P. Jones, G. V. Blessing, and C. R. Robbins, "Dry-Coupled Ultrasonic Elasticity Measurements of Sintered Ceramics and Their Green States," *Materials Evaluation*, **44**, 859-62 (1986).

61. H. A. Kearsey and L. E. Gill, “A Study of the Sedimentation of Flocculated Thoria Slurries using a Gamma-Ray Technique,” *Trans. Inst. Chem. Engrs.*, **41**, 296-306 (1963).
62. N. Kulkarni, B. Moudgil, and M. Bhardwaj, “Ultrasonic Characterization of Green and Sintered Ceramics; I, Time Domain,” *Bull. Am. Ceram. Soc.* **73** (6) 146-53 (1994).
63. N. Kulkarni, B. Moudgil, and M. Bhardwaj, “Ultrasonic Characterization of Green and Sintered Ceramics; II, Frequency Domain,” *Bull. Am. Ceram. Soc.* **73** (7) 83-85 (1994).
64. D. C. Kunerth, K. L. Telschow, and J. B. Walter, “Characterization of Porosity Distributions in Advanced Ceramics: A Comparison of Ultrasonic Methods,” *Materials Evaluation* 47, May 1989, pp. 571-5.
65. D. S. Kupperman and J. P. Karplus, “Ultrasonic Wave Propagation Characteristics of Green Ceramics,” *Bull. Am. Ceram. Soc.* **63** (12) 1505-09 (1984).
66. D. S. Kupperman, H. B. Karplus, R. B. Poeppel, W. A. Ellingson, H. Berger, C. Robbins, and E. Fuller, “Application of NDE Methods to Green Ceramics: Initial Results,” in *Nondestructive Evaluation: Applications to Materials Processing*, edited by O. Buck and S. W. Wolf (The Minerals, Metals, and Materials, Society, Warrendale, Pennsylvania, 1983), pp. 89-101.
67. L. A. Lott and D. C. Kunerth, “NDE Testing and Inspection,” in *Engineered Materials Handbook, Ceramics and Glasses, Volume 4* (American Society of Metals, Metals Park, Ohio 1991), pp. 617-27.
68. M. Maisl, H. Reiter, and P. Holler, “Micro-Radiography and Tomography for High Resolution NDT of Advanced Materials and Microstructural Components, *A.S.M.E. Trans.*. **112**, April 1990, pp. 223-26.
69. L. P. Martin and M. Rosen, “Analysis of Ultrasonic Measurements on Sintering Zinc Oxide,” *Ultrasonics* **35** (1) 65-71 (1997).
70. L. P. Martin and M. Rosen, “Correlation Between Surface Area Reduction and Ultrasonic Velocity in Sintered Zinc Oxide Powders,” *J. Am. Ceram. Soc.* **80** (4) 839-44 (1997).
71. J. F. Moore, “Evolution of Computed Tomography,” *Materials Evaluation*, **48**, May 1990, pp. 630-40.
72. K. K. Phani and S. Niyogi, “Porosity Dependence of Ultrasonic Velocity Modulus in Sintered Uranium Dioxide - A Discussion,” *J. Mater. Sci. Lett.* **5**, 427-30 (1986).
73. D. H. Phillips and J. J. Lannutti, “ X-ray Computed Tomography for the Testing and Evaluation of Ceramic Processes,” *Am. Ceram. Soc. Bull.* **72** (11) 69-75 (1993).

74. J. M. Rebello, and J. Charlier, "Some Aspects of NDE Applied to Ceramics," *European J. NDT* **2**, 150-8 (1993).
75. R. Roberts, "A Dry-Contact Coupling Technique for Ultrasonic Nondestructive Evaluation of Green State Ceramics," *Mater. Eval.* **46**, May, 758-66 (1988).
76. J. G . Saxton, "Resonant Inspection of Production Parts," *Am. Ceram. Soc. Bull.* **75** (10) 48-50 (1996).
77. T. Taylor, W. A. Ellingson, and W. D. Koenigsberg, "Evaluation of Engineering Ceramics by Gamma-Ray Computed Tomography," *Ceram. Eng. Sci. Proc.* **7** (7-8) 772-83 (1986).
78. J. Winterfield, H. D. Tietz, and T. Palm, "Non-destructive Testing of Volume Defects in Structural Ceramics," *Ceram. Forum Int.* **71**, 96-100 (1994).
79. C. Zhang, Z. Guo. Y. Dong, H. Zhang, and J. Tian, "Testing Advanced Structural Ceramics," *Am. Ceram. Soc. Bull.* **75** (1) 64-66 (1996).
80. *NDT of High Performance Ceramics, Conference Proceedings*, August 25-27, 1987, Boston, Massachusetts (American Ceramic Society, Westerville, Ohio, 1987).
81. *Proceedings of the Conference on Nondestructive Analysis of Modern Ceramics,* July 9-12, 1990, Columbus, Ohio (American Society for NonDestructive Testing, Columbus, Ohio, and the American Ceramic Society, Westerville, Ohio, 1990).
82. D. H. Phillips and J. J. Lannutti, "Measuring Physical Density with X-ray Computed Tomography," *NDT&E International* **30** (6) 339-350 (1997).
83. J. K. Beddow, "Morphological Analysis," in *Advances in Ceramics, Vol. 21: Ceramic Powder Science* (American Ceramic Society, Westerville, Ohio 1987), pp. 747-757.
84. J. K. Beddow and D. W. Luerkens, "Effects of Powder Production and of Material Processing on Powder Particle Morphology of Adipic Acid," *Powder Technology,* **68**, 163-173 (1991).
85. D. Luerkens, Business Systems International, Iowa City, Iowa, personal communication (1997).
86. G. Thurston, "Characterization of Polymer Solutions in Flow Through Porous Media," *Advances in Rheology,* **4**, 385-91 (1984).
87. R. D. Costley, W. M. Ingham, and J. A. Simpson, "Ultrasonic Velocity Measurement of Molten Liquids," published in this proceedings.
88. G. Levin, Brimrose Corporation, Baltimore, Maryland, personal communication (1997).
89. J. Szymanski, Zeltex Incorporated, Hagerstown, Maryland, personal communication (1997).

90. P. Krygsman, Bruker Spectrospin Limited, Milton, Ontario, Canada, personal communication (1997).
91. J. Watkins, Epsilon Industrial Incorporated, Austin, Texas, personal communication (1997).
92. J. Schwedes and D. Schulze, "Measurement of Flow Properties of Bulk Solids," *Powder Technology*, **61**, 59-68 (1990).
93. C. P. Wroth and G. T. Houlsby, "Applications of Soil Mechanics Theory to Ceramics Processing," in *Ultrastructure Processing of Ceramics, Glasses, and Composites,* edited by L. L. Hench and D. R. Ulrich (Wiley, New York, 1984), pp. 448-63.
94. J. G. Conley, B. Moran, and J. N. Gray, "A New Paradigm for the Design of Safety Critical Castings," in *Proceedings of the Symposium on Aluminum in Automotive Applications, SP-1350*, Society of Automotive Engineers, February 1998, pp. 25-38.
95. J. S. Abel, G. C. Stangle, C. H. Schilling, and I. A. Aksay, "Sedimentation in Flocculating Colloidal Suspensions," *J. Mater. Res.* **9** (2) 451-61 (1994).
96. T. Aizawa, "Granular Modeling for Computer Aided Design of Colloidal Processing," in *Science, Technology, and Applications of Colloidal Suspensions,* edited by J. H. Adair, J. A. Casey, C. A. Randall, and S. Venigalla (American Ceramic Society, Westerville, Ohio, 1995), pp. 251-269.
97. T. Watanabe, T. Aizawa, and J. Kihara, "Ultrasonic Array Sensing for In-situ NDE in Extrusion Process for Green Ceramics," published in this proceedings.
98. R. Challis and J. Tebbutt, "Ultrasonic Wave Interactions with Colloidal Mixtures," published in this proceedings.
99. V. Garcia and C. H. Schilling, "Models Relating Ultrasonic Measurements to the Microstructures of Granular Media: A Review," published in this proceedings.
100. R. Buscall and L. R. White, "The Consolidation of Concentrated Suspensions, Part 1: The Theory of Sedimenation," *J. Chem. Soc., Faraday Trans. I,* **83**, 873-91 (1987).
101. A. P. Gast, W. B. Russel, and C. K. Hall, "An Experimental and Theoretical Study of Phase Transitions in the Polystyrene Latex and Hydroxymethylcellulose System," *J. Colloid Interface Sci.* **109** (1) 161-71 (1986).
102. J. H. D. Hampton, S. B. Savage, and R. A. L. Drew, "The Clogging Effects in Slip Casting," *Ceramic Transactions, Volume 1, Ceramic Powder Science II, B,* edited by G. L. Messing, E. R. Fuller, and H. Hausner (American Ceramic Society, Westerville, Ohio, 1988), pp. 749-57.

103. C. W. Hong and P. Greil, "Discreet Element Modelling of Colloidal Powder Processing," in *Science, Technology, and Applications of Colloidal Suspensions,* edited by J. H. Adair, J. A. Casey, C. A. Randall, and S. Venigalla (American Ceramic Society, Westerville, Ohio, 1995), pp. 235-249.
104. M. A. Janney and G. Y. Onoda, Jr., "Particulate Mechanics of Highly Loaded Ceramic Systems," in *Advances in Ceramics, Vol. 21: Ceramic Powder Science,* edited by G. L. Messing, K. S. Mazdiyasni, J. W. McCauley, and R. A. Haber (American Ceramic Society, Westerville, Ohio, 1987), pp. 615-26.
105. B. J. Kellett and C. Y. Lin, "Mechanics of Constant-Rate Filter Pressing of Highly Flocculated Slurries," *J. Am. Ceram. Soc.* **80** (2) 381-93 (1997).
106. L. T. Kuhn, R. M. McMeeking, and F. F. Lange, "Modeling Powder Consolidation," in *Powders and Grains*, edited by J. Biarez and R. Gourves (Balkema, Rotterdam, 1989), pp. 331-38.
107. M. R. Kuhn and J. K. Mitchell, "The Modeling of Soil Creep with the Discreet Element Method," *Proc. First U.S. Conference on Discreet Element Methods,"* edited by G. W. Mustoe, M. Hendricksen, and H. P. Huttelmaier (CSM Press, Golden, Colorado, 1989).
108. W.-H. Shih, W. Y. Shih, S. I. Kim, J. Liu, and I. A. Aksay, "Scaling Behavior of the Elastic Properties of Colloidal Gels," *Phys. Rev. A.,* **42** (8) 4772-79 (1990).
109. W. Y. Shih, I. A. Aksay, and R. Kikuchi, "Reversible Growth Model: Cluster-Cluster Aggregation with Finite Binding Energies," *Phys. Rev. A.,* **36** (10) 5015-19 (1987).
110. F. M. Tiller and C. D. Tsai, "Theory of Filtration of Ceramics: I, Slip Casting," *J. Am. Ceram. Soc.* **69** (12) 882-87 (1986).
111. J. Zheng, W. B. Carlson, and J. S. Reed, "Flow Mechanics on Extrusion through a Square-Entry Die," *J. Am. Ceram. Soc.* **75** (11) 3011-16 (1992).
112. R. P. Behringer and J. T. Jenkins, editors, *Powder and Grains 97* (Balkema, Rotterdam, 1997).

Suspension and Green-Body Characterization

ULTRASONIC WAVE INTERACTIONS WITH COLLOIDAL MIXTURES

Richard Challis and John Tebbutt
Ultrasonics and Digital Signal Processing Laboratory
Keele University
Staffordshire
ST5 5BG UK

INTRODUCTION

A colloid is a mixture of materials in which small particles, which may be solid or liquid, are suspended in a liquid. It is generally accepted that 'small particles' are of a size such that they would be subject to Brownian motion, with diameters between 0.1 μm and 1 μm. Very many commercially significant materials are either colloids, or will have passed through a colloidal state during processing. For many reasons which include product quality and process economy there is a need to determine the physical state of colloids both in the laboratory and at plant level. What is required here are reliable estimates of particle concentration, and particle size and its distribution. In addition, it may be necessary to gain measures of flocculation and/or network formation between particles. These are difficult to define in a general way, since flocs and networks can take many forms. There are many techniques that have potential for the characterisation of colloids, examples being optical scattering or turbidity measures, sedimentation tests, neutron and x-ray scattering, electric conductivity and/or permittivity, electroacoustic methods, and acoustic (ultrasonic) methods alone. Ultrasonic methods are presently attracting considerable interest since they complement other possibilities, are *potentially* rugged and cheap, and can be used on mixtures of high concentration that would be too opaque for optical methods. This paper provides a brief summary of the measurement techniques involved and then reviews the physics of acoustic interactions with colloids, and some theoretical formulations, with examples of results of recent laboratory studies.

SUMMARY OF ULTRASONIC TECHNIQUES

The majority of techniques involve propagating near-plane compression waves across a test medium in order to estimate the wave absorption coefficient $\alpha(f)$ (Nepers m^{-1}) and phase velocity $c(f)$ in the test medium as functions of frequency, f, in the range 1 MHz $< f <$ 100 MHz. Both $\alpha(f)$ and $c(f)$ are complicated functions of particle size and concentration as well as the physico-thermal properties of all constituents of the mixture. The physics of these relationships is reasonably well-established where the particles remain separated, but is not yet significantly developed where there is flocculation and/or network formation. Ultrasonic characterisation of colloids generally involves the fitting of a theoretical propagation model to measured data to yield estimates of particle size and concentration. It is to be noted that a measure of α and c at a single spot frequency may not correspond to a unique combination of concentration and particle size, let alone size distribution. What is required is absorption and velocity spectra measured over wide frequency bandwidths. The physical means of achieving this may take many forms. Transducers for the transmission and reception of ultrasonic waves may operate in pairs, one transmitting and one receiving, or may operate singly where the same device is used for both transmission and reception - the pulse-echo configuration. The operating bandwidth of transducers is generally of the same order as the centre frequency of their main resonance, and so multiple transducers may be required to achieve required operating bandwidths. The propagation distance between transducers may be fixed or variable, and many different geometries are possible for acoustic test cells. Signals transmitted between transducers may take many forms: a series of single frequency measurements, short pulse measurements combined with Fourier analysis(1) which yields $\alpha(f)$ and $c(f)$ from modulus and phase spectra respectively, or pseudo-random signal schemes(2) where low signal amplitudes are required in situations of fire or explosion hazard.

PHYSICS OF ACOUSTIC INTERACTIONS WITH COLLOIDS

There are seven *principal* effects associated with acoustic wave interactions with colloids, all of which contribute to $\alpha(f)$ and $c(f)$:

(i) The wave may be attenuated by viscosity, and/or chemical relaxation in the continuous phase.

(ii) The wave "spends some time" in the suspended particles and will exhibit modified attenuation and phase velocity if the properties of the two phases differ.

(iii) Where there is a contrast between the densities of the two phases, the particles will undergo oscillatory motion with respect to the continuous phase in response to oscillatory pressure gradients associated with the wave. Frictional losses between the phases and dipole-form reradiation from the moving particle result. These are known as viscoinertial phenomena.

(iv) Where there is a contrast between the compressibilities of the particles and the continuous phase then the particles will, in response to excitation by the passing wave, undergo a 'breathing' motion with respect to the continuous phase. This brings about monopole-form reradiation.

(v) There will be coupling between pressure cycles in the wave and local temperatures in the two phases. Significant temperature gradients can lead to heat flux between the phases.

(vi) If the local temperature changes combined with the thermal expansivities of the two phases lead to differential expansion between the two phases in response to the passing wave, then the particle will reradiate as a monopole.

(vii) At wave frequencies that are high enough to excite resonances in the particles, complex reradiation patterns develop around the particles, and these remove energy from the principal axis of propagation. This corresponds to quadripole or higher order reradiation.

THEORETICAL FORMULATIONS

Interest in acoustic interactions with particulate mixtures goes back around 100 years, and the literature is correspondingly voluminous. The choice and validity of any given theory depends on the strengths of the contrasts in physical properties between phases that underpin the seven phenomena outlined above. Due to constraints of space a full review is not possible here; we take an engineer's view and focus our discussion on those theories that are in current use, or are most likely to contribute to the development of instruments for colloidal characterisation. In situations where all contrasts between phases, except compressibility, are low, simple effective medium concepts may be appropriate(3). Where the principal contrast is density, then the hydrodynamic model of Harker and Temple(4) may be employed. The model of greatest generality includes scattering, and is generally attributed to Epstein and Carhart(5) and Allegra and Hawley(6).

Simple Equivalent Medium Theory(3)

Compression wave propagation speed is predicted on the basis of effective density ρ_{eff} and effective compressibility β_{eff} in the medium, thus

$$c_0 = \frac{1}{\sqrt{\rho_{eff}\,\beta_{eff}}} \tag{1}$$

where

$$\rho_{eff} = \phi\rho_D + (1 - \phi)\rho_C \tag{2}$$

$$\beta_{eff} = \phi\beta_D + (1 - \phi)\beta_C \tag{3}$$

ϕ is the volume fraction of the dispersed phase and subscripts D and C refer to dispersed and continuous phases respectively. The theory can only be considered in the long wavelength limit and gives unrealistic results as contrasts between thermal properties of the phases and/or densities increase. However, there are situations in which it is applicable, in that it represents a lower sound velocity limit in dilute suspensions. Urick and Ament[7] later published a more comprehensive formulation which included primary scattering and viscosity but which was applicable to very weak scatterers in very dilute systems. It extended the concepts to the calculation of attenuation but has found little application in practice.

A Hydrodynamical Theory

Harker and Temple[4] proposed this theory which is based on four differential equations - continuity for each phase, an equation expressing drag on one phase by the other, and conservation of momentum for the mixture. In forming the wave equation the rate of change of momentum was related to local pressure gradients, a Stokes drag, and an extra force that would be required to accelerate the *induced mass* of the particle. This latter quantity depends on the particle concentration and shape and accounts for fluid carried with the particle as it moves. The formulation used the effective density and compressibility as given by eqns 2 and 3, and an effective viscosity, η_{eff}, intended to account for interactions between moving particles, thus

$$\eta_{eff} = \eta(1 + 2.5\phi + 7.35\phi^2 \ldots) \qquad \textbf{(4)}$$

η is the viscosity of the pure continuous phase. Coupling of the motions of the two phases was incorporated in a frequency dependent term $M(\omega,\phi,R,\eta_{eff})$, where R is the particle radius and $\omega = 2\pi f$. Their complex wave-number $k = \omega/c(\omega) + i\alpha(\omega)$, came out to be

$$k^2 = \beta_{eff}\,\omega^2\left[\frac{\rho_c[\rho_D + (\phi\rho_D + (1 - \phi)\rho_c)M(\omega)]}{(\rho_c + (1 - \phi)\rho_D) + \rho_c M(\omega)}\right] \qquad \textbf{(5)}$$

This theory has recently been extended by Evans and Attenborough[8] to include thermal transport between phases, but comprehensive comparisons with other approaches have not yet been completed.

"Comprehensive" Theories

A significant step forward in theoretical formulation occurred with the work of Epstein and Carhart[5], later extended by Allegra and Hawley[6]. These works are generally recognised as a cornerstone in theoretical development and have come to be known as the ECAH formulation; it is used, sometimes with significant extensions, by many research workers worldwide, and recently it has formed the basis for particle sizing instrumentation[9]. Because of its growing significance it will be reviewed here in more detail than other formulations. The theory assumes that suspensions were relatively dilute (see below), that there were no interactions between particles, and that the effects of individual particles were additive. It included particle resonance, scattering, and a rigorous description of thermal transport and viscous mechanisms. Indeed, if a complex wavenumber for the continuous phase is included, the formulation incorporates all of the seven effects listed earlier.

It is assumed that a plane compression wave incident on a spherical particle results in compression waves scattered from the particle and refracted into the particle; in addition, mode conversion at the particle surface gives rise to thermal waves and transverse waves being scattered away from and into the particle (fig 1); the transverse waves were evanescent in liquid media.

Figure 1 Mode converted waves at particle surfaces after Epstein and Carhart[5] and Allegra and Hawley[6] models

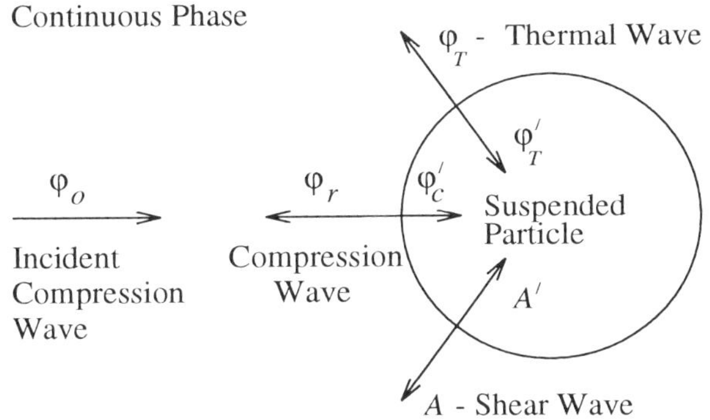

Continuity at the particle surface for stress, displacement (or velocity), temperature and heat flux gave a 6x6 set of equations for potentials for the three types of wave, inside and outside the particle. Of interest is the scattered compression wave field since this is (indirectly) observed through measurements of attenuation and phase velocity in the bulk medium. The relevant potential is

$$\phi_r = \sum_{n=0}^{\infty} i^n (2n+1)\ A_n h_n(k_c r) P_n(\cos\theta) \qquad \textbf{(6)}$$

Here A_n are partial wave coefficients that link the scattered field to the basic physical phenomena at the particle boundary; h_n is the spherical Hankel function, k_c the compression wavenumber and r radial distance from the particle centre. The angular distribution of the scattered field is given by P_n, the ordinary Legendre polynomial of order n. The six boundary equations are not easily soluble analytically, and many approximations exist for various contrasts between the physical properties of the particles and continuous phase, generally in the long wavelength limit. Inversion of the boundary equations to obtain A_n by means of a computer is fraught with problems of underflow and overflow associated, at least in part, with the need to evaluate Bessel and Hankel functions of small and large order and with small and large complex arguments. However, computer solutions in lieu of analytical approximations are attractive on account of their potential for generality in relation to the many different combinations of contrast between the physical properties of both phases of the mixture. We have made considerable inroads into this problem (hitherto unpublished) and calculations given later result from these methods.

The coefficient A_0 incorporates many physical effects that are dominant when there is a low density contrast between the phases - emulsions or dispersed polymers for example. The most basic effect is the difference between the compression wave attenuations and velocities between the phases. It also incorporates the difference in volume change between the particle and the surrounding fluid, due to different compressibilities, in response to pressure fluctuations in the wave, leading to monopole reradiation. There are two thermal contributions to A_0: where coupling between pressure cycles in the wave and local temperatures on either side of the particle boundary lead to a temperature differential, heat flux between the phases occurs and this may not remain in phase with the wave, the result being increased attenuation. Finally, thermal scattering, again monopolar, contributes to A_0 when the thermal expansion coefficients of the particles and the dispersed phase are significantly different. The coefficient A_1 includes the effects of motions of the particles with respect to the continuous phase which result in viscous drag losses and dipolar reradiation; these effects become dominant where the difference in densities between the phases is the principal contrast. The quadripole ($n = 2$) and higher order A_n coefficients correspond to increasingly complex particle resonances, which occur as wavelengths of various wave modes diminish to become comparable with particle dimension. Whilst they become increasingly difficult to calculate as order and frequency increase, their contributions may become significant in recently developed instrumentation aimed at particle sizing by means of ultrasound. The A_n coefficients are also functions of frequency and particle size. They describe the wave interaction with a single particle, and the effect of the scattered waves on the wavenumber of the bulk mixture is achieved by summing over all particles in the medium. This is generally done under the assumption of a homogeneous random distribution of non-interacting particles in the long-wavelength far-field limit, the result being the complex wavenumber for the mixture β:

$$\left(\frac{\beta}{k_c}\right)^2 = 1 + \frac{3\phi}{ik_c^3R^3}\sum_{n=0}^{\infty}(2n+1)A_n \tag{7}$$

where

$$\beta = \frac{\omega}{c(\omega)} + i\alpha(\omega) \tag{8}$$

When the RHS of eqn 7 is close to unity, β is approximated easily; for higher values then calculation by de Moivre's theorem is appropriate. Extensions to eqn 7 exist which approximate the effects of multiple scattering[10,11], and distributions of particle radius R can be approximated by further summations over $\phi(R)$.

COMPARISON OF THEORIES

We have carried out a comprehensive study[12] of the theories of Urick[3], Urick and Ament[7], Harker and Temple[4], and Epstein and Carhart[5] and Allegra and Hawley[6], comparing the expected frequency dependent absorption and phase velocity between theories, and with experiment. Experimental results were obtained using a short pulse transmission spectrometer[1] based on thick transducers set at fixed range, and which yielded $\alpha(f)$ and $c(f)$ over frequency ranges that depended on the signal to noise conditions resulting from the actual values of $\alpha(f)$; the maximum range for measurements was 1 MHz to 70 MHz.

The aim in this comparison was to challenge the theories with mixtures whose physical contrasts fall into four categories, Table I.

Where appropriate the comparison involved calculations based on multiple scattering concepts[10,11] for higher contrasts and concentrations, and incorporated particle size polydispersity (TiO_2).

Table I. Dispersed mixture types, showing principal physical contrasts

Mixture Type	Contrast Between Phases Density	Thermal Props	Experimental Mixtures
A	small	large	Polystyrene/water Hexadecane/water Silica/water
B	large	small	TiO_2/water
C	large	large	Iron/water Iron/silicone oil
D	small	small	No sample

RESULTS AND DISCUSSION

Due to constraints of space only a small sample of our results is presented in detail here. Figs 2(a)-(c) show the comparisons for polystyrene in water, silica in water and TiO_2 in water, respectively. The polystyrene result indicates that for type A mixtures the ECAH model alone is applicable; this is expected due to the dominance of thermal phenomena over viscoinertial effects. The results for silica show that ECAH is the most applicable, for the same reason, but that models based on viscoinertial effects do better than they did in the case of polystyrene; this is expected due to the higher density contrast compared to polystyrene. The TiO_2 result indicates convergence of the models, again expected, but this time due to the dominance of viscoinertial effects; the contribution of the A_1 coefficient to absorption in the ECAH formulation is around sixty times that of A_0 or A_2 at 10 MHz for this mixture.

Results for phase velocity, not shown here, indicated that, in most cases, the models of Urick, Urick and Ament, and Harker and Temple, tended to be in agreement over the frequency ranges used for experiment, They showed convergence to the Urick frequency - independent value at low frequency for dilute suspensions. However, for type A mixtures the ECAH model gave close agreement with experiment. The other models, whilst giving consistent results with each other, gave poor agreement with experiment.

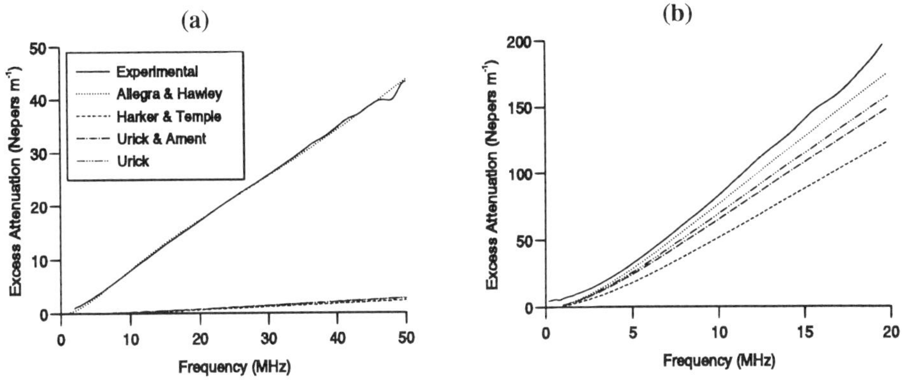

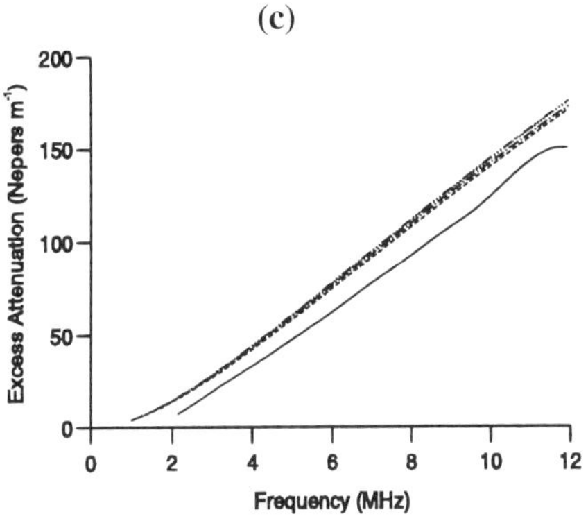

Figure 2 Comparisons between experimentally measured absorption spectra and the simulations of Allegra and Hawley[6], Harker and Temple[4], Urick and Ament[7] and Urick[3] for (a) 200 nm diameter polystyrene particles, 5.6% v/v in water; (b) 320 nm diameter silica particles, 8% v/v in water; and (c) 260 nm (Stokes) diameter titanium dioxide particles, 1% v/v in water. NOTE: differing abscissae.

Experimental data for polydisperse mixtures was compared to the ECAH model with wavenumber calculated on the basis of the median particle size using eqn 7, and also with eqn 7 adapted for polydispersity. For a hexadecane (median particle diameter 1 μm) in water emulsion the two formulations agreed with experiment up to around 10 MHz, but there was divergence between the models at higher frequencies (10% at 35 MHz), the polydisperse model being the best predictor. Similar behaviour was observed for the TiO_2 suspension (median particle diameter 260 nm) but significant divergence began at 4 MHz, rising to 25% at 13 MHz.

For more concentrated suspensions the predictors based on eqn 7 were compared with two alternatives that incorporate multiple scattering[10,11]. For a 30% v/v suspension of polystyrene particles (diameter 270 nm), the three approaches gave near identical results, generally agreeing with experiment, as would be expected due to the dominance of thermal contrast. For mixtures involving higher density and elastic contrasts the three methods of calculation gave significantly different results for both attenuation and velocity, but we do not yet have experimental data for comparison.

CONCLUDING REMARKS

It is clear that the ECAH model when used to predict absorption and phase velocity in liquid-based suspensions has the greatest generality across different physical contrasts between the dispersed and continuous phases. However, for dispersions of relatively dense particles, other models may have some application. In order to apply the ECAH model to a wide range of mixtures, analytical approximations are to be avoided, since so many different approaches would be required. This implies that effort applied to stabilisation of computer solutions to the diffraction problem would be worthwhile.

To preserve generality over mixtures that are polydisperse and/or are of high concentration, then calculations of the mixture wavenumber will require extensions to eqn 7. It is to be noted here that the ECAH model is <u>not</u> self-consistent and that there will be a limit to the dispersed phase concentration to which it is applicable, although this may be greater than 30% v/v in some cases. Recent developments in instrumentation indicate that the ECAH model will be used as the basis for the measurement of dispersed phase particle size by means of ultrasonic absorption and velocity measurements. It is to be noted that whilst the forward prediction of $\alpha(f)$ and $c(f)$ on the basis of particle size and concentration (and a knowledge of physical properties) is reasonably tractable, the inverse calculation to obtain particle parameters is not trivial and is potentially ill-conditioned.

REFERENCES

[1] R.E. Challis et al, *J. Acoust. Soc. Am*, **90** 730-740 (1992).

[2] R.E. Challis, "Developments in Acoustics and Ultrasonics", 67-88, Edited by M.J.M. Povey and D.J. McClements, IOPP, UK, 1992.

[3] R.J. Urick, *J. Appl. Phys.*, **18** 983-987 (1947).

[4] A.H. Harker and J.A.G. Temple, *J. Phys. D*, **21** 1576-1588 (1988).

[5] P.S. Epstein and R.R. Carhart, *J. Acoust. Soc. Am.*, **25** 553-565 (1953).

[6] J.R. Allegra and S.A. Hawley, *J. Acoust. Soc. Am.*, **51** 1545-1564 (1972).

[7] R.J. Urick and W.S. Ament, *J. Acoust. Soc. Am.*, **21** 115-119 (1949).

[8] J.M. Evans and K. Attenborough, *Proc. 15th Int. Con. Acoustics*, June 1995.

[9] Malvern Instruments Ltd, UK. Personal Communication, 1996.

[10] P.C. Waterman and R. Truell, *J. Math. Phys.*, **2** 512-537 (1961).

[11] P. Lloyd and M.V. Berry, *Proc. Phys. Soc.*, **91** 678-688 (1967).

[12] J.S. Tebbutt, PhD Thesis (1996) Keele University, UK.

ULTRASONIC ARRAY SENSING FOR IN-SITU NDE IN EXTRUSION PROCESS OF GREEN CERAMICS

T.Watanabe, T.Aizawa, and J.Kihara

Department of Metallurgy, University of Tokyo, Tokyo, 113, Japan
+81-3-3812-2111

ABSTRACT

On-process quantitative nondestructive evaluation is necessary to promote more advanced extrusion forming of green ceramic. New ultrasonic sensing is proposed both for mechanical characterization and nondestructive inspection of extruded green body. A through-transmission transducer system with the fixed frequency of 10MHz was housed in the forward extrusion former to make on-process measurement both in the normal and the 45°-inclined directions to a cavity channel. The acoustic velocity in the specimen is 776 m/s by off-line measurement; it varies by the change of mechanical properties due to high pressure in forming. The measured ultrasonic signals were found to have sufficient sensitivity and S/N ratio to detect the artificial defects only by comparison of the peak intensity to the full-dense greens. Furthermore, the size and position of pores inside the cavity channel can be predicted even in the forming.

INTRODUCTION

In the literature, quantitative nondestructive evaluation (QNDE) methods have been studied for quality control of ceramic materials and products: the X-ray methods[1,2] and the ultrasonic testing [3]. In the former method, since X-ray penetrates the specimen up to 10 to 100μm in thickness, the obtained information is limited to the mechanical conditions near the surface.

Most research in the latter method are off-process one; the testing was not done in the forming process. Especially in the successive forming process, it is difficult to pick up the sample for the check. Hence a testing method in the forming process is needed. However a lot of analysis can be utilized for the evaluation in off-process methods, in most ultrasonic research in the forming

process, it is utilized only whether the ultrasonic wave can be received through the specimen or not[4].
We developed and examined the experimental instrument to utilize the velocity and attenuation for the evaluation and to aim the phase analysis and on-process quantitative nondestructive evaluation in extrusion process.

EXPERIMENTAL METHOD

Our developed instrument for the extrusion forming is depicted in **Fig. 1**. Here, the container, the mouthpiece, the base and the sensor attachment are

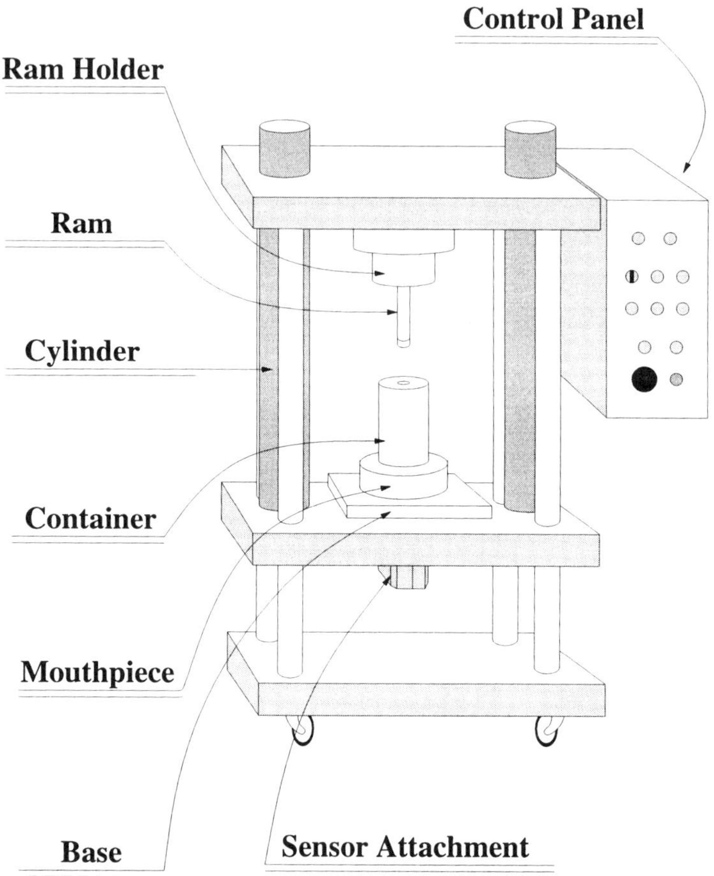

Fig. 1: Diagram of Experimental Instrument

carried up by the cylinder, and then the green ceramics situated in the container was extruded by the ram. In this process, the maximum load is designed as 10 tons.

In the present paper, the green ceramic consisted of the PZT powder, binder, plasticible material, dispersant agent, and water.

The schematic view of the extrusion forming process in the present instrument is shown in **Fig. 2**. The diagram of the mouthpiece are shown in **Fig. 3**. In

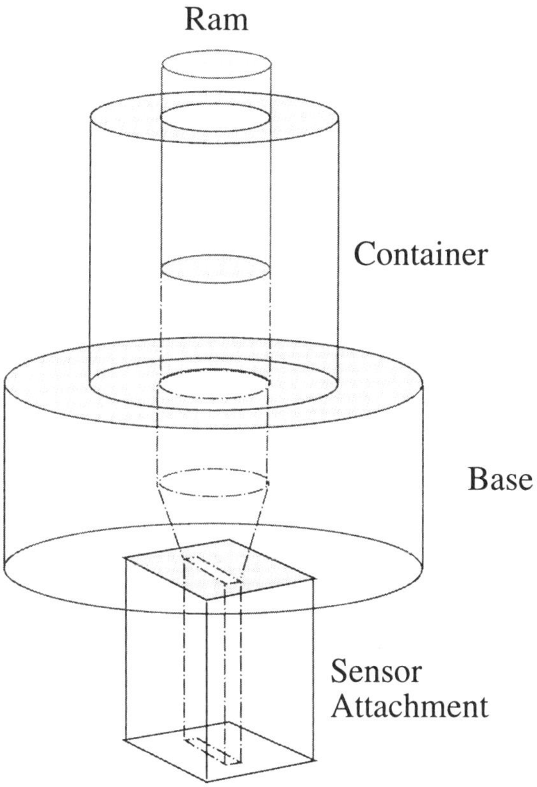

Fig. 2: Diagram of Experimental Forming Process

the present study, two type of the mouthpieces are provided; the cross section of the cavity channels are 1 mm × 10mm and 5 mm × 10 mm. The extruded green ceramics are carried through the mouthpiece and formed as the products like a belt. And then the products are flowed through the cavity of the sensor attachment, and measured the kinetic properties with the ultrasonic sensors.

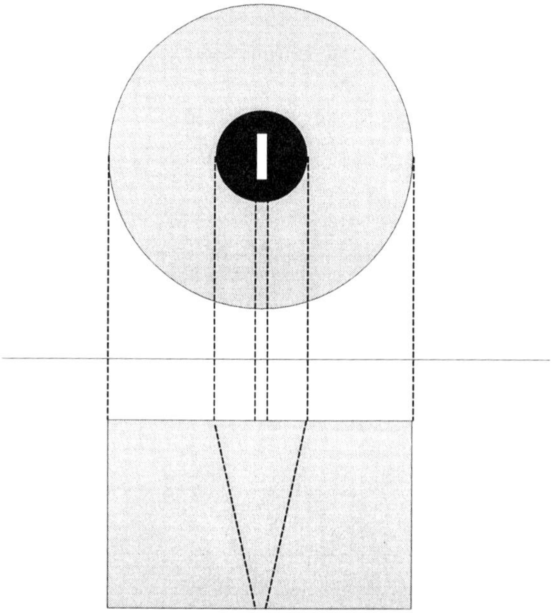

Fig. 3: Diagram of the Mouthpiece

EXPERIMENTAL TRANSDUCERS AND THE SENSOR ATTACHMENT

Various transducers were listed as the candidate for the sensors in the present experimental instrument. In this instrument, following points are required for the sensor,

1. Improving the coupling of the impedance between the transducer and the sensor attachment,

2. Increasing the pulse amplitude to be supplied for the transducers as much as possible.

Examined some types of transducers[5,6], we prepared the PZT type transducer as shown **Photo. 1** as the sensor. It has a diameter of 10 mm and one of 7 mm as the effective transmitting area, This transducer is designed to have the resonant frequency in the range of 10 MHz.
For obtaining the clear transmitted signal, since there is no multi–reflected signal between the transducer and the specimen in the transmitted signal, it is the best that the transducers are directly contacted to the specimen. Nevertheless, during the forming process, the specimen is loaded the pressure

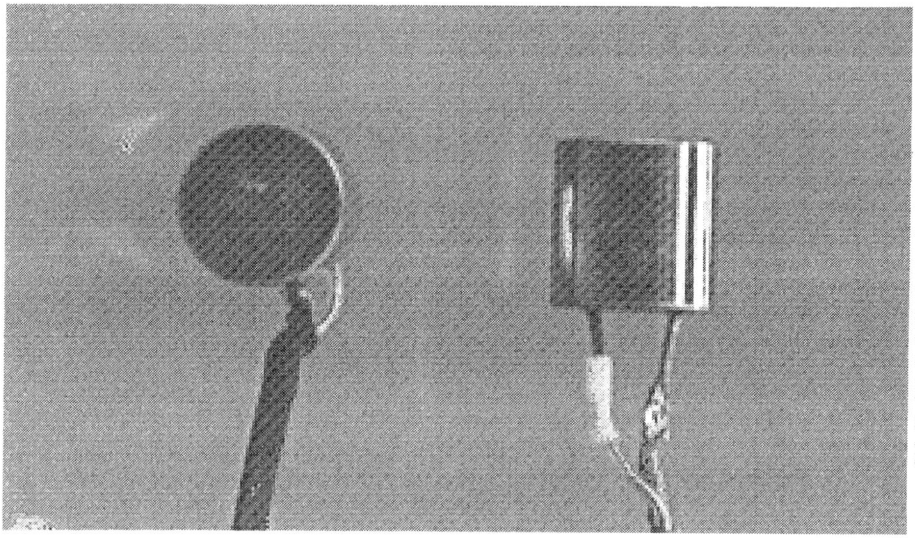

Photo. 1: Ultrasonic Transducer

caused from the load of 10 tons. Then, it is not desirable that the pressure loads the transmitting area of the transducers, because there is possibility that the transducer is deformed or moves away in the sensor attachment by the pressure. Hence to examine which system is more suitable for this evaluating system, we measured the specimen with two types of alignment of the transducers as shown in **Fig. 4(A)**, **(B)**.

On the other hand, to assemble the arrayed sensors in the future, the measurement in a diagonal direction must be achieved. Then in the present research, a pair of transducers were fixed as shown in **Fig. 4(C)** and we examined the obtained wave profile.

Here, in the direct contact type(**Fig. 4(A)**), the transducers are fixed by adhesive agent (epoxy resin), and in the others (**Fig. 4(B)**, **(C)**) are fixed with the way shown in **Fig. 5**. In the present method, for the purpose of improving coupling of the impedance between the transmitting area of the transducer and the tool steel, the thin layer of the ethanol between them is added.

EXPERIMENTAL DATA ACQUISITION SYSTEM

The frequency, the amplitude of supplied voltage, the gain level and the low and high pass filter are adjusted by RAM–10000 (RITEC Co.). The obtained transmitted signal is shown on the oscilloscope and translated to the personal

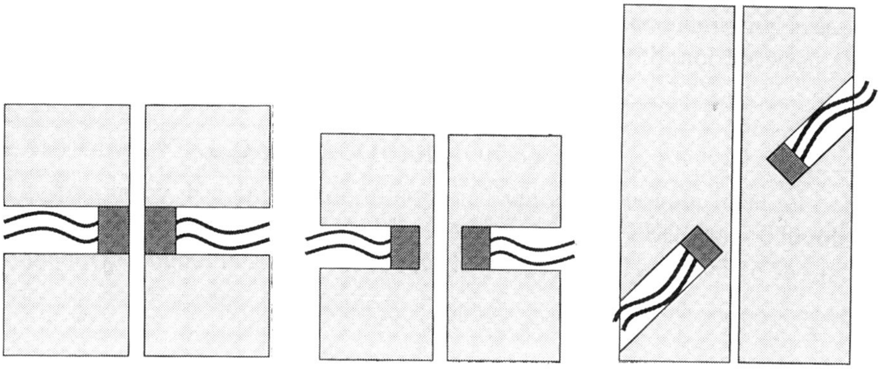

(A) Direct Contact (B) Indirect Contact (C) Direction of 45°

Fig. 4: Diagram of the Sensor Attachment

computer as the text format. The experimental system is shown in **Fig. 6**.

EXPERIMENTAL RESULT

TRANSMITTED WAVE PROFILE

In the experiment with the direct contact type, several times, the transducer was pushed back by the pressure from the specimen. But the surface of the transducers was not injured by the friction of the specimen. It was found that fixing the transducers with a back–up which prevents the transducer from moving back, the measurement can be done with the direct contact type attachment.

The transmitted wave profiles with each type of the sensor attachment are shown in **Fig. 7**. In the transmitted wave profile with the indirect contact attachment as shown in **Fig. 7(B)**, in comparison with **Fig. 7(A)**, peaks of multi–reflection in the space of the tool steel of the sensor attachment between the transducer and the specimen are observed.

However the first transmitted pulse rose up sharply, then the measurement of the ultrasonic velocity in the specimen is not influenced by the multi–reflected peaks. With this wave profile, the velocity in the specimen is measured as 776 m/s. Furthermore the multi–reflected peaks stand periodically, it means that the signal is only a superposition of the waves of same profiles, and then the analysis of the frequency and the phase can be done easily.

Secondly, the amplitude of the transmitted signal with the indirect contact attachment as shown in **Fig. 7(B)** was smaller than the one with the attachment

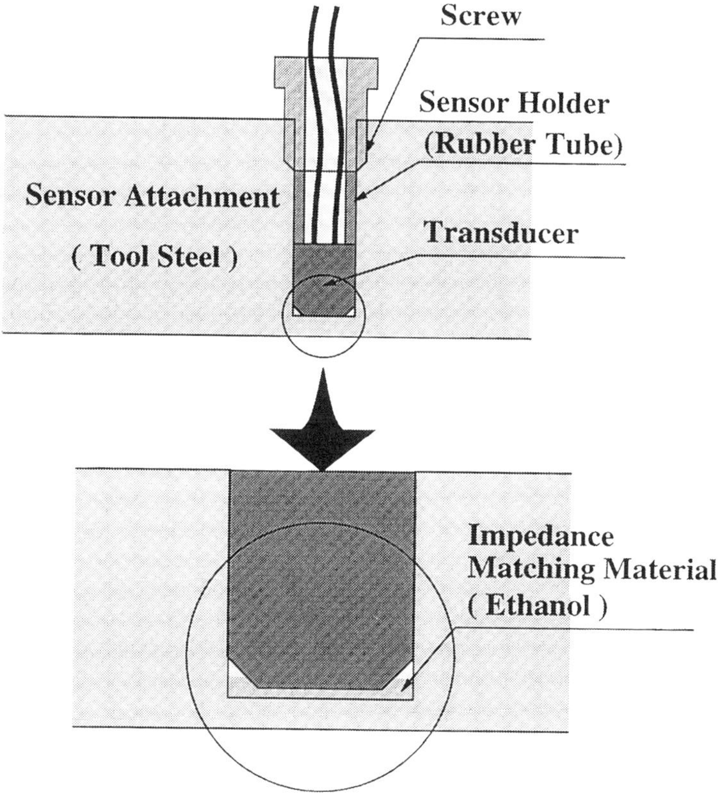

Fig. 5: Schematic View of the Fixing Method

in diagonal direction as shown in **Fig. 7(C)**. It is caused that the transmitted signal reflected much at the interface between the tool steel of the sensor attachment and the specimen. Nevertheless the first transmitted peak rose up sharply, and it can be confirmed that the transit time corresponds to the distance between the transmitted area of transducers. From the above, it is found that the specimen will be able to be measured with the arrayed sensors.

FAULT DETECTION

In the previous section, the transmitted wave profile was measured obviously with the direct contact sensor attachment. Hence we checked the detectability of the faults embedded in the specimen with this method. First of all by including sufficient amount of air into the specimen, we produced the product

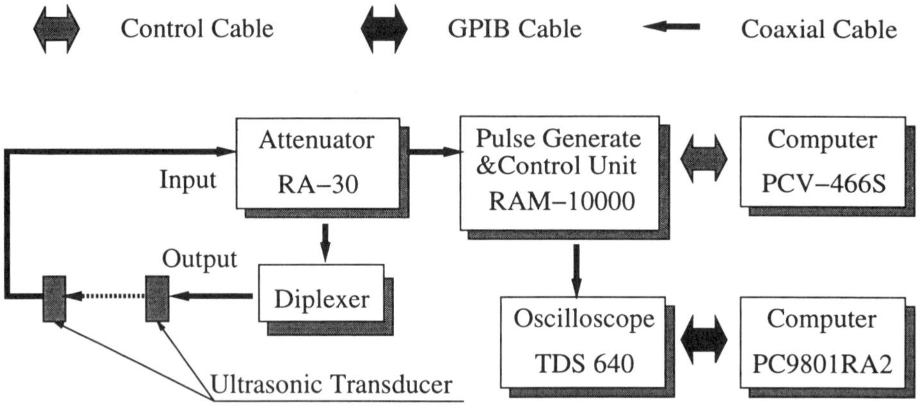

Fig. 6: Experiment System

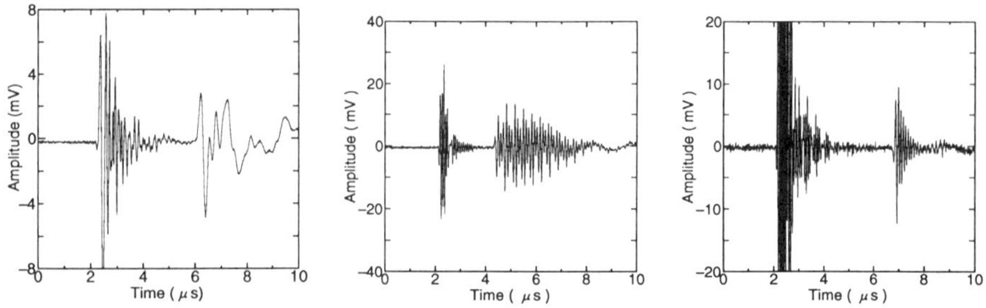

(A) Direct Contact (B) Indirect Contact (C) Direction of 45°

Fig. 7: Transmitted Wave Profile

with artificial faults. Both the product with and without faults are shown in **Photo. 2**. In comparison with the successful (faults free) product, the faults caused by the pressed air which expands when the forming pressure is unloaded can be observed.

Secondly the transmitted wave profiles at the part of specimen with and without faults in **Fig. 8**, respectively. It can be observed that the amplitude of the received peak with the faults is smaller than the one without faults.

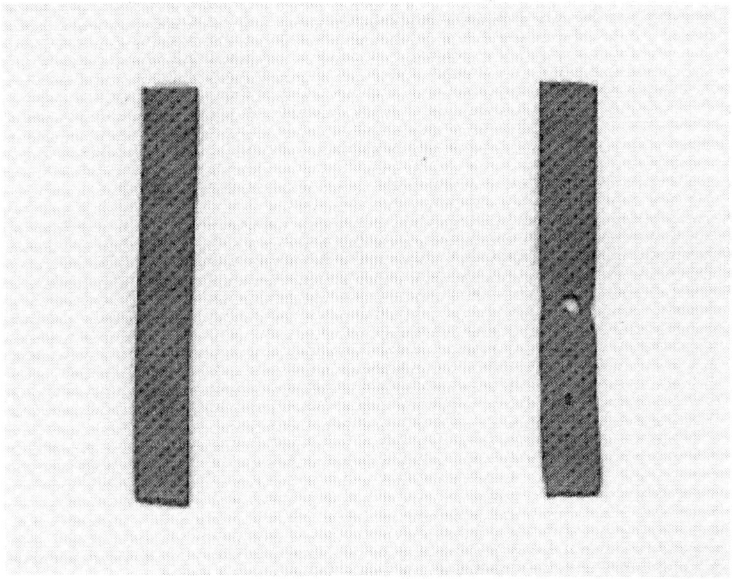

(A) Successful (B) With Faults

Photo. 2: Products

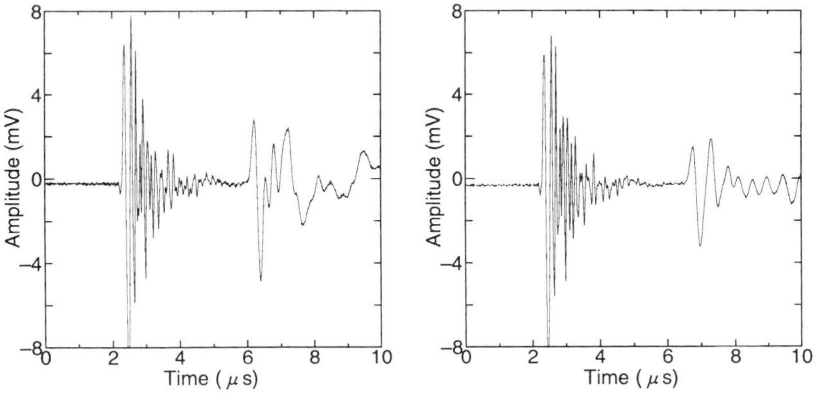

(A) Successful (B) With Faults

Fig. 8: Transmitted Wave Profiles

DISCUSSION

In the transmitted wave profile, it can be observed that the amplitude of the received signal through the specimen with the faults is smaller than the one without faults. It is known that the ultrasonic ray incident on the pore embedded in the material is divided the reflected ray and the creeping ray around the pore as shown in **Fig. 9**[7]. Hurst et al. reported that the velocity of the creeping ray was smaller than the one of the ultrasonic ray incident on the specimen. Hence, in the transmitted wave profile, the creeping ray appears

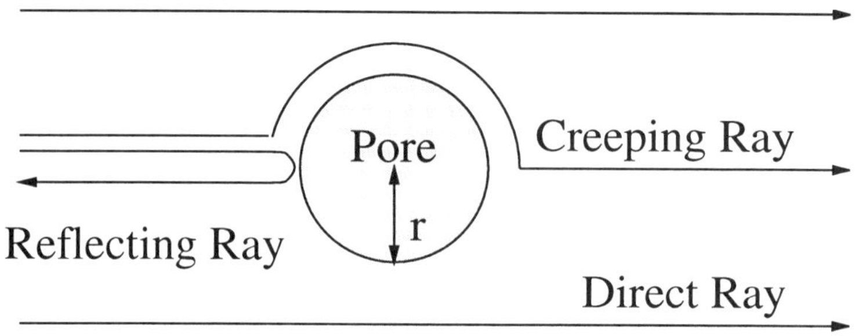

Fig. 9: The Effect of Pore on Transmitted Ray

as the delayed transmitted peak, but does not influence the amplitude of the first transmitted peak. Therefore the difference of the amplitude of the first transmitted peak is caused by the reflection at the interface between the pore and the specimen as shown in **Fig. 10**. With assumption that the ultrasonic ray reflected as shown in **Fig. 10**, the amplitude of the first transmitted peak is proportional to the projected area in the incident direction. And then assuming that the pore is sphere with radius of r, the following equation is held;

$$\frac{A_1}{A_0} = 1 - \frac{\pi r^2}{S} \ . \tag{1}$$

Here, the amplitude of the first transmitted peak of the specimen without faults, the one of the specimen with faults, and the effective transmitting area of the transducer are A_0, A_1, and S respectively.
If we take account of the diffraction of the ray, the distance between the pore and the transducer becomes the variable and it is difficult to decide the pore size. In this experiment, the thickness of the specimen is sufficiently smaller than the size of the transmitting area of the transducer, then we assumed that the ray transmits without diffraction in the **Eq. (1)**.

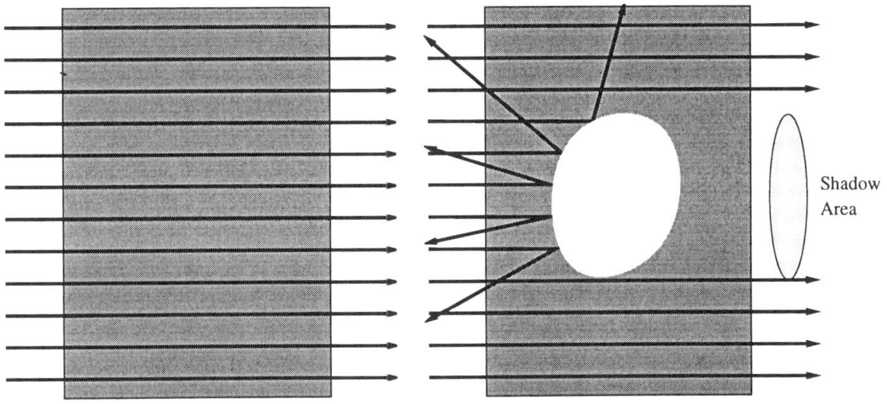

(A) Successful (B) With Faults

Fig. 10: Attenuation with a Fault

Accordingly it is found that by the monitoring of the amplitude of the first transmitted peak, the faults including invisible ones from the surface of the products can be detected quantitatively with the information of the size and the position in the longitudinal direction of the product, even in the forming,

CONCLUSION

With our developed instrument, we ensure followings in the extrusion forming process;

1. the transmitted wave profile can be obtained stably and clearly both in the normal and 45°–inclined direction to a cavity channel of the sensor attachment,
2. since the transmitted wave can be obtained in diagonal direction, the arrayed sensor will be able to be assembled,
3. the ultrasonic velocity can be obtained accurately with the clear transmitted wave profile,
4. measuring the relationship between the density of the specimen and the ultrasonic velocity in the specimen previously, it will be possible to evaluate the density of the green ceramic specimen quantitatively in the forming process,
5. the faults can be detected with the information of the size and the position by monitoring the amplitude of the first transmitted peak.

Furthermore, to be noted that obtaining the clear transmitted wave profile means that the analysis of the phase, which is related to the distribution of faults and binder, can be done.
Then the present method has a potential to be used not only to detect the faults, but also to evaluate the density of the products and the distribution of the binder or faults in the forming process.

REFERENCES

1) Yasushi Ymauchi et al. : Microtomography Using Conventional X-ray Sources : Nondestructive Characterization of Materials VI : 129–136 : Plenum Press, New York(1994)

2) J.Muller et al. : Quantitative Nondestructive Evaluation of Density of Green State Compressed Products : Nondestructive Characterization of Materials VI : 175–182 : Plenum Press, New York(1994)

3) Y.Higo et al. : Nondestructive Evaluation of Micromechanism of Deformation Process During Fatigue Testing on Polymer by ETFuM (Elastic–Wave Transfer Function Method) : Nondestructive Characterization of Materials VI : 223–230 : Plenum Press, New York(1994)

4) Derrick D. Hongerholt et al. : ULTRASONIC SENSORS FOR POWDER INJECTION MOLDING PROCESS MONITORING : Advances in Powder Metallurgy & Particulate Materials –1996 : Vol.5, 19-249 – 19-257, METAL POWDER INDUSTRIES FEDERATION(1996), USA

5) T.Watanabe et al. : On–process Quantitative Evaluation of Ceramic Green Body in Extrusion Process, : The Proceedings of the 46th Japanese Joint Conference for the Technology of Plasticity : 199–200, THE JAPAN SOCIETY FOR TECHNOLOGY OF PLASTICITY(1995–9),

6) T.Watanabe et al. : Effect of ultrasonic frequency on on–process quantitative evaluation of ceramic greens in extrusion process : THE PROCEEDINGS OF THE 1996 JAPANESE SPRING CONFERENCE FOR THE TECHNOLOGY OF PLASTICITY : 478–479, THE JAPAN SOCIETY FOR TECHNOLOGY OF PLASTICITY(1996–5)

7) Junji Takatsubo : The development of quantitative nondestructive evaluation system with AE technique : Ph.D. Paper at the University of Tokyo : the University of Tokyo, Tokyo

RECENT ADVANCES IN X-RAY CHARACTERIZATION OF COMPOSITES AND TAPE CAST CERAMICS

S. P. Huss,[1,2] C. H. Schilling,[1] J. N. Gray,[2,3] and C. W. Maranville[1,2]

[1] Ames Laboratory[+] and Department of Materials Science and Engineering
Iowa State University, Ames, Iowa 50010

[2] Center for Nondestructive Evaluation
Iowa State University, Ames, Iowa 50010

[3] Department of Mechanical Engineering
Iowa State University, Ames, Iowa 50010

ABSTRACT

We have recently developed two prototype x-ray techniques for the nondestructive evaluation of ceramic tapes and composites: (*i*) x-ray radiography of tape casting defects, including tape thickness variations, surface pits, cracks, pores, bubbles, and density gradients, and (*ii*) energy-dispersive absorption spectroscopy for simultaneous mapping of composition and porosity in ceramic composites. These methods can both function as two-dimensional area detectors (rather than conventional point- or line-detectors), providing rapid, statistical information on microstructure heterogeneities in flat sheets of ceramic material. Both techniques demonstrate the potential to serve as process development tools capable of inspecting both green and sintered materials.

I. BACKGROUND

Cracks and warpage that arise during drying, debinding, and sintering of ceramics are exacerbated by green-microstructure heterogeneities, e.g., agglomerates, packing-density gradients, composition gradients, isolated pores and bubbles, and so on. Earlier in this proceedings, we reviewed major problems associated with traditional methods of controlling these heterogeneities in ceramic

[+] Ames Laboratory is operated by Iowa State University under the contract number W-7405-eng-82 with the U.S. Department of Energy.

production: (*i*) they provide insufficient information on the spatial- and temporal-evolution of green-microstructure defects from one unit operation to the next, (*ii*) they usually involve slow and costly trial-and-error procedures that are off-line and destructive in nature, and (*iii*) they also have limited value when attempting to sort through and optimize so many different processing variables that typically affect microstructure-defect evolution in a given production operation.[1] To overcome these barriers, advances in nondestructive (NDE) instrumentation are needed, advances not only for on-line inspection purposes but also for off-line process development.

The goals of this paper are to briefly summarize two prototype techniques that we have recently developed for characterizing green and sintered ceramics. These techniques include: (*i*) x-ray radiography of tape casting defects, including tape thickness variations, surface pits, cracks, pores, bubbles, and density gradients, and (*ii*) energy-dispersive absorption spectroscopy for simultaneous mapping of composition and porosity in ceramic composites. In the following sections we discuss design details, performance characteristics, and relevant background information for each technique.

II. AREA DETECTION OF TAPE-CASTING DEFECTS

Tape casting of microelectronic substrates and capacitors is a prime example of a high-value-added, ceramic-manufacturing process where new, on-line methods of defect sensing and process control are needed. Ceramic tapes are typically characterized by off-line techniques, such as visual inspections, transmission optical microscopy, reflection optical microscopy, and single-point thickness measurements. There are a few examples of nondestructive, on-line inspection of ceramic tapes, including surface analysis by the use of optical microscopy, and thickness gauging using measurements of x-ray transmission with a solid-state point detector[2].

We took advantage of recent improvements in image processing and x-ray microfocus radiography to develop a prototype apparatus that would allow area detection of thickness variations, surface pits, pores, bubbles, and density gradients in tape cast ceramics[3,4]. The apparatus was designed to incorporate several benefits over traditional characterization routes: (*i*) the system functions as a two-dimensional area detector (rather than a conventional point- or line-detector), providing rapid, statistical information on two-dimensional maps of microstructure heterogeneities in ceramic tapes; (*ii*) it is capable of detecting small bubbles, pores and surface pits on the order of 10 microns in size; (*iii*) it can also detect tape thickness variations on the order of 10 microns. Using standard equipment, it is also amenable to scale up for quantitative process monitoring in an industrial setting.

The operating principle entails irradiating a ceramic tape with a microfocus x-ray generator and digitizing the projected image, followed by image processing. The projected image can be obtained using x-ray film or an amorphous silicon array detector. We developed an off-line, prototype technique, i.e., it was applied to stationary pieces of ceramic tape that were removed from a tape casting apparatus. Figure 1 shows a schematic illustration of a scaled-up, on-line system that could be developed based on our prototype. We can envision use of a scaled-up, on-line system to monitor tape defects in a moving tape as a function of time and position immediately after casting and also during subsequent debinding and sintering.

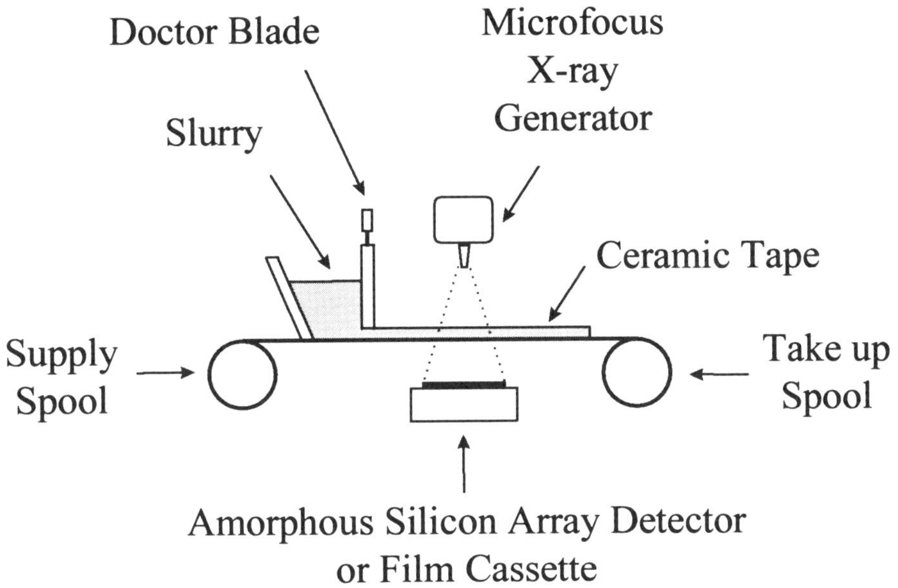

Figure 1. Schematic illustration of an on-line monitoring device for area detection of tape casting defects.

We recently benchmarked the performance of a film-based, prototype system in a study of tape-cast alumina[3,4]. In that study, our goal was to achieve maximum detectability of small pores, density gradients, and small tape-thickness variations through the proper selection of exposure parameters. Results of the radiography data were correlated with optical and scanning electron microscopy. First, we developed a protocol to quantitatively analyze image-blurring effects that are inherently produced by grains in the x-ray film. This protocol involved use of a charge-coupled-device (CCD) camera to digitize exposed pieces of x-ray film, followed by computerized image analysis. We also employed a bremsstrahlung x-ray beam having a maximum voltage of 30 kV. This relatively low voltage increased the detectability of small pores, density gradients, and small tape-thickness variations, as a result of the higher degree of attenuation that naturally occurs with low-energy photons.

Figure 2 illustrates the application of this prototype to a 200 μm-thick piece of a green tape that was cast using α-alumina powder (0.4 μm average particle size). The first thing to notice is the original radiograph (a) in Figure 2, which appears brighter when passing from the upper left to the lower right corner. This increase in brightness signals an increase in the tape thickness. Thickness variations can easily be determined by digitizing the radiograph and correlating the results with separate, digitized radiographs of controlled-thickness specimens. Also shown in Figure 2, the digitized image of radiograph (a) was normalized in order to mask the effects of tape-thickness variations and in turn improve the resolution of discreet, isolated features, including bubbles, pores, and surface pits. The first part of this normalization process involved calculating an average, background grayscale level in radiograph (a), which was then subtracted from all pixels in this image to produce the normalized image in radiograph (b). As shown in Figure 2, one-dimensional, grayscale profiles (line scans) are easily generated from the digitized images, profiles that can be useful in an industrial setting to analyze isolated defects. Grayscale levels in all pixels of this resulting image were then thresholded to allow viewing of pixel clusters that represented isolated regions where there was a loss of alumina (Figure 3). We then calculated a pixel cluster size distribution in Figure 3, which collectively represents thresholded data from radiographs of several different regions of the same alumina tape (not shown). This type of analysis illustrates the technique's potential for statistical process control.

Figure 4 illustrates another example of this technique which was used to study the effects of the average particle size of alumina on tape microstructures before and after sintering. As expected, the larger-sized powder results in larger-sized defects. Figure 4 illustrates how bubbles that are introduced in the green tape (0.4 micron particle size) are also visible in the radiograph after sintering.

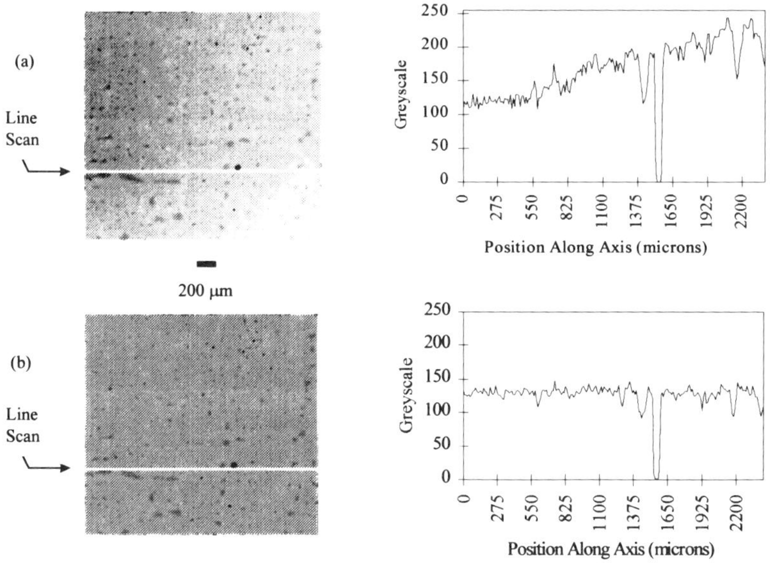

Figure 2. Radiographs (a) and (b) respectively show the same region of an alumina tape before and after background normalization. Line-scans along the same path in both radiographs (right) show how this normalization can enhance the detection of isolated spots where there is loss of alumina (pores, bubbles, or surface pits).

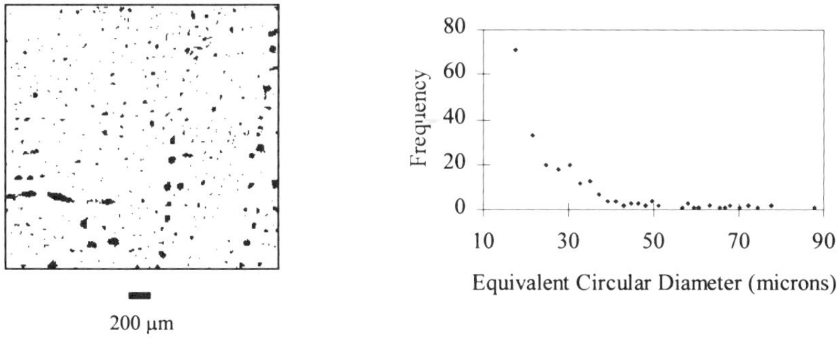

Figure 3. Radiograph (b) of Figure 2 is thresholded on the left. Black pixels indicate regions where there is a loss of alumina. A distribution of pixel-cluster sizes (right) was obtained by analyzing thresholded images from several radiographs of different regions of the same alumina tape.

Of course, one of the drawbacks of this method is that it does not reveal information on the location of a given defect in a direction normal to the surface of the tape. In other words, it is uncertain whether a given pixel cluster signals the occurrence of a surface pit or a sub-surface pore. We are currently exploring the use of stereography to overcome this problem.

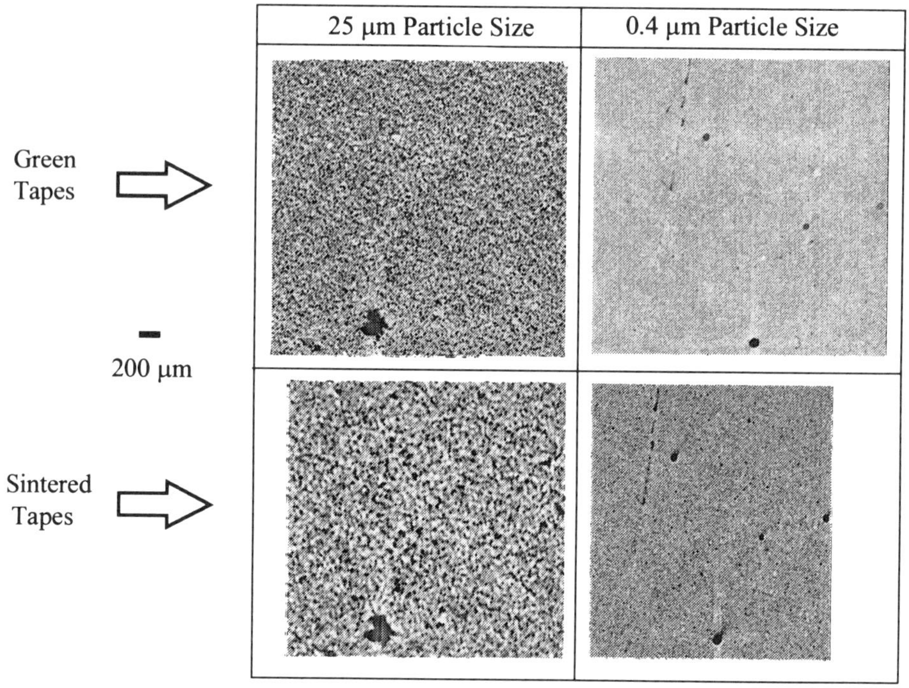

Figure 4. Radiographs of tape-cast alumina showing effects of the average powder size. Images corresponding to each powder size were obtained at approximately the same region of each tape before and after sintering to 1550 °C.

III. COMPOSITION & POROSITY MAPPING WITH X-RAY ABSORPTION SPECTROSCOPY

The Center for Nondestructive Evaluation (CNDE) at Iowa State University has developed an x-ray absorption spectroscopy technique allowing quantitative, simultaneous mapping of both the composition and porosity of a broad range of ceramic, metal, and polymer composite materials. Much like the radiographic technique described above for tape casting applications, this technique involves a two-dimensional projection of a flat test specimen. It uses a single, solid-state detector or an array of detectors to capture the transmitted intensity. Coupled with a multi-channel analyzer, this configuration measures the transmitted intensity as a function of the x-ray photon energy.

Our procedure takes advantage of the fact that each constituent phase of a composite material tends to exhibit a unique x-ray attenuation coefficient as a function of the photon energy. This can be understood by recognizing that the x-ray attenuation coefficient primarily describes the degree of photon scattering by electrons and by the photoelectric effect in an irradiated material. In turn, this coefficient strongly depends on the energy of the incident photons and the electron density (or mass number) of the irradiated material. In order to obtain reliable maps of composition and porosity, is essential to make an accurate determination of the x-ray attenuation coefficient as a function of the photon energy for each constituent phase in the composite. Theoretical calculations are useful in this regard, however, they require detailed information regarding the composition and mass density of each constituent phase[5]. Laboratory measurements of x-ray attenuation coefficients are easily performed by exposing calibration specimens of each constituent phase with white radiation and then measuring forward attenuation with an energy-discriminating detector. Calibration specimens can be in monolithic, fibrous, or powder forms[6]. Figure 5 illustrates a set of such measurements for α-alumina powder and SiC - based fiber.

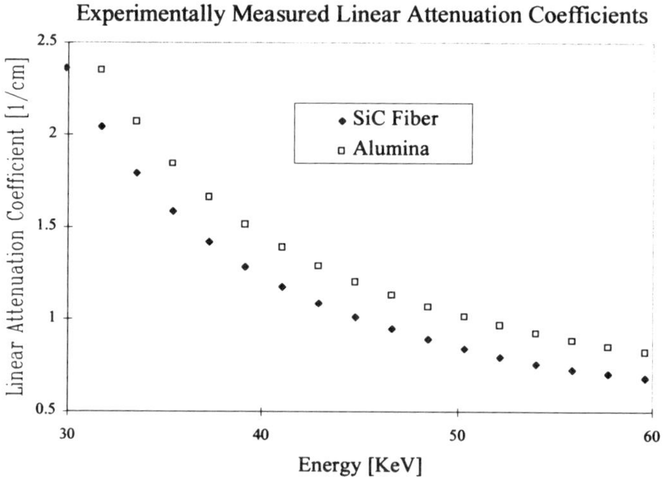

Figure 5. Experimentally-measured x-ray attenuation coefficients (μ) of alumina powder and SiC - based fiber.

Once the attenuation coefficient as a function of the photon energy is known for each constituent phase, the composite specimen is exposed to white radiation at various locations, and the incident and transmission spectra are recorded. Maps of composition and porosity are then calculated by substituting the incident and transmitted spectral intensities along with measured attenuation coefficients into the Beer-Lambert law and mass conservation equations:

$$I(E) = I_0(E)\exp\left[-\sum_i x_i \mu_i(E)\right] \tag{1}$$

In equation (1), E is the photon energy, $I(E)$ and $I_0(E)$ are the energy-dependent intensities of the transmitted and incident x-ray beams, respectively, x_i is the thickness of constituent i in the path of the x-ray beam, and μ_i is the linear x-ray attenuation coefficient of constituent i. For a two-constituent composite (e.g., alumina + SiC fiber), Equation (1) can be solved knowing x-ray attenuation coefficients and x-ray intensity data at two photon energies. The thickness of each constituent, x_1 and x_2, can then be used together with an independent measure of total sample thickness, x_m, to calculate the fraction of porosity, f_p, the fraction of component 1, f_1, and the fraction of component 2, f_2, as shown in equation (2).

$$f_p = 1 - \frac{x_1 + x_2}{x_m}, \quad f_1 = \frac{x_1}{x_1 + x_2}, \quad f_2 = \frac{x_2}{x_1 + x_2} \qquad (2)$$

Researchers in soil chemistry initially reported a so-called dual-energy technique, where they simultaneously irradiated soil mixtures with two mono-energetic radioisotopes to measure the bulk density and water content of soils[7]. Instead of using radioactive isotopes with fixed energies, we employ a brehmstrahlung x-ray source and an energy discriminating, solid-state detector, which allows us to choose a wide range of photon energies. An advantage of having a brehmstrahlung source is that we can enhance detectability by optimizing the choice of energies based on the accuracies of measured attenuation coefficients[8]. Early work by Ting, et al. demonstrated this optimized energy dispersive technique on carbon-prepreg composites to measure variations in the concentrations of the fiber and resin.[8,9] Ting, et al. also used this method to measure the thickness of alumina corrosion films on aluminum.[8] However, a concern with this early work is that it required long acquisition times (up to 24 hours per point) to achieve sufficient counting statistics needed for good measurement sensitivity. We have since improved the technique by analyzing a larger portion of the energy spectrum, which in turn increases the counting statistics and significantly reduces the measurement time (2.5 minutes per point).

In Figure 6, the technique is used to map spatial variations of porosity in a SiC-fiber-reinforced mullite-matrix composite that was prepared from different shape-forming techniques in our laboratory: uniaxial pressing of a dry powder and pressure filtration of a strongly flocculated suspension. The Figure illustrates a 0.4 inch by 0.75 inch area that was sampled at an increment of 0.02 inches (508 microns). Both the scans and their respective histograms (Figure 7) reveal that while the overall porosity ranges are similar in both samples (a mean porosity value, ρ, of 0.41), more severe spatial variations of porosity are present in the dry pressed sample. These differences highlight the potential of this technique to be used as a process development tool by analyzing the effects of different processing variables (in this case, the shape forming technique) on the porosity distribution in the green state.

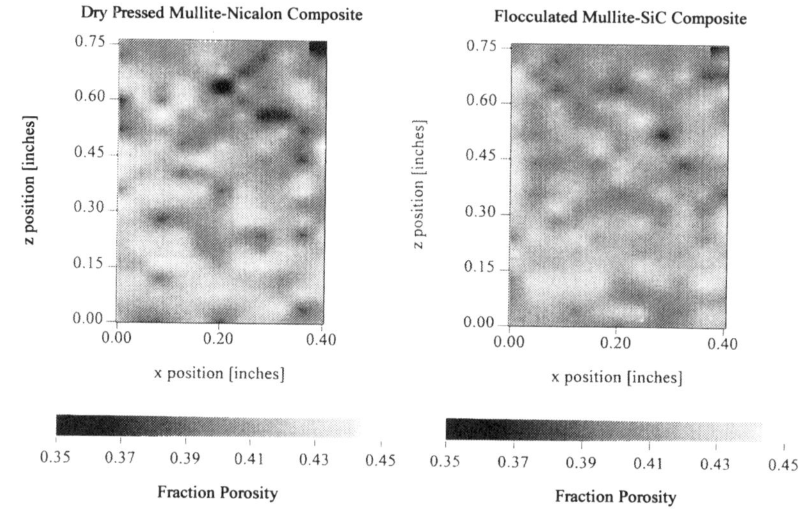

Figure 6. 2D porosity distributions of green ceramic samples obtained through energy dispersive x-ray measurements.

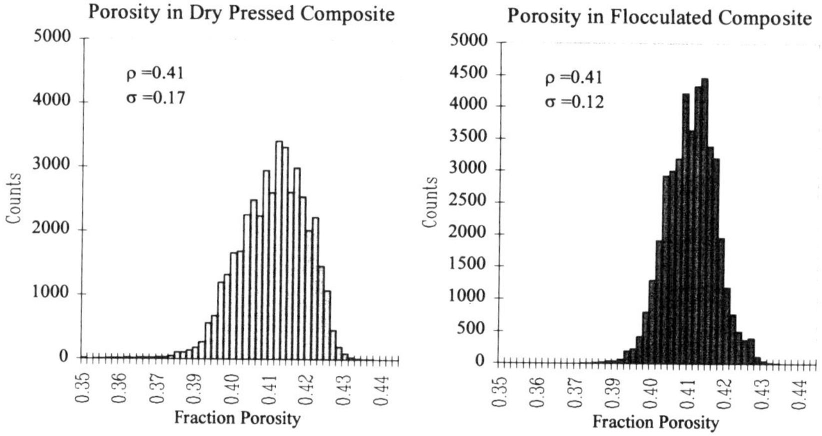

Figure 7. Histograms of the above porosity distributions show the same mean porosity value, ρ, for both samples. A larger standard deviation, σ, indicating a wider porosity range is observed the dry pressed sample.

Figure 8 shows maps of composition and porosity of an alumina-matrix composite containing woven fibers of SiC. In this specimen, a 0.4 inch by 0.5 inch area was sampled at an increment of 0.025 inch (635 microns). The porosity and compositional data are plotted against one another in Figure 9 illustrating a strong positive correlation between regions of high porosity and regions of high alumina concentration in this sample. This type of analysis illustrates the technique's potential for statistical process control.

In the above examples, fairly coarse sample-positioning intervals were used (508 and 635 microns) in the interest of reducing the overall scan time. Much finer spatial resolution is possible by reducing the width of the detector collimator. We frequently demonstrate 30 micron spatial resolution with our system. With a single point detector, mapping out a reasonably sized area of 1 cm^2 at 30 micron intervals is overly time-consuming. However, for on-line applications, this trade-off between acquisition time and spatial resolution can be overcome by implementing 1D and 2D detector arrays.

The accuracy of the technique depends strongly upon the acquisition time and the accuracy of the attenuation coefficients of the constituent materials. While a full analysis is beyond the scope of this paper, we believe the overall accuracy of the data in the above examples is acceptable (2% for porosity and 5% for composition) based on a comparison of mean values of porosity and composition that we obtained by independent measurements.

IV. SUMMARY

Two prototype x-ray techniques have been developed for the nondestructive evaluation of ceramic tapes and composites: (*i*) x-ray film radiography of tape casting defects allowing detection of tape thickness variations, surface pits, cracks, pores, bubbles, and density gradients and (*ii*) energy-dispersive x-ray spectroscopy which provides quantitative maps of both porosity and composition in green and sintered composites. We have used both techniques to relate processing variables such as starting powder size and shaping method to the green and sintered microstructures in ceramic tapes and composites. In addition, we are currently exploring the potential of examining multi-constituent (3 and up) ceramic composites using the energy dispersive technique.

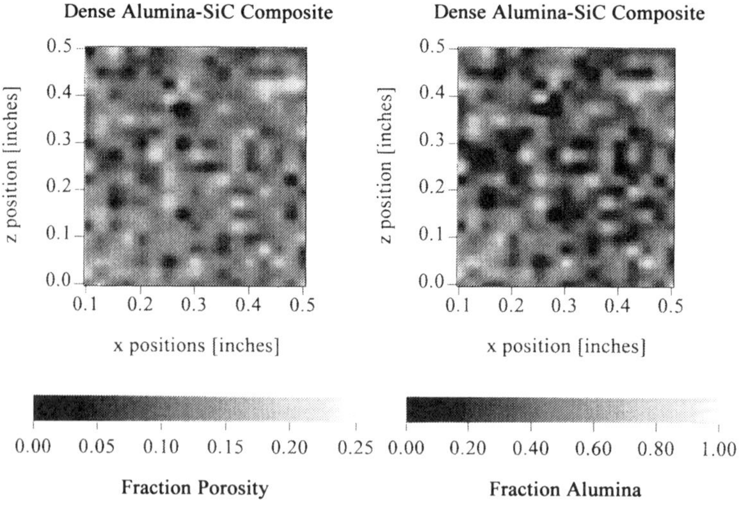

Figure 8. 2D porosity and alumina distributions in a dense alumina-SiC fiber composite obtained through energy dispersive measurements.

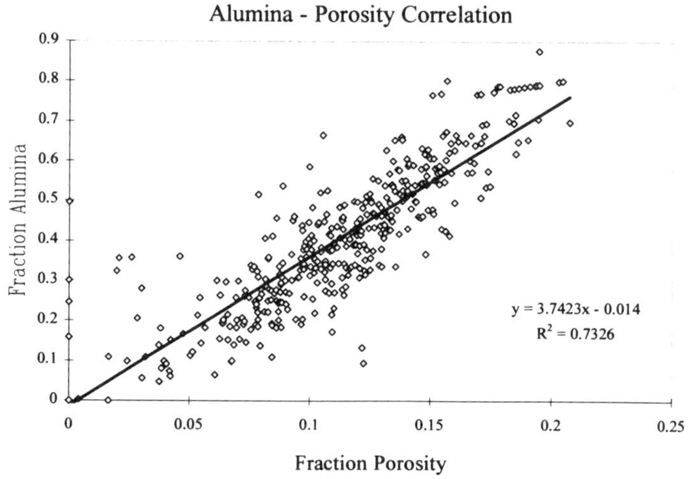

Figure 9. Porosity plotted against alumina fraction from Figure 8 illustrates a strong positive correlation between regions of high porosity and regions of high relative alumina content.

ACKNOWLEDGEMENTS

This research is supported by the Office of Basic Energy Sciences at the U.S. Department of Energy and by the NSF Industry/University Cooperative research program.

REFERENCES:

[1]C. H. Schilling and J. N. Gray, "Needs and Opportunities for NDE in Ceramic Processing," in *Ceramic transactions. Volume 89: Nondestructive Evaluation of Ceramics*, edited by C. H. Schilling and J. N. Gray (American Ceramic Society, Westerville, OH, 1998).

[2]E. Twiname, Richard E. Mistler, Inc., Morrisville, Pennsylvania, personal communication (1997).

[3]C.W. Maranville, "Radiographic Imaging of Microstructural Defects in Ceramic Tapes," M.Sc. Thesis, Iowa State University, 1996.

[4]C.W. Maranville, J. N. Gray, and C H. Schilling, "Novel X-ray Imaging Method for Evaluating Defect Evolution in Ceramic Tapes," in *Review of Progress in Quantitative Nondestructive Evaluation*, edited by D. O. Thompson and D. E. Chimenti (Plenum Press Co., New York and London, 1996), Vol. 15b, pp. 2331-2338.

[5]E. B. Saloman, J. H. Hubbell, and J. H. Scofield, "Attenuation Cross Sections for Energies 100eV to 100keV and Elements Z = 1 to 92," *At. Data Nucl. Data Tables 38*, 1-197 (1988).

[6]S. P. Huss, "Processing and Characterization of SiC Fiber Reinforced Mullite," M.Sc. Thesis, Iowa State University, 1994.

[7]B. J. Bridge and N. Collis-George, "A Dual Source Gamma Ray Traversing mechanism Suitable for the Non-Destructive Simultaneous Measurement of Bulk Density and Water Content in Columns of Swelling Soils," *Australian Journal of Soil Research* **11**, 83-92 (1972).

[8]J. Ting, T. Jensen, and J. N. Gray, "Quantitative Determination of Material Composition of Composites Using Energy Sensitive Measurements," in *Review of Progress in Quantitative Nondestructive Evaluation*, edited by D. O. Thompson and D. E. Chimenti (Plenum Press, New York, 1991), Vol. 11B, pp. 1499-1505.

[9]J. Ting, "Quantitative Evaluation of Material Composition in Composites Using X-Ray Energy-Dispersive NDE Technique," M.Sc. Thesis, Iowa State University, 1993.

FINE SCALE RESOLUTION OF CERAMIC MICROSTRUCTURES BY X-RAY COMPUTED TOMOGRAPHY: AN ECONOMICAL APPROACH

S. P. Huss, [1,2] J. N. Gray,[2,3] C. H. Schilling[1,2]

[1] Ames Laboratory[+] and Department of Materials Science and Engineering
Iowa State University, Ames, Iowa 50010

[2] Center for Nondestructive Evaluation
Iowa State University, Ames, Iowa 50010

[3] Department of Mechanical Engineering
Iowa State University, Ames, Iowa 50010

ABSTRACT

High-magnification, 3-dimensional imaging of ceramics by x-ray computed tomography (CT) typically requires use of a synchrotron and / or a massive array of solid-state detectors. The high cost of this research equipment generally limits its use as a process-development tool by ceramic manufacturers. In this paper, we demonstrate a more economical approach which produces image quality comparable to that of standard, high performance equipment. In this method we utilized an inexpensive image intensifier and applied techniques of frame averaging, gray-scale sub-ranging and other noise reduction and artifact removal methods in order to improve contrast sensitivity while maintaining low cost. After discussing critical factors affecting image quality, we discuss cost / performance tradeoffs associated with different CT equipment options.

I. BACKGROUND

In an earlier paper in this proceedings[1], we described two x-ray techniques capable of two dimensional imaging of green-body defects in ceramics. We also discussed the importance of gaining knowledge of the specific location and shape

---

[+] Ames Laboratory is operated by Iowa State University under the contract number W-7405-eng-82 with the U.S. Department of Energy.

of these defects in order to improve our understanding of relationships between manufacturing variables and defect evolution[2]. Microstructural features ranging from micron-sized pores to millimeter-scale density gradients are all critical to this understanding. While no technique yet exists to nondestructively image individual, submicron ceramic particles in a green body, current, state-of-the-art x-ray CT approaches this goal with an imaging capability of 10 to 100 micron sized features.

High Resolution CT of Advanced Ceramics: Previous Research

In the past decade, several researchers have used high performance x-ray computed tomography to study microstructure evolution during ceramic processing. For example, Sawicka et al. combined a high energy, mono-energetic x-ray source with a single, solid-state detector and observed density variations as small as 1 volume % in sintered alumina specimens[3] having an overall size of 10 cm. The spatial resolution of this system was limited by the collimation of the incident beam, reported at approximately 0.7 mm. Since then, significant increases in spatial resolution and significant reductions in acquisition time have been made possible by the introduction of two-dimensional, solid-state diode detector arrays in conjunction with either a microfocus x-ray generator or a synchrotron photon source. For example, Phillips et al. used a commercially available CT system consisting of a high energy x-ray source and a solid state detector array to observe density gradient evolution during dry powder compaction[4]. Features as small as 200 microns were observed in compacts of ZnO, measuring 0.5 inches in diameter. Higher resolutions were reported by Kinney et al. using a synchrotron based x-ray CT microscope[5]. In that study, the microstructure evolution of chemical vapor infiltrated composite samples was analyzed at different stages of the fabrication process at a voxel resolution of 15.8 microns. We should also mention the work of Shen et. al who used a microfocus x-ray source and an array of solid-state detectors to observe time-dependent microstructure evolution in consolidating silica gels[6]. Of his more notable findings is the evidence for macroscopic fracturing during the consolidation of stiff gels.

It is important to note that the above work entails use of either a synchrotron or an array of solid-state detectors. The high cost of this research equipment generally limits its use as a process-development tool by ceramic manufacturers. In this paper we wish to discuss a more economical approach where the costs can be reduced by an order of magnitude while still maintaining acceptable levels of image quality.

Critical Factors Affecting Image Resolution

It is necessary to appreciate that CT image resolution can be separated into two major components: spatial resolution and contrast sensitivity. Spatial resolution represents the smallest detectable length-increment of an object perpendicular to the x-ray beam, and is determined by the spatial sampling frequency of the field of view. For example, if a 1 cm specimen is magnified and imaged across a detector array with 500 elements, the spatial resolution is nominally equal to 1 cm divided by 500, or 20 microns. There are several ways this limit can be improved by moving the specimen or detector array to increase sampling frequency, however, this is at the expense of acquisition time. Contrast sensitivity represents the smallest detectable length-increment of an object parallel to the x-ray beam and is governed by the x-ray counting statistics and the noise characteristics and dynamic range of the detector. For example, if a one-cm-thick ceramic specimen is examined by an image intensifier system having a typical contrast sensitivity of 5%, the smallest detectable change in the thickness of the specimen would be 5% of one cm or 500 microns.

Current Equipment Choices

Let us consider the current limits and equipment choices regarding spatial resolution. Prior to the availability of microfocus x-ray generators in the 1980's, high spatial resolution CT (i.e., capable of resolving features finer than 100 microns) was only possible at synchrotron ports where a stream of parallel x-rays could be used to image an object[7]. In this parallel-beam configuration, the spatial resolution is limited by the pixel width of the detector surface. In the early 1980's, standard industrial x-ray generators possessed large spot sizes (>1 mm) prohibiting object magnification without severe image blurring. Since then, minifocus and microfocus x-ray tubes with respective spot sizes of less than 500 and 10 microns have enabled object magnification in the fan beam or cone beam configuration, thereby allowing previously untenable spatial resolution (down to a few microns).

One approach to reduce the cost of an x-ray CT system while retaining the ability to resolve fine-scale features is to implement a minifocus source rather than a microfocus source (or a synchrotron). Minifocus sources are available with a wide range of x-ray spot sizes and cost approximately half that of high-performance microfocus sources. We can demonstrate the magnification capability of a minifocus source by imaging a standard resolution gauge, which consists of a thin sheet of plastic coated with a thin, metal pattern (e.g., Au or W) of converging lines. Figure 1 illustrates digitized images of projection radiographs of such a resolution gauge at 3x magnification using both a microfocus (5 micron spot size) and minifocus (200 micron spot size) source. The images reveal that the lines of the gauge are clearly resolvable up to at least 10 line pairs per millimeter, or 50 micron increments. As a result, one could expect to resolve 50 micron features in computed tomography using such a minifocus source.

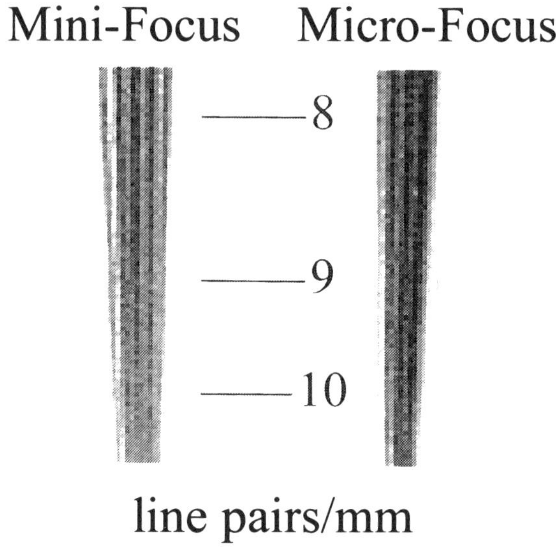

Figure 1 - digitized radiographic images of a standard resolution gauge for both the minifocus (~200 micron spot size) and microfocus (~5 micron spot size) x-ray source.

Let us now consider the current limits and equipment choices regarding contrast sensitivity. While image intensifiers and linear arrays of charge-coupled device (CCD) detectors are fast and relatively inexpensive, the contrast sensitivity is low, typically between 1 and 5%. Solid-state detectors provide much better contrast sensitivity (typically 0.1%), however, they are much more expensive, and they require prohibitively long acquisition times. Of course, the acquisition time can be reduced by implementing an array of solid-state detectors rather than a single detector, however, this significantly raises the equipment cost.

II. IMAGE-INTENSIFIER APPROACH

In order to overcome these problems our approach uses a microfocus generator and an inexpensive image intensifier with an 8-bit digital CCD camera. Using techniques of frame averaging, gray-scale sub-ranging and other noise reduction and artifact removal methods[8,9] we significantly improved the contrast sensitivity to below 1%. For example, shown in Figure 2 are projection images of a single layer of 300 micron diameter soda-lime silicate spheres that were positioned on a quarter-inch-thick plate of the same material. The diameter of each sphere is approximately 3% of the thickness of the glass plate. Therefore, a contrast sensitivity of better than 3% is required to resolve this thickness change and image a single sphere. The left side of Figure 2 illustrates use of a standard acquisition scheme (i.e., no adjustment to the 8 bit signal) where the glass spheres are simply not visible. However, the spheres begin to take shape in the image on the right side of Figure 2, which was obtained by integrating multiple frames of an 11-bit subranged signal.

The effect of this frame accumulation and subranging is further demonstrated in Figure 3 by the CT images of 300 micron diameter soda-lime silicate spheres packed inside a 1.5 cm inside-diameter glass vial. The pixel size in this image is approximately 20 microns. It is apparent that these signal processing techniques produce a major improvement in the quality of the image, allowing resolution of the individual 300 micron spheres.

Finally, we would like to mention that the current limits of spatial resolution depend not only on the x-ray equipment but also on the data handling capacity of the computers working with the images. It is important to consider that the storage capacity requirements for a given image are inversely proportional to the voxel volume. For example, an object measuring 1 cm^3 would require approximately 64 megabytes of memory if sampled at a voxel size of 25 microns and 64 gigabytes if sampled at a voxel size of 2.5 microns. It is only recently that advances in computer technology have made it possible to capture, display, and store these large images conveniently.

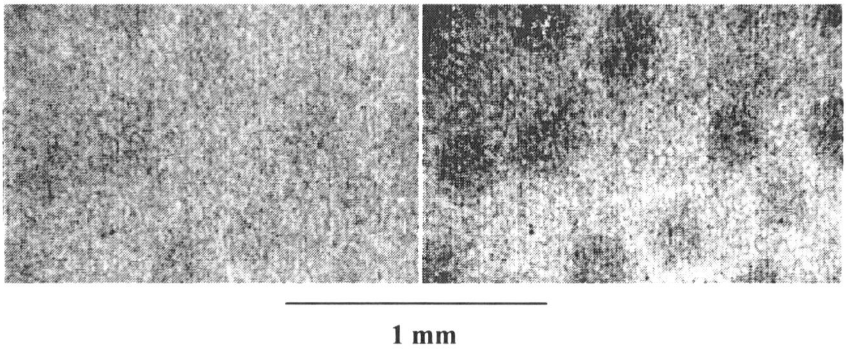

Figure 2 - Digital x-ray projection images of 300 micron diameter soda-lime silicate spheres. The image on the left obtained from an unprocessed 8-bit signal from the CCD camera. The image on the right was obtained by accumulating 60 frames of an 11-bit sub-ranged signal.

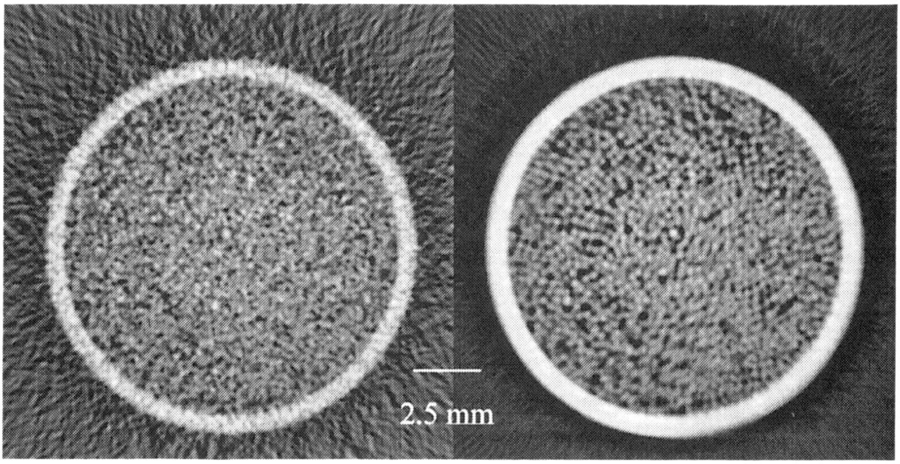

Figure 3 - X-ray CT images of 300 micron soda-lime silicate spheres in a 1.5 cm glass vial. The image on the left was obtained from an unprocessed 8-bit signal from the CCD camera. The image on the right was obtained by accumulating 60 frames of an 11-bit sub-ranged signal.

IV. SUMMARY

We augmented a relatively inexpensive image intensifier based CT system to reap the high spatial resolution inherent in a microfocus x-ray source. In turn, we have a practical system capable of achieving image quality comparable to that of the high performance CT systems (i.e., synchrotron or massive solid state detector array systems) without the high cost.

ACKNOWLEDGEMENTS

This research is supported by the Office of Basic Energy Sciences at the U.S. Department of Energy and by the NSF Industry/University Cooperative Research program.

REFERENCES:

[1]S. P. Huss, C. H. Schilling, J. N. Gray, "Recent Advances in X-ray Characterization of Composites and Tape Cast Ceramics," in *Ceramic Transactions, Vol. 89: Nondestructive Evaluation of Ceramics*, edited by C. H. Schilling and J. N. Gray (American Ceramic Society, Westerville, OH, 1998).

[2]C. H. Schilling and J. N. Gray, "Needs and Opportunities for NDE in Ceramic Processing," in *Ceramic Transactions, Vol. 89: Nondestructive Evaluation of Ceramics*, edited by C. H. Schilling and J. N. Gray (American Ceramic Society, Westerville, OH, 1998).

[3]B. D. Sawicka, "Photon CT Scanning of Advanced Ceramic Materials," in *Review of Quantitative Nondestructive Evaluation*, edited by Donald O. Thompson and Dale E. Chimenti (Plenum Press, New York, 1987), Vol. 6a, pp. 421-432.

[4]D. H. Phillips and J. J. Lannutti, "Measuring Physical Density with X-ray Computed Tomography," *NDT&E International* **30**, 339 (1997).

[5]J. H. Kinney, T. M. Breunig, T. L. Starr, "X-ray Tomographic Study of Chemical Vapor Infiltration Processing of Ceramic Composites," *Science* **260** (May 7), 789-792 (1993).

[6]C. Shen, W. Russel, and F. M. Auzerais, "Colloidal Gel Filtration: Experiment and Theory," *AIChE Journal* **40** (11), 1876-1891 (1994).

[7]R. W. Parish, "Microfocus X-ray Technology - A Review of Developments and Application," in *Review of Progress in Quantitative Nondestructive Evaluation*, edited by D. O. Thompson and D. E. Chimenti (Plenum Press Co., New York and London, 1985), Vol. 4a, pp. 1-20.

[8]S. Kakummanu and J Gray, "Development of a Model for the X-ray Image Intensifier-Camera Imaging System and Examples of Its Use," in *Review of Progress in Quantitative Nondestructive Evaluation*, edited by D. O. Thompson and D. E. Chimenti (Plenum Press Co., New York and London, 1994), Vol. 13a, pp. 685-92.

[9]V. Kini and J Gray, "Real Time X-ray Radiography and its Use in Economic and Fast Tomographic Reconstruction," in *Review of Progress in Quantitative Nondestructive Evaluation*, edited by D. O. Thompson and D. E. Chimenti (Plenum Press Co., New York and London, 1994), Vol. 13a, pp. 541-6.

MODELS RELATING ULTRASONIC VELOCITY TO THE MICROSTRUCTURES OF GREEN AND SINTERED CERAMICS: A REVIEW

V.J. García and C.H. Schilling

Ames Laboratory* and Department of Materials Science and Engineering
Iowa State University, Ames, Iowa 50011

ABSTRACT

We review theoretical models which can be used in a practical setting to relate measurements of ultrasonic velocity to microstructural features of green and sintered ceramics. Features that are explicitly related include the particle coordination number, the specific surface area, and the packing density. We have rewritten some of the models and combined them with experimental results to obtain a general, power-law relation between the ultrasonic velocity and the reduction of specific surface area during sintering.

INTRODUCTION

Ultrasonic measurements are commonly used to evaluate porosity, cracks, thickness variations, and inclusions in ceramic- or metal-powder compacts [1-9]. Some of the more exotic applications of ultrasound in densely-packed, particulate media include the following: (*i*) studies of spatial variations of packing-density in green ceramics [10], (*ii*) remote photoacoustic testing of ceramics in the course of sintering [11] (*iii*) ultrasonically-inferred measurements of specific surface area during sintering [12,13], (*iv*) ultrasonic velocity and attenuation measurements as a function of the sintering temperature and the pH of slip-cast specimens of colloidal alumina [14,15], and (*v*) use of ultrasonic attenuation to probe mechanical energy losses in lunar and terrestrial (sedimentary) rocks, energy

* Ames Laboratory is operated by Iowa State University under the contract number W-7405-eng-82 with the U.S. Department of Energy.

losses that strongly and reversibly depended on the sorption-desorption of various gases [16,17].

A deeper understanding of the relationship between ultrasonic velocity and the distinct microstructural changes that occur within a powder compact at various stages of processing is needed in order to realize the full potential of ultrasonic measurements for characterizing green and sintered ceramics. Significant advances in the theoretical framework relating ultrasonic measurements to explicit microstructural parameters (e.g., the particle coordination number, specific surface area, pore size) have been recently reported. Given the recent interest in the use of NDE for defect characterization in ceramics, we believe a review of such models is timely. The main purpose of this paper is to review such models, emphasizing those which can be used in a practical setting to relate measurements of ultrasonic velocity to microstructural features of green and sintered powder compacts. The review specifically pertains to the so-called long-wavelength regime, where scattering attenuation is minimal because the ultrasonic wavelength λ is much greater than the particle size. We will not focus on ultrasonic attenuation in this paper. The reader can refer to several excellent publications in this regard [9,18-22].

In this review paper, we will consider four regimes of ultrasonic wave propagation in ceramics, each regime corresponding to the following type of specimen: (*i*) fully dense polycrystalline ceramics, (*ii*) green bodies with non-adhesive particles, in which case dry particles, with clean surfaces, contact each other (without chemical adhesion) by the action of an external, mechanical force, (*iii*) green bodies with adhesive particles, in which case dry particles are packed together with chemical bonds attaching the particle surfaces together, and (*iv*) partially-sintered powder compacts. In case (*iii*), interparticle adhesion may arise from van der Waals attraction forces or by the existence of a binder coating on the particle surfaces.

FULLY DENSE, POLYCRYSTALLINE CERAMICS

The longitudinal sound velocity V_B in a fully-dense, polycrystalline ceramic is conveniently expressed as follows [23,24]:

$$V_B^2 = \frac{(1-v)E}{(1+v)(1-2v)\rho} \tag{1}$$

In this model, the sound velocity depends on the Young's modulus (E), the bulk density (ρ), and the Poisson ratio (v) of the specimen. This model applies to a broad range of isotropic materials and is derived from elasticity theory. Schreiber et al. [23] presented an extensive review of elastic constants and their measurement.

GREEN BODIES WITH NON-ADHESIVE PARTICLE CONTACTS

It is generally understood that ultrasonic wave propagation through a granular medium strongly depends on micro-mechanical and microstructural features such as the interparticle contact area, the particle coordination number, solid volume fraction, the stiffness of a contact, and so on. Several researchers have modeled the elastic behavior of assemblies of spherical particles by describing the micromechanics of each particle-particle contact using the Hertz model [25-29]. Goddard [30] describes some interesting anomalies with this behavior.

We shall begin by reviewing Brettell's model [31] for elastic wave propagation through a linear chain of spherical particles, each sphere having a radius of r. A static compressive force is first exerted along the length of the chain, producing a circular contact area between particles. The contact radius, a, of this circular area is the same at all contacts along the chain. The chain is then subjected to a simple harmonic compression wave such that the displacement of each sphere is parallel to the length of the chain. The deformation of each sphere is elastic and described by Hertz's law. The longitudinal (compression) wave velocity, V_P, is expressed as

$$\frac{V_P}{V_B} = f(v)\left(\frac{a}{r}\right)^{1/2} \tag{2}$$

$$f(v) = 3^{1/2} g(v) \qquad \text{and} \qquad g(v) = \left[\frac{(1+v)(1-2v)}{\pi(1-v)(1-v^2)}\right]^{1/2} \tag{3}$$

In these expressions, v is the Poisson ratio of the dense material comprising each spherical particle. An important result of this model is that the ultrasonic velocity is proportional to the square root of a/r.

Let us now consider the case of a three-dimensional packing of spherical particles. Belyaeva et al. [32] developed a general model of the macroscopic, elastic stress-strain behavior for a range of packing structures, including random packing, cubic packing, and so on:

$$\sigma = \frac{b\langle c\rangle\phi\, E}{\pi\left(1 - v^2\right)}\varepsilon^{3/2} \tag{4}$$

In this case σ is a hydrostatic or uniaxial stress applied to a powder compact, ε is the resulting macroscopic strain in the compact, ϕ is the volume fraction of particles in the compact, $\langle c\rangle$ is the average coordination number of the particles, v is the Poisson ratio of the dense material comprising each spherical particle, and b is an empirical constant. This constant is equal to 1 if the powder compact is subjected to a hydrostatic stress, and it is equal to 1/3 if the compact undergoes uniaxial compression. This model was derived based on an energy balance of a powder compact when it is subjected to external forces. The central assumption of the model is that the Poisson ratio of the particulate skeleton is approximately zero. As discussed by Belyaeva [33], this is a valid assumption considering that point contacts between spherical particles are much more compliant than the bulk material within a given sphere.

In general, the longitudinal sound velocity of an isotropic medium, V_P, can be expressed as a function of the Young's modulus and the bulk density, ρ, using the following relation [34]:

$$V_P^2 = \frac{1}{\rho}\frac{d\sigma}{d\varepsilon} \tag{5}$$

We can substitute Equation (4) into Equation (5) to provide a general expression for the longitudinal sound velocity:

$$\frac{V_P}{V_B} = \langle c\rangle^{1/2} f(v,\phi)\left(\frac{a}{r}\right)^{1/2} \tag{6}$$

$$f(v,\phi) = \phi^{1/2}(3b)^{1/4}\, g(v) \tag{7}$$

Note that the expression for ultrasonic velocity in Equation (6) is similar to that of the Brettell model for a linear chain of spherical particles in Equation (2). Both of these equations express the sound velocity as proportional to the square root of a/r. An important aspect of Equation (6) is that the velocity is also proportional to the square root of the average coordination number.

Digby [28] reported a model for the elastic properties of randomly-packed mono-size spherical particles, Equation (8):

$$\frac{V_P}{V_B} = \langle c \rangle^{1/2} h(v) \left(\frac{a}{r} \right)^{1/2} \tag{8}$$

$$h(v) = \left[\frac{1}{10\pi} \left(\frac{3}{1-v} + \frac{4}{2-v} \right) \frac{1-2v}{1-v} \right]^{1/2} \tag{9}$$

Much like the Belyaeva model, the Digby model in Equation (8) expresses the ultrasonic velocity as proportional to the square root of both the particle coordination number and a/r. We have used the Belyaeva model in Figure 1 to illustrate the general trend of the ratio (V_p/V_B) for different packing structures: cubic, random and tetragonal. For these calculations we assumed that the powder compacts are subjected to uniaxial compression (b = 1/3), the Poisson ratio is equal to 0.22 (for polycrystalline Al_2O_3) and the specific values of coordination number and packing density shown in Table I. Figure 1 illustrates that significant increases in velocity can be produced as a function of the particle coordination number, packing density, and the static stress. Although the Digby model was derived from a different approach, Figure 1 shows that the Digby model is in good agreement with the Belyaeva approach.

Table I. Values of coordination number and packing density used in Fig. 1

Microstructure	Average coordination number $\langle c \rangle$	Solid Volume Fraction ϕ	Reference
Cubic	6	~0.52	[35]
Random	8.84	~0.60	[28]
Tetragonal	10	~0.70	[35]

GREEN BODIES WITH ADHESIVE PARTICLE CONTACTS

Several authors have modeled the elastic properties of powder compacts consisting of randomly-packed arrangements of adhesive spherical particles [28,29,36,37]. The models are generally similar and consistent with each other. The following general assumptions apply for each case: (*i*) powder specimens are homogeneous, isotropic, and consist of linear elastic, spherical particles,

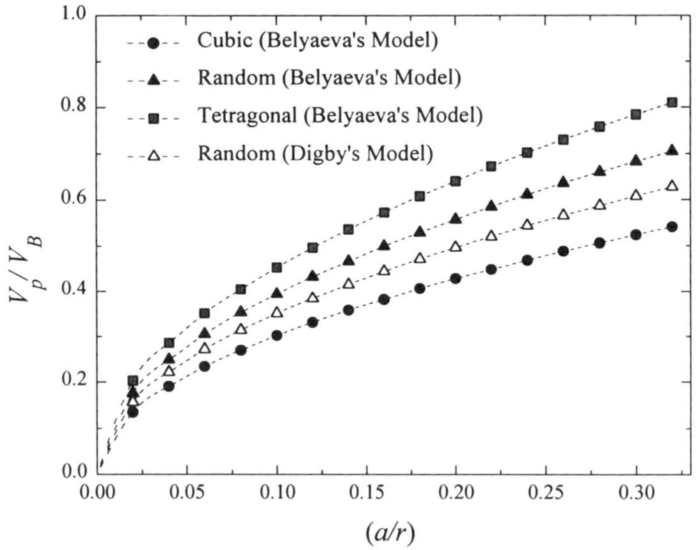

Figure 1. Calculated values of the ratio $\left(V_p/V_B\right)$ for cubic, random, and tetragonal structures as a function of the ratio (a/r).

(*ii*) elastic stress-strain parameters are volume-averaged, global descriptors of local microstructural variables, (*iii*) particles do not undergo rotation, rolling, or torsion, and (*iv*) the displacements of the particle centers are consistent with the uniform displacement field applied to a specimen. The total thickness of adhesive between two contacting particles is 2δ. The compression-wave velocity (V_p) and the shear-wave velocity (V_s) are then expressed as

$$V_p^2 = \frac{3}{20}\frac{\Gamma}{\rho}\left(k_n + \tfrac{2}{3}k_t\right) \tag{10}$$

$$V_s^2 = \frac{1}{20}\frac{\Gamma}{\rho}\left(k_n + \tfrac{3}{2}k_t\right) \tag{11}$$

$$\Gamma = \frac{\langle c \rangle}{\pi r\left(1 + \delta/r\right)} \tag{12}$$

In these equations, k_n and k_t are respectively equal to the average normal and tangential stiffness of a particle-particle contact. In order to obtain potentially useful microstructural parameters from ultrasonic measurements, we can rearrange these equations to express the contact stiffnesses in terms of the measured velocities:

$$k_n = \frac{4\rho}{\Gamma}\left(3V_p^2 - 4V_s^2\right), \quad \text{and} \quad k_t = \frac{4\rho}{\Gamma}\left(6V_s^2 - 2V_p^2\right) \tag{13}$$

We recently utilized these equations to estimate the role of specimen processing variables (binder chemistry) on the interparticle-contact stiffness [19]. In that study, we were able to correlate the normal stiffness with the fracture strength of alumina green-bodies.

PARTIALLY-SINTERED CERAMICS

Jernot et al. [38-41] performed extensive image analysis on partially sintered specimens made of spherical metallic powder and developed an empirical correlation between the ultrasonic velocity and both the volume-fraction solids and the average coordination number:

$$\frac{V_P}{V_B} = \left(\langle c\rangle - \frac{\langle c\rangle - 1}{\phi^{1/3}}\right)^{1/2} \tag{14}$$

Gurland defined *contiguity* as the proportion of the surface of a particle that is in contact with other particles [42-44]. Jernot's image analysis on metallic specimens involved use of this term and stereology equations that were originally derived by Saltikov [43,45]. Jernot describes the reduction of the specific surface area in terms of the coordination number, $\langle c\rangle$, the radius of the contact area, a, and the particle radius, r:

$$\frac{\Delta S}{S_o} = \frac{\langle c\rangle}{4}\left(\frac{a}{r}\right)^2 \tag{15}$$

In this expression, S_o is equal to the initial, specific surface area of the powder compact prior to sintering and $\Delta S = S_o - S$, where S is the specific surface area of the partially-sintered specimen. Using this expression, we can rewrite the Belyaeva model (Equation 6) in terms of the reduction of specific surface area:

$$\frac{V_P}{V_B} = 2^{1/2}\langle c\rangle^{1/4} f(v,\phi)\left(\frac{\Delta S}{S_o}\right)^{1/4} \tag{16}$$

In this expression $f(v,\phi)$ is given by Equation (7). It can be observed that this theoretical expression is more convenient than the empirical relation by Jernot, as it takes into account the specific surface area reduction, ΔS.

In addition, Arzt and co-workers [46-49] have shown experimentally and theoretically that the coordination number can be related to the volume fraction solids by the following convenient relation: $\langle c\rangle = 12\phi$. When we assume that $\langle c\rangle = 12\phi$, $b = 1/3$, and that the solid volume fraction is proportional to the specific surface area reduction $\phi \propto \left(\Delta S/S_o\right)$, Equation (16) can be rewritten as such:

$$\frac{V_P}{V_B} \propto g(v)\left(\frac{\Delta S}{S_o}\right) \tag{17}$$

In this expression, $g(v)$ is a factor that depends on the Poisson ratio of the material and is given by Equation (3). We should emphasize that Equation (17) is based on a particular case of the Belyaeva model for ideal, randomly packed, mono-size spherical particles. Notice that Equation (17) predicts a linear dependence of the ultrasonic velocity on the reduction of the specific surface area.

Martin and Rosen first observed a linear relationship between the ultrasonic velocity and the specific surface area over broad range of porosity in sintered specimens of ZnO and Al_2O_3 [12,13]:

$$S = b_1 V_P + b_2 \tag{18}$$

In this expression, b_1 and b_2 are empirical constants. This result leads to the conclusion that changes in the ultrasonic velocity, ΔV_P, are proportional to variations in the reduction of the specific surface area:

$$\Delta V_P = k\Delta S \tag{19}$$

In this expression, k is a constant of proportionality. It is possible to rewrite Equation (19) if we assume that the specific surface area upon sintering to a near-fully-dense condition is extremely small relative to the specific surface area of the powder compact prior to any sintering:

$$\frac{V_P}{V_B} = \frac{V_{P_O}}{V_B} + \left(1 - \frac{V_{P_O}}{V_B}\right)\frac{\Delta S}{S_o} \tag{20}$$

In this expression, V_{p_o} is the compression wave velocity of the green body prior to sintering.

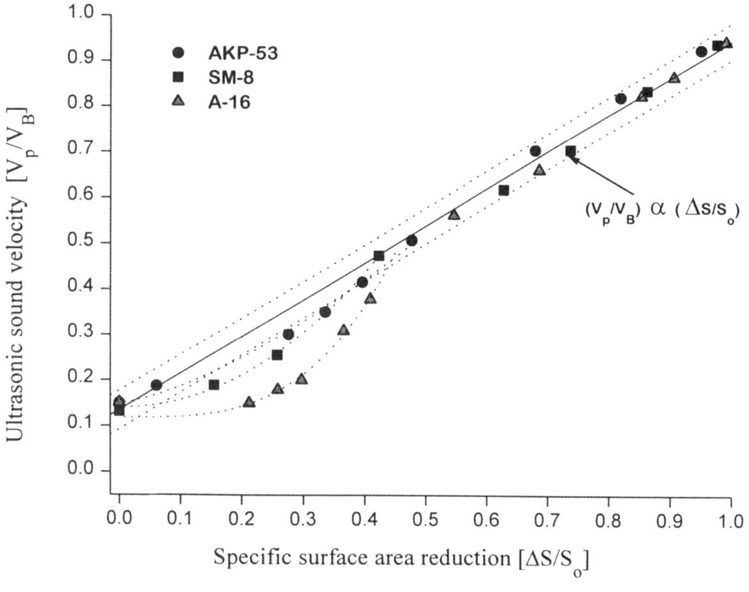

Figure 2 experimental values of ultrasonic velocity and specific surface area reduction reported by Martin et al. [50] in three different commercial alpha alumina powder compacts that were partially sintered to different pore fractions; AKP-53, Sumitomo Chemical America, New York, NY; SM-8, Baikowski International, Charlotte, NC; A-16SG, Alcoa Industrial Chemicals, Bauxite, AK.

Martin et al. [50] recently reported deviations from the linear behavior in Equation (18) and attributed these deviations to a distribution of particle sizes. Their data were reported as ultrasonic velocity as a function of the specific surface area. As shown in Figure 2, we have replotted his data in a slightly different form, using ultrasonic velocity as a function of the *reduction* in specific surface area. We chose this alternate form because it more clearly shows the deviation from linearity. These data represent specimens of three different commercial alpha alumina powders partially sintered to different pore fractions. Using linear regression, we included in Figure 2 a prediction band for a 95% confidence level. The correlation coefficient, *R*, was 0.99806 and Prob>F = 0.0001.

Let us focus on the deviation from linearity in Figure 2 during the early stages of sintering ($\Delta S / S_o < 0.5$). Martin states that this deviation is attributed to local differences in sintering kinetics inherent with a broad distribution of particle sizes in a given specimen (small particles sinter faster than big particles) [50,51]. Martin states that, during the initial stages of sintering, it is possible to have a substantial reduction of specific surface area without a major change in the sound velocity. This would be possible if a percolated network of large particles is the main contributor to the elastic properties of a powder compact throughout initial-stage sintering.

We propose the following general relationship between the longitudinal sound velocity and the reduction of specific surface area:

$$\frac{V_P}{V_B} \propto \left(\frac{\Delta S}{S_o} \right)^x \tag{21}$$

At this point three scenarios would be possible:

1) Ideal packing of mono-size, spherical particles, where the average coordination number remains constant during sintering. In this case we would expect $x \leq 1$ according to Belyaeva model in Equation (16).
2) The average coordination number changes upon sintering. In this case, a linear correlation between the longitudinal sound velocity and the reduction of specific surface area would be expected ($x = 1$) according to Equations (17) and (20). This scenario is not necessarily restricted to ideal packing of mono-size, spherical particles.

3) With a distribution of particle sizes, we can expect a nonlinear relation between the longitudinal sound velocity and the reduction of specific surface area ($x \geq 1$) because of the different sintering rates discussed earlier.

Tables II and III summarize all of the models in this paper relating ultrasonic velocity to the microstructures of green and sintered ceramics.

Table II. Summary of models relating ultrasonic velocity to the microstructures of green ceramics.

1. ***Green Ceramics (Non-Adhesive Particle Contacts)***

- *Linear chain, Brettell* [31];

$$\frac{V_P}{V_B} = f(v)\left(\frac{a}{r}\right)^{1/2} \; ; \; f(v) = 3^{1/2} g(v) \; ; \; g(v) = \left[\frac{(1+v)(1-2v)}{\pi(1-v)\left(1-v^2\right)}\right]^{1/2}$$

- *3-D structures, Belyaeva* [32];

$$\frac{V_P}{V_B} = \langle c\rangle^{1/2} f(v,\phi)\left(\frac{a}{r}\right)^{1/2} \; ; \; f(v,\phi) = \phi^{1/2}(3b)^{1/4} g(v)$$

- *Random packing, Digby* [28];

$$\frac{V_P}{V_B} = \langle c\rangle^{1/2} h(v)\left(\frac{a}{r}\right)^{1/2} \; ; \; h(v) = \left[\frac{1}{10\pi}\left(\frac{3}{1-v} + \frac{4}{2-v}\right)\frac{1-2v}{1-v}\right]^{1/2}$$

2. ***Green Ceramics (Adhesive Particle Contacts);***

Elata and Dvorkin, Dvorkin, and Winkler [28,29,36,37];

$$V_p^2 = \frac{3}{20}\frac{\Gamma}{\rho}\left(k_n + \tfrac{2}{3}k_t\right) \; ; \; V_s^2 = \frac{1}{20}\frac{\Gamma}{\rho}\left(k_n + \tfrac{3}{2}k_t\right) \; ; \; \Gamma = \frac{\langle c\rangle}{\pi r(1+\delta/r)}$$

Table III. Summary of models relating ultrasonic velocity to the microstructures of partially sintered and fully dense ceramics.

1. ***Partially-Sintered Ceramics***

Jerno [38];
$$\frac{V_P}{V_B} = \left(\langle c \rangle - \frac{\langle c \rangle - 1}{\phi^{1/3}} \right)^{1/2}$$

Derived from Belyaeva's model;
$$\frac{V_P}{V_B} = 2^{1/2} \langle c \rangle^{1/4} f(v, \phi) \left(\frac{\Delta S}{S_o} \right)^{1/4}$$

Experimentally observed, Martin and Rosen [12,13];

$$S = b_1 V_P + b_2 \rightarrow \Delta V_P = k \Delta S$$

Derived from Martin and Rosen [12,13];

$$\frac{V_P}{V_B} = \frac{V_{P_o}}{V_B} + \left(1 - \frac{V_{P_o}}{V_B} \right) \frac{\Delta S}{S_o}$$

Derived from Belyaeva's model when $\langle c \rangle = 12\phi$, $b = 1/3$, *and* $\phi \propto (\Delta S / S_o)$;

$$\frac{V_P}{V_B} \propto g(v) \left(\frac{\Delta S}{S_o} \right)$$

General power-law relation;

$$\frac{V_P}{V_B} \propto \left(\frac{\Delta S}{S_o} \right)^x$$

2. ***Fully dense, polycrystalline ceramics;***

Schreiber [23] *and Birch* [24];

$$V_B^2 = \frac{(1 - v) E}{(1 + v)(1 - 2v) \rho}$$

ACKNOWLEDGMENTS

- Office of Basic Energy Sciences at the U.S. Department of Energy.
- Center for Nondestructive Evaluation at Iowa State University.
- Ames Laboratory and Department of Materials Science at Iowa State University.

REFERENCES

1. D. S. Kupperman and H. B. Karplus, "Ultrasonic Wave Propagation Characteristics of Green Ceramics," *Am. Ceram. Soc. Bull.* **63** (12), 1505-09 (1984).
2. D. S. Kupperman, H. B. Karplus, R. B. Poeppel, W. A. Ellingson, H. Berger, C. Robbins, and E. Fuller, "Application of NDE Methods to Green Ceramics: Initial Results," in *Nondestructive Evaluation: Applications to Materials Processing*, edited by O. Buck and S. W. Wolf, TMS Fall Meeting (Philadelphia, PA, Oct. 3-4, 1983), pp. 89-101.
3. M. P. Jones, G. V. Blessing, and C. R. Robbins, "Dry-Coupled Ultrasonic Elasticity Measurements of Sintered Ceramics and Their Green States," *Materials Evaluation* **44**, 859-62 (1986).
4. E. P. Papadakis and B. W. Petersen, "Longitudinal Velocity as a Predictor of Density in Sintered Powder Metal Parts," *Materials Evaluation*, 76-80 (April, 1979).
5. J. M. Rebello and J. Charlier, "Some Aspect of NDT Applied to Ceramics," *European J. NDT* **2**, 150-58 (1993).
6. R. W. Davidge, "Defects in Ceramics- The Target for NDT," *Br. Ceram. Trans. J.* **88**, 113-16 (1988).
7. J. Winterfeld, H. D. Tietz, and T. Palm, "Non-destructive Testing of Volume Defects in Structural Ceramics," *Ceram. Forum. Int.* **71**, 96-100 (1994).
8. K. K. Phani and S. Niyogi, "Porosity Dependence of Ultrasonic Velocity Modulus in Sintered Uranium Dioxide- A Discussion," *J. Mater. Sci. Lett.* **5**, 427-30 (1986).
9. M. C. Bhardwaj, "Fundamental Developments in Ultrasonics for Advanced NDC," Conference Proceedings, NDT of High-Performance Ceramics (Boston, Mass., Aug 25-27, 1987) (American Ceramic Society, Westerville, Ohio, 1988), pp. 472-527.
10. R. A. Roberts, "A Dry-Contact Coupling Technique for Ultrasonic Nondestructive Evaluation of Green-State Ceramics," *Materials Evaluation* **46**, 758-66 (May, 1988).

11. G. Busse, L. Lyamshev, and J. Stanullo, "Remote Photoacoustic Testing of Ceramics in the Course of Sintering," *Acoustical Physics* **42** (1), 35-39 (1996).
12. L. P. Martin and M. Rosen, "Analysis of Ultrasonic Velocity Measurements on Sintering Zinc Oxide," *Ultrasonics* **35** (1), 65-71 (1997).
13. L. P. Martin and M. Rosen, "Correlation Between Surface Area Reduction and Ultrasonic Velocity in Sintered Zinc Oxide Powders," *J. Am. Ceram. Soc.* **80** (4), 839-44 (1997).
14. N. Kulkarni, B. Moudgil, and M. Bhardwaj, "Ultrasonic Characterization of Green and Sintered Ceramics: I, Time Domain," *Am. Ceram. Soc. Bull.* **73** (6), 146-53 (1994).
15. N. Kulkarni, B. Moudgil, and M. Bhardwaj, "Ultrasonic Characterization of Green and Sintered Ceramics: II, Frequency Domain," *Am. Ceram. Soc. Bull.* **73** (7), 83-85 (1994).
16. B. R. Tittman, "Possible Mechanism for Seismic Attenuation in Rocks Containing Small Amounts of Volatiles," *J. Geophysical Research* **85** (B10), 5199-5208 (1980).
17. V. A. Clark, B. R. Tittmann, and T. W. Spencer, "Effect of Volatiles on Attenuation (Q^{-1}) and Velocity in Sedimentary Rocks," *J. Geophysical Research* **85** (B10), 5190-98 (1980).
18. R. B. Thompson, W. A Spitzig, and T. A. Gray, "Relative Effects of Porosity and Grain Size on Ultrasonic Wave Propagation in Iron Compacts," in *Review of Progress in Quantitative Nondestructive Evaluation*, edited by D. O. Thompson and D. E. Chimenti (Plenum Press, New York, 1986), pp. 1643-53.
19. C. H. Schilling, V. J. Garcia, R. M. Smith, and R. A. Roberts, "Ultrasonic- and Mechanical Behavior of Green- and Partially-Sintered Alumina: Effects of Slurry Consolidation Chemistry," *J. Am. Ceram. Soc. in press* (1998).
20. V. K. Varadan, Y. Ma, and V. V. Varadan, "A Multiple Scattering Theory for Elastic Wave Propagation in Discreet Random Media," *J. Acoust. Soc. Am.* **77** (2), 375-85 (February, 1985).
21. F. E. Stanke and G. S. Kino, "A Unified Theory for Elastic Wave Propagation in Polycrystalline Materials," *J. Acoust. Soc. Am.* **75** (3), 665-81 (March, 1984).
22. A. G. Evans, B. R. Tittman, L. Ahlberg, B. T. Khuri-Yakub, and G. S. Kino, "Ultrasonic Attenuation in Ceramics," *J. Appl. Phys.* **49** (5), 2669-79 (1978).
23. E. Schreiber, O. L. Anderson, and N. Soga, *Elastic Constants and Their Measurement* (McGraw-Hill Inc., New York, 1973).

24. F. Birch, "The Velocity of Compressional Waves in Rocks to 10 Kilobars, Part 2," *Journal of Geophysical Research* **66** (7), 2199-224 (July, 1961).
25. H. Hertz, "Uber die Beruhrung Fester Elastisher Korper (On the Contact of Elastic Solids)," *J. reine und angewandte Mathematik* **92**, 156-171 (1882).
26. R. D. Mindlin, "Compliance of Elastic Bodies in Contact," *Trans. ASME* **71**, A-259 (1949).
27. K. Walton, "The Oblique Compression of two Elastic Spheres," *J. Mech. Phys. Solids* **26**, 139-150 (1978).
28. P. J. Digby, "The Effective Elastic Moduli of Porous Granular Rocks," *J. Appl. Mech.* **48**, 803-808 (1981).
29. D. Elata and J. Dvorkin, "Pressure Sensitivity of Cemented Granular Materials," *Mech. of Mater.* **23** (1996).
30. J. D. Goddard, "Nonlinear Elastic and Pressure-Dependent Wave Speeds in Granular Media," *Proc. R. Soc. Lond. A* **430**, 105-131 (1990).
31. J. M. Brettell, "Elastic Wave Propagation in Powders," *Acustica* **65**, 49-51 (1987).
32. I. Yu. Belyaeva, V Zaitsev, Yu., and L. A. Ostrovsky, "Nonlinear Acoustoelastic Properties of Granular Media," *Acoust. Phys.* **39**, 11-15 (1993).
33. I. Yu. Belyaeva, L. A. Ostrovsky, and V. Yu. Zaitsev, "Nonlinear Acoustoelastic Properties of Media with Complex Structure," in *Research Trends in Physics: Nonlllinear Acoustics*, edited by K. A. Naugol'nykh and L. A. Ostrovsky (American Institute of Physics, L a Jolla International School of Physics, 1994), pp. 1-27.
34. A. N. Lazaridi and V. F. Nesterenko, "Observation of a New Type of Solitary Waves in a One-Dimensional Granular Medium," *Zh. Prikl. Mekh. Tekh. Fiz.*3, 115-118 (1985).
35. H. Deresiewicz, "A Review of Some Recents Studies of the Mechanical Behavior of Granular Media," *Appl. Mech. Rev.* **11**, 259-261 (1958).
36. J. Dvorkin, A. Nur, and H. Yin, "Effective Properties of Cemented Granular Materials," *Mech. of Mater.* **18**, 351-66 (1994).
37. K. W. Winkler, "Contact Stiffness in Granular Porous Materials: Comparison between Theory and Experiment," *Geophys. Res. Lett.* **10**, 1073-76 (1983).
38. J. P. Jernot, M. Coster, and J.L. Chermant, "Model to Describe the Elastic Modulus of Sintered Materials," *Phys. Stat. Sol. (a)* **72** (325-32, 1982).
39. J. P. Jernot, J. L. Chermant, and M. Coster, "A New Model to Describe the Variation of Electrical Conductivity in Materials Sintered in Solid Phase," *Phys. Stat. Sol. (a)* **74**, 475-83 (1982).

40. J. P. Jernot, J. L. Chermant, and M. Cooster, "The Mean Free Path in the Porous Phase of Sintered Materials," *Powder Technology* **30**, 31-35 (1981).
41. J. P. Jernot, M. Coster, and J. L. Chermant, "Model of Variation of the Specific Surface Area During Sintering," *Powder Technology* **30**, 21-29 (1981).
42. J. Gurland, "The Measurement of Grain Contiguity in Two-Phase Alloys," *Transactions of The Metallurgical Society of AIME* **212**, 452-455 (1958).
43. L. Oger and J. P. Jernot, "Different Methods to Analyze the Structure of Granular Media," in *Disorder and Granular Media*, edited by D. Bideau and A. Hansen (North-Holland, Amsterdan, 1993), pp. 165-198.
44. R. Cytermann, "A New Way to Investigate the Dependence of Elastic Moduli on the Microstructure of Porous Materials," *Powder Metallurgy International* **19** (4), 27-30 (1987).
45. S.A. Saltykov, *Sterometric Metallography* (Second Edition, Metallurgizdat, Moscow, 1958).
46. E. Arzt, "The Influence of an Increasing Particle Co-ordination on the Densification of Spherical Powders," *Acta Metallurgica* **30**, 1883 (1982).
47. E. Arzt, M. F. Ashby, and K. E. Easterling, "Practical Applications of Hot-Isostatic Pressing Diagrams: Four Case Studies," *Metallurgical Transactions A* **14A**, 221 (1983).
48. H. F. Fischmeister and E. Arzt, "Densification of Powders by Particle Deformation," *Powder Metallurgy* **26**, 82 (1983).
49. R. M. McMeeking, "Constitutive Laws for Sintering and Pressing of Powders," in *Mechanics of Granular Materials and Powder Systems, MD-Vol. 37*, edited by M. M. Mehrabadi, The Winter annual meeting of the American Society of Mechanical Engineers (ASME, Anaheim, California, November 8-13, 1992), pp. 51-61.
50. L. P. Martin, D. Nagle, and M Rosen, "Relationship between Specific Surface Area and Ultrasonic Sound Wave Velocity in Sintered Ceramic Powders," in *Ceramic Transactions, Volume 89: Nondestructive Evaluation of Ceramics,* edited by C. H. Schilling and J. N. Gray (American Ceramic Society, Westerville, Ohio, 1998).
51. F. F. Lange, "Sinterability of Powder Agglomerates," *J. Am. Ceram. Soc.* **67** (2), 83-89 (1984).

Characterization of Sintered Ceramics and Composites

IN SITU MEASUREMENT OF ULTRASONIC SOUND WAVE VELOCITY DURING SINTERING OF CERAMIC POWDERS

L. P. Martin, D. Dadon and M. Rosen
The Johns Hopkins University, Department of Materials Science and Engineering, Baltimore, MD 21218

ABSTRACT

A technique is presented for *in situ* measurement of the longitudinal wave velocity in ceramic powder compacts undergoing sintering. The technique uses laser generation and contact detection of the ultrasonic signal. Results are presented for sintering zinc oxide and alumina. The data demonstrate the capability for performing *in situ* longitudinal wave velocity measurements at high temperatures, and in attenuative materials such ceramic powder compacts. In addition, the data demonstrate the sensitivity of ultrasonic velocity measurements to changes within the sintering compact which are not detectable by dimension measurement alone. A novel interpretation of the ultrasonic velocity data is also presented. This interpretation is based upon analysis of the longitudinal wave velocity data in the context of existing models treating surface area reduction during sintering. A previously observed correlation between the longitudinal wave ultrasonic velocity and the specific surface area facilitates this analysis. Activation energies for surface area reduction during sintering of ZnO which were determined by this technique compare well with values reported in the literature.

INTRODUCTION

During the initial stages of sintering in ceramic powder compacts, the ultrasonic velocity can be observed to increase significantly with little or no change in the overall pore volume fraction [1,2]. This is attributable to changes in pore geometry associated with sintering mechanisms, such as surface diffusion, which provide material transport without significant shrinkage. Thus, ultrasonic velocity measurements, which can be made *in situ* during sintering, may provide information about the state of the sintering material not available from more common *in situ* measurements, such as dilatometry. During the intermediate and final stages of sintering, the ultrasonic velocity continues to increase and is accompanied by a concurrent reduction in the pore volume fraction. In order to relate the measured velocities to existing sintering or ultrasonic models, numerous simplifying assumptions as to the pore/particle geometry are often made. These are generally unrealistic assumptions when considering real powder systems, which tend towards non-uniform particle size and shape. A correlation between the ultrasonic velocity and the specific surface area reduction in partially sintered zinc oxide (ZnO) and alumina (Al_2O_3) powder compacts has recently been reported [1,2]. This correlation provides a means of gathering data (sintering kinetics via evaluation of the specific surface area) which have previously not been available by measurements performed *in situ* during the sintering cycle.

A technique is presented here for accurate measurement of the longitudinal wave velocity up to a temperature $T = 1500\ ^{\circ}C$ in sintering powder compacts. Representative results are presented for heating cycles performed upon sintering ZnO and Al_2O_3 powder compacts. A previously observed correlation between the longitudinal wave ultrasonic velocity and the specific surface area is applied towards interpretation of the ultrasonic velocity data in ZnO in terms of existing models treating surface area reduction during sintering.

SAMPLE PREPARATION AND CHARACTERIZATION

ZnO powder compacts were fabricated from a commercial zinc oxide powder (Z-1012, Cerac, Milwaukee, WI). Specific surface area (3.10 m^2/g) and SEM analysis indicated highly agglomerated particles with an average grain size on the order of ~0.3 μm. The samples were uniaxially pressed, without binder, to 37 MPa or cold isostatic pressed (CIP'ed) to 300 MPa. These green samples were cylindrical, with a diameter of 31 mm, thickness of approximately 16 mm. The CIP'ed samples had an average pore volume of P~0.31, and the uniaxially pressed samples had an average of P~0.47.

Alumina (Al_20_3) samples were prepared by uniaxial pressing 15-20 g of powder, without binder, to15 MPa, followed by CIP'ing to 150 MPa. The CIP'ing was performed in order to minimize the elastic anisotropy associated with uniaxial pressing. The powder (AKP-53, Sumitomo Chemical America, Inc. New York, NY) was specified by the manufacturer as >99.99 % pure alpha alumina. The measured specific surface area was 13.19 m^2/g, with typical green sample specifications as follows: diameter 30.3 mm, thickness approximately 11.7 mm, and pore volume fraction P=0.42. All ZnO and Al_2O_3 samples were pre-dried at 105 °C for 12 hours. Bulk density was determined from dimensional measurement and weighing. Pore volume fractions P were determined from

$$P = 1 - \frac{\rho}{\rho_0} \tag{1}$$

where ρ represents the bulk density, as determined experimentally from mass and volume, and ρ_o represents the theoretical (X-ray) density.

Room temperature ultrasonic velocity measurements were performed using commercial 5 or 3.5 MHz longitudinal wave transducers. The transducers were excited by timed pulses from an ultrasonic pulser-receiver (5052PR, Panametrics, Waltham, MA). Time of flight measurements for the ultrasonic signals were performed using the pulse-echo overlap method on a digital oscilloscope (9450A, LeCroy, Chestnut Ridge, NY). Contamination of the porous samples with couplant was prevented by use of a polymer barrier between the sample and the transducer/couplant when making room temperature time of flight measurements.

HIGH TEMPERATURE ULTRASONIC VELOCITY MEASUREMENT

Two slightly different systems were used for performing the *in situ* ultrasonic velocity measurements. The details of the measurement systems are as follows. For the ZnO data, the ultrasonic signal was generated on one face of the cylindrical sample using a Nd:YAG pulsed laser source (MK-367, Kigre, Hilton Head, SC) delivering 17.5±2 mJ per 4 ns pulse at a wavelength of 1064 nm. The laser was focused to a spot size of approximately 0.5 mm. Signal detection was performed using a contact piezoelectric transducer situated outside the furnace at the end of a 30 cm fused quartz acoustic rod, 1.50 cm in diameter. In order to prevent excessive heating of the transducer by infrared radiation (IR) transmitted down the length of the acoustic rod, a reflective Al foil layer was placed at the end of the rod. Since scattering of the IR from the end of the rod resulted in localized heating at that location, it was required that the transducer be further removed by placing a second short (1.5 cm) length of fused quartz rod between the foil and the transducer. In this configuration, the transducer faceplate temperature was maintained below approximately 35 °C. Henceforth, for brevity, this entire fixture of long rod/Al foil/short rod will be referred to simply as the 'buffer rod'. Pressure coupling was used to achieve transmission of the ultrasonic signal from the sample to the buffer rod. Repeatable pressure was achieved by spring loading the sliding stage upon which the buffer rod was fixed. Since the application of pressure can significantly enhance the sintering process, the spring-loaded buffer rod was placed in contact with the sintering sample only during the brief period of each measurement (< 5 sec/measurement). At all other times, it was removed so that the end of the buffer rod was approximately 2-3 mm from the sample surface. The pulsed laser used to excite the ultrasonic signal was fired down the length of a fused quartz tube aligned with the buffer rod. Changes in sample dimension were monitored using a dial gauge reading the displacement of the sliding stage supporting the buffer rod. Calibration runs using fully dense ZnO samples were performed in order to allow dimension changes due to expansion in the quartz rods and tube, and thermal expansion in the samples, to be accounted for. This apparatus was determined to be capable of measuring the longitudinal wave ultrasonic velocity over a temperature range from room temperature to approximately 1100 °C.

The temperature limitation of the apparatus described above was principally a result of the fused quartz components. In the second system, used for measurement of the alumina samples, these components were replaced by high purity alumina, and the length of each of these components was reduced by the use of a small box furnace rather than the long tube furnace used in the original system. The use of the alumina buffer rod alleviated the problem of IR transmission down the length of the rod, and increased the effective temperature range to approximately 1500 °C. However, due to the increased thermal conductivity and reduced length, it was necessary to affix a water cooling jacket to the end of the alumina buffer rod to prevent excessive heating of the ultrasonic transducer. Also, in the second system a 450 mJ, frequency doubled Nd:YAG pulsed laser (Surelite I, Continuum, Santa Clara, CA) was used for generation. The pulse width was 4-6 ns at a wavelength of 523 nm. The maximum energy output at the 532 nm wavelength was approximately 230 mJ/pulse, however the beam energy was attenuated significantly in order to prevent ablation damage to the sample. Finally, in the second system the data collection (via computer interface) was somewhat more sophisticated than in the original system. Here a digital dial gauge was directly interfaced via Labview to allow direct recording of the displacement associated with the sample and rod expansion/shrinkage. In the future, the entire system will be automated. A schematic of the experimental apparatus is shown in Figure 1.

The ultrasonic time of flight can be determined using the experimental systems described above by subtracting the one-half the time of flight of a pulse echo signal in the buffer rod from the time of flight of the laser generated signal through the sample and buffer rod. Sidewall reflections in the buffer rod, however, hindered interpretation of the pulse echo ultrasonic signal. This problem was ameliorated by the use of a higher frequency for the pulse echo measurement. At the higher frequencies where the pulse echo measurements were easily made upon the buffer rod (>10 MHZ) the attenuation in the green samples was excessive. For this reason, the laser generated signal was detected using a conventional contact piezoelectric transducer with a fundamental frequency of 5 or 3.5 MHZ. This transducer was then driven at a higher harmonic, approximately 15-18 MHZ, in order to perform the pulse echo measurement on the buffer rod assembly.

The ultrasonic time of flight in the heated sample was measured by a differential measurement technique. Room temperature pulse echo measurements were performed on the sample, outside the high temperature measurement apparatus, to accurately determine time of flight t_s^o and the sample thickness l_o, where the velocity is given by t_s^o/l_o. The sample was then placed in the high temperature measurement apparatus. A through transmission waveform, representing the room temperature time of flight of the laser generated signal through the sample and buffer, was generated and stored using a personal computer (PC) interfaced with a digital oscilloscope. Also, a pulse echo waveform was generated and stored. This waveform represents the time of flight of the pulse echo signal in the buffer rod assembly at room temperature. Since the time of flight in the sample is known at this temperature, all subsequent measurements were performed by overlapping with these two waveforms. Thus, at some temperature T the laser is manually fired to initiate the measurement process. The waveform representing the laser generated signal transmitted through the sample and the buffer rod is gathered on the digital oscilloscope and automatically stored on the PC. The PC then automatically initiates a procedure whereby the waveform from the pulse echo signal in the buffer rod is collected and stored. At a later time, the waveforms collected at temperature T are compared via overlap on the oscilloscope with the waveforms collected at room temperature. Since the time of flight in the sample at room temperature, t_s^o, was known, the time of flight at any time-temperature during a heating run, t_s, can be subsequently determined by

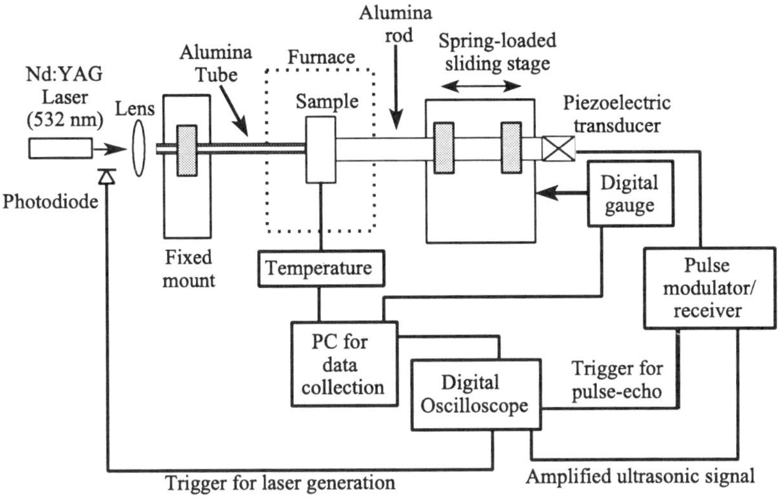

Figure 1: Schematic of high temperature ultrasonic velocity measurement system.

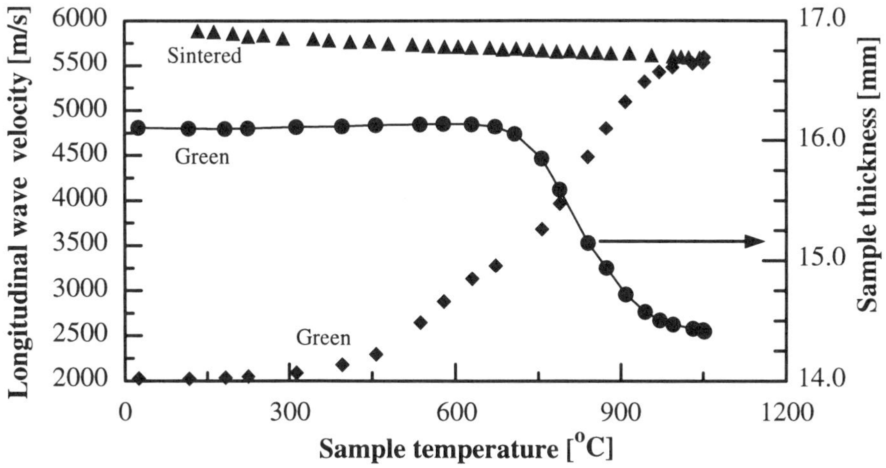

Figure 2: Longitudinal wave velocity and sample thickness versus temperature during sintering of a green ZnO sample. Also shown, longitudinal wave velocity versus temperature during heating of a near full density (sintered) ZnO sample.

$$t_s = t_s^0 + (\Delta t_1 - \frac{1}{2}\Delta t_2) = t_s^0 + \Delta t_s. \quad (2)$$

where Δt_1 represents the change in the time of flight of the laser generated signal through the sample and the buffer rod , and Δt_2 represents the change in the pulse echo time of flight in the buffer rod. To determine velocity from time of flight, thickness changes determined by the dial gauge (reading the sliding stage displacement) were corrected for expansion in the rods to determine the change in sample size Δl as a function of time and temperature. Thus, the longitudinal wave velocity at temperature T, $V_L(T)$, is determined by this differential technique to be

$$V_L(T) = \frac{l_o + \Delta l}{t_s^o + \Delta t_s}. \quad (3)$$

Automation of the data collection using the PC makes the acquisition very rapid, and allows the data to be analyzed in near real time, or after completion of the sintering run. In the present investigation, data analysis was performed after the sintering run was complete.

RESULTS AND DISCUSSION

The methodology described above was applied to the determination of the longitudinal wave velocity in sintering ZnO and Al_2O_3 powder compacts. The results of a typical sintering run, with a heating rate of 5 °C/min, during which an isostatically pressed (green) ZnO sample was sintered to near full density are shown in Figure 2. During this sintering run, it can be seen that the ultrasonic (longitudinal wave) velocity begins to increase significantly at a temperature well below that at which sintering shrinkage begins to occur. In fact, the velocity can be seen to increase by over 60 %, to approximately 3250 m/s from the green value of approximately 2000 m/s, over the temperature range 25 to 625 °C. This indicates that the ultrasonic velocity is affected by changes in the pore structure which are not detectable by more conventional techniques such as dilatometry. When shrinkage begins to occur at approximately 625 °C, there is a noticeable change in the rate at which the ultrasonic velocity increases with temperature. In this sintering run, the heating was stopped at 1050 °C, and the sample held at this temperature for 15 minutes before a second velocity measurement was made. The increase in the ultrasonic velocity during this dwell is shown by the pair of points at 1050 °C. Figure 2 also shows the longitudinal wave velocity measured during a second heating of the same sample. During this run there was no additional densification of the sample as near full density was reached during the initial sintering run. Over the temperature range from 25 to 1050 °C the longitudinal wave velocity in the dense ZnO decreases by approximately 5%. This observation is consistent with that of previous investigators [3].

In Figure 3, the longitudinal wave velocity is shown for two separate heating cycles, also at 5 °C/min, of an Al_2O_3 sample. The first cycle represents the sintering of the green sample, which reached ~0.99 of the theoretical density (<1% porosity) during the sintering treatment. Sample dimension during the sintering is also shown, and from these data it is apparent that there is approximately a 150 % increase in the sound wave velocity before the onset of densification, at approximately 1000 °C. In addition, there is a noticeable change in the slope of the sound wave velocity as a function of temperature when densification begins. These effects are qualitatively similar to those exhibited by the ZnO system. The reduction in velocity between 200 and 400 °C is attributable to the desorption of surface species, for example hydroxyl groups [2]. Also shown in Figure 3 is the longitudinal wave velocity versus temperature for the same sample during a second heating, again at 5 °C/min. During this run there was no additional densification of the sample as near full density was reached during the initial sintering run. These data demonstrate that the temperature coefficient of the longitudinal wave velocity in alumina is small in the temperature range from room temperature to 1500 °C, approximately 0.25 m/s/°C. The scatter in the data are attributed to experimental error associated with measurement of the very small time of flight of the ultrasonic signal in the sintered alumina sample (t_s<0.9 μs). This level of scatter is expected to be improved upon as the measurement methodology is refined during future work.

The data presented in Figures 2 and 3 demonstrate the capability for making *in situ* longitudinal wave velocity measurements at high temperatures, and in attenuative materials such as sintering ceramic powder compacts. In addition, the data demonstrate the sensitivity of the ultrasonic sound wave velocity measurements to changes within the sintering compact which are not detected by dimension measurement alone. However, the quantitative interpretation of such data in porous materials is not straightforward [4-8]. For this reason, ultrasonic velocity measurements have experienced only limited use as a parameter for monitoring sintering processes.

A novel application of ultrasonic velocity measurements performed *in situ* during initial stage sintering of ZnO is presented below. This application is based upon previous observations of the linear the variation of longitudinal wave velocity with specific surface area in partially sintered porous samples prepared from the present material system [1]. Figure 4 presents longitudinal wave velocity data as a function of time during initial stage sintering of a uniaxially pressed sample of ZnO. The initial pore fraction in the sample was P~0.47. The time-temperature profile used consists of isothermal dwells at four temperatures: 234, 366, 433 and 507 °C. The sample dimensions measured at room temperature were identical before and after the sintering run, indicating that there was no sintering shrinkage involved. Despite this, it is apparent that there is a large increase in the longitudinal wave velocity. The longitudinal wave velocity changes by 0.4, 2.8, 8.0, and 17.8 % during dwells of approximately 100 minutes at each of the respective temperatures. In the calculation of the ultrasonic velocities from the measured times of flight during this sintering run, the thermal expansion in the sample was assumed to be negligible. This is supported by the small thermal expansion coefficient of dense ZnO, $<7.5\times10^{-6}$ K^{-1} for T<500 °C [9], and by the high pore fraction in this sample, P~0.47. Thus, the sample thickness was assumed to be constant at the room temperature value of 17.1 mm throughout the

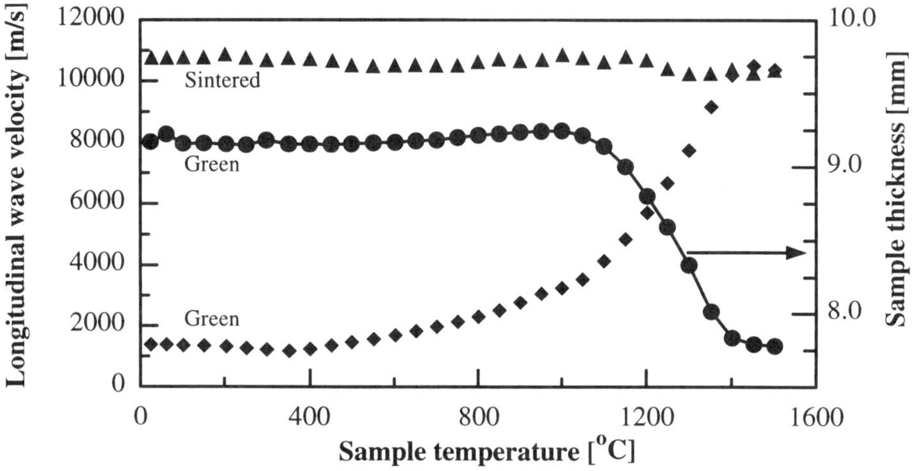

Figure 3: Longitudinal wave velocity and sample thickness versus temperature during sintering of a green alumina sample. Also shown, longitudinal wave velocity versus temperature during heating of a near full density (sintered) alumina sample.

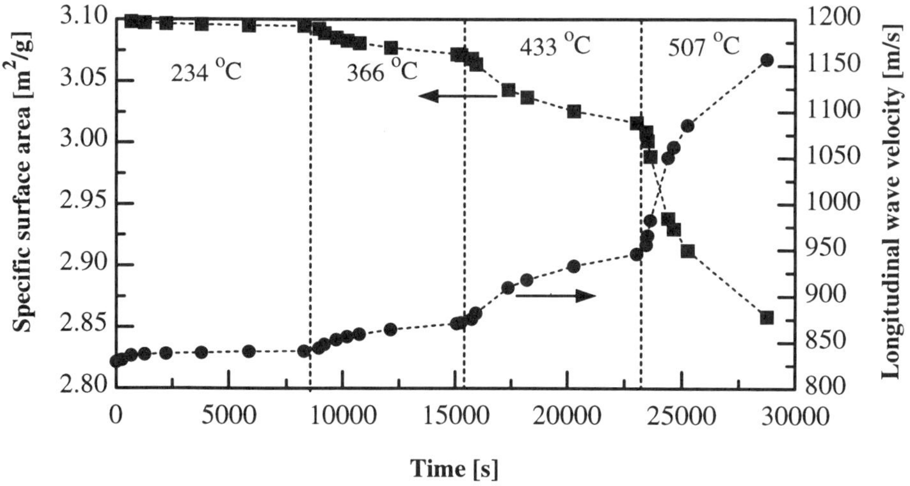

Figure 4: Specific surface area and longitudinal wave velocity versus time for a green ZnO sample subject to several consecutive isothermal dwells. Surface area was inferred from the velocity data using equation (4). Labels indicate dwell temperatures.

sintering treatment shown in Figure 4. This results in an error of no more than approximately 0.1% at 200 °C and 0.3% at 530 °C.

The increase in the ultrasonic velocity at constant pore fraction, i.e. without densification, shown in Figure 4 can be attributed to the changing morphology of the particle contact regions in the powder compact. Such changes in morphology are manifested in the surface area of the porosity as well as in the ultrasonic velocity. It has been observed experimentally that in certain material systems the longitudinal wave velocity varies linearly with specific surface area reduction over a wide range of porosity [1,2]. The relation between specific surface area, as measured by Brunnauer, Emmet and Teller (BET) analysis of Kr gas adsorption at 78 K, and the longitudinal wave velocity in ZnO samples identical to that shown in Figure 4 has been determined to be

$$S = 3.72 - 7.45x10^{-3} V_L, \qquad [m^2/g] \qquad (4)$$

were S is the surface are expressed in m^2/g and V_L is the longitudinal velocity in m/s [1]. The observed linear relation between the increase in the longitudinal wave velocity and the reduction in specific surface area in porous powder compacts allows interpretation of *in situ* measurements, such as those presented in Figure 4, in the context of existing models treating surface area reduction during initial stage sintering.

For comparison with the ultrasonic velocity, the specific surface area, as inferred from the ultrasonic velocity data using equation (4), is also shown in Figure 4. The reduction in the specific surface area S with time t during the initial stage of isothermal sintering may be expressed as

$$\left(\frac{\Delta S}{S_0}\right)^{\gamma} = \left(\frac{Ct}{T}\right) \exp(-Q/RT), \qquad (5)$$

where S_0 is the initial specific surface area, ΔS is the incremental surface area reduction S_0-S, C is a constant determined by the geometry, surface energy and transport mechanism [9,10]. The exponential term arises from the diffusivity $D=D_0 exp(-Q/RT)$ [10]. The value of γ is primarily determined by the dominant sintering mechanism, with geometry (particle coordination and interparticle neck size) playing a secondary role. If isothermal treatment is considered, it is useful to restate equation (5) as

$$S = S_0 \left(1 - \left(\frac{C}{T}\right)^{\frac{1}{\gamma}} \exp\left(\frac{-Q}{RT\gamma}\right) t^{1/\gamma} \right). \qquad (6)$$

In Figure 5 the surface area, as inferred from the ultrasonic velocity data, are shown with lines indicating the best fit of equation (6) to each isothermal segment of the data.

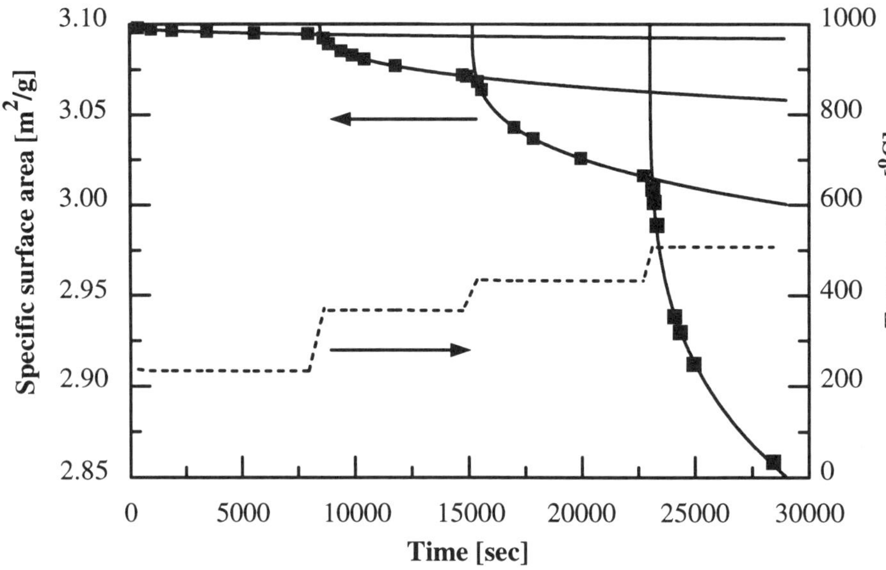

Figure 5: Specific surface area and isothermal dwell temperature versus time during sintering. Curves indicate fit of equation (6) to surface area data.

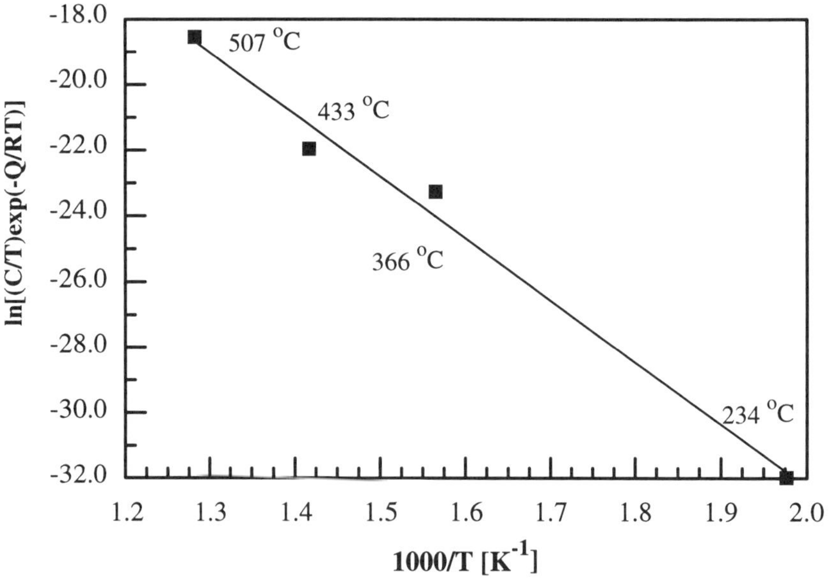

Figure 6: ln[(C/T)exp(-Q/RT)], determined at each isothermal dwell temperature, versus 1000/T. The slope, -Q/R, yields the activation energy for the surface area reduction process during the initial stage of sintering.

From the values of γ and *(C/T)exp(-Q/RT)* determined from fitting equation (6) to the surface area data for each isothermal segment shown in Figure 5, it is possible to evaluate the activation energy Q as follows. Consider the equality

$$\ln\left[\left(\frac{C}{T}\right)\exp\left(\frac{-Q}{RT}\right)\right] = \ln\left(\frac{C}{T}\right) - \left(\frac{Q}{R}\right)\left(\frac{1}{T}\right). \qquad (7)$$

Since the temperature dependence in the logarithm is small compared to the term *(1/T)*, a plot of *ln[(C/T)exp(-Q/RT)]* versus *1/T* yields a straight line with a slope *-Q/R*. This is shown in Figure 6, where the slope yields Q=155$\pm$15 kJ/mol over the range of isothermal dwell temperatures T~230-500 °C. The correlation coefficient for this linear fit, 0.994, indicates a statistically significant correlation. At each of the temperatures investigated, the optimum value of γ was within the range 3.1-3.9. The values of Q and γ determined by this technique compare favorably with those reported in the literature: Q ranging from 118-147 kJ/mol, γ ranging from 3.7-4.1 [1,12,13]. It should be noted that the previously reported experimental determinations of Q and γ were based upon post sintering measurements performed upon many samples, each sintered for a specified time at a single temperature. That the present analysis was performed upon *in situ* data gathered from a single sample demonstrates the usefulness of this ultrasonics based technique.

SUMMARY

A technique for accurate measurement of the longitudinal wave velocity up to 1500 °C in sintering ceramic powder compacts was presented. The technique uses laser generation and contact detection of the ultrasonic signal. Results were presented for heating cycles performed upon sintering ZnO and Al_2O_3 powder compacts, and upon a pre-sintered dense samples. The data demonstrate the sensitivity of ultrasonic velocity measurements to changes within the sintering compact which are not detected by dimension measurement alone. These changes are attributable to changes in pore geometry associated with sintering mechanisms, such as surface diffusion, which provide material transport without significant shrinkage. Thus, *in situ* ultrasonic velocity measurements during sintering may provide information about the state of the sintering material not available from more common *in situ* measurements, such as dilatometry. Furthermore, a novel application of ultrasonic velocity measurements performed *in situ* on sintering ZnO was presented. A previously observed correlation between the longitudinal wave ultrasonic velocity and the specific surface area facilitated interpretation of *in situ* velocity measurements. The analysis yielded an activation energy for surface area reduction in sintering ZnO which compares well with values reported in the literature. This analysis of ultrasonic velocity data in the context of surface area reduction provides a unique method for *in situ* evaluation of sintering mechanism and kinetics.

ACKNOWLEDGMENTS

The authors wish to express their appreciation for financial support from the U.S. Army Research Laboratory under award No. DAAL019620047, and from the. U.S. Department of Energy, Division of Advanced Energy Projects, under contract No. DE-FG02-94ER-12140.

REFERENCES

1. L. P. Martin and M. Rosen, "Correlation Between Surface Area Reduction and Ultrasonic Velocity in Sintered Zinc Oxide Powders," *J. Am. Ceram. Soc.*, **80** [4], 839-846, (1997).

2. L. P. Martin, D. Nagle and M. Rosen, "Relationship Between Specific Surface Area and Ultrasonic Sound Wave Velocity in Sintered Ceramic Powders," submitted to this *Proceedngs.*

3. K. L. Telschow, J. B. Walter and G. V. Garcia, "Laser Ultrasonic Monitoring of Ceramic Sintering," *J. Appl. Phys.*, **68** [12], 6077-6082 (1990).

4. S. Spinner, F. P. Knudsen and L. Stone, "Elastic Constant-porosity Relations for Polycrystalline Thoria," *J. Res. Natl. Bur. Std.., Sect. C*, **67** [1], 39-46 (1963).

5. E. A. Dean and J. A. Lopez, "Empirical Dependence of Elastic Moduli on Porosity for Ceramic Materials," *J. Am. Ceram. Soc.*, **66** [5], 366-370 (1983).

6. L. P. Martin, D. Dadon and M. Rosen, "Evaluation of Ultrasonically Determined Elasticity-Porosity Relations in Zinc Oxide," *J. Am. Ceram. Soc.*, **79** [5], 1281-1289 (1996).

7. H. Ledbetter, M. Dunn, S. Kim and R. Fields, "Void Shape in Sintered Titanium," in *Review of Quantitative Nondestructive Evaluation, Vol. 14*, D. O. Thompson and D. E. Chimenti, eds. (Plenum Press, NY, 1995), pp. 1633-1639.

8. R. M. Spriggs, "Expression for Effect of Porosity on Elastic Modulus of Polycrystalline Refractory Materials, Particularly Aluminum Oxide," *J. Am. Ceram. Soc.*, **44** [12], 628-629 (1961).

9. *Thermophysical Properties of Matter: Volume 13*, Y.S. Touloukian, R. K. Kirby, R. E. Taylor and T. Y. R. Lee, eds., (Plenum Press, New York, 1977), pp. 444-447.

10. R. M. German, "Interpretation of Isochronal Temperature-dependent Alumina Sintering Data," *Powder Tech.*, **17**, 287 - 289 (1977).

11. R. M. German, "A Sintering Parameter for Submicron Powders," *Sci. of Sintering*, **10** [1], 11 - 25 (1978).

12. S. K. Moon, "Diffusion in and on Solid Oxides," *J. KIChE.*, **11** [3], 157-163 (1973).

13. L. F. Norris and G. Parravano, "Sintering of Zinc Oxide," *J. Am. Ceram. Soc.*, **46** [9], 449-452 (1963).

RELATIONSHIP BETWEEN SPECIFIC SURFACE AREA AND ULTRASONIC SOUND WAVE VELOCITY IN SINTERED CERAMIC POWDERS

L. P. Martin, D. Nagle and M. Rosen
The Johns Hopkins University, Department of Materials Science and Engineering, Baltimore, MD, 21218

ABSTRACT

Specific surface area and ultrasonic velocity were measured in zinc oxide and alumina powder compacts partially sintered to different pore fractions. A correlation between the reduction in surface area and the increase in ultrasonic velocity with degree of sintering was observed. In addition, the particle size distribution was observed to have a strong effect upon the relative evolution of the elastic properties and the surface area reduction during sintering of alumina. The results indicate that, in the case of a broad particle size distribution, the initial surface area reduction is dominated by the sintering of the smaller particles in the compact, and can occur with relatively insignificant increases in the elastic moduli (i.e. ultrasonic velocity). In contrast, the increase in the elastic moduli, as characterized by the ultrasonic velocity, appears to be dominated by the sintering of the larger particles. When the particle size distribution is sufficiently narrow, the surface area reduction occurs concurrently with increased ultrasonic velocity over nearly the entire range of sintering. These observations contribute to the understanding of the dynamic evolution of the elastic properties and particle contact morphology during sintering of ceramic powder compacts. In addition, the observed correlation between ultrasonic velocity and surface area reduction allows analysis of ultrasonic data in the context of existing sintering models, thus facilitating the use of ultrasonic velocity measurements as a real-time tool for monitoring and controlling sintering processes.

INTRODUCTION

Sensitive techniques for *in situ* monitoring of the sintering process can provide a better understanding of the interrelationship between the evolution of particle contact morphology and the physical properties, such as the elastic moduli. This is particularly true for systems in which the sintering behavior is not well understood, such as in nano-powder compacts. Also, the capability to detect the effects of changes in morphology attributable to the different stages of sintering may permit optimization of sintering cycles to provide tailored porosity/moduli/surface area characteristics. Further, in some processes two or more initial powders are mixed together and then reaction sintered. Consequently, changes in the properties of the powder compact result from both the densification and chemical reaction between the initial powders. In this case, *in situ* diagnostic techniques may provide valuable information regarding not only the morphological changes in the compact, but also the extent of the reaction between the initial species.

Specific surface area has been shown to be a useful parameter for the evaluation of sintering kinetics [1-6]. Analysis of the morphology evolution during the initial stages of sintering spheres has shown the specific surface area S to be dependent upon interparticle neck radius x by

$$\left(\frac{\Delta S}{S_0}\right) = K_1\left(\frac{x}{a}\right)^m \tag{1}$$

where ΔS is the loss in surface area $(S-S_0)$, S_0 is the initial surface area (in the unsintered sample), x/a is the neck radius divided by the particle radius a, and K_1 and m are constants dependent upon the particle packing, geometry and transport mechanism [7,8]. Various models predicting the reduction in surface area during initial stage sintering are well developed in the literature [1-5]. Much of this work is based upon that of Kuczynski, in which mass transport is assumed to be driven predominantly by the high curvature gradients associated with the regions of interparticle contact [9]. Surface area has many advantages as a parameter for characterization of the sintering powder compact [3]. In particular, specific surface area is strongly affected (due to the direct dependence upon interparticle neck diameter) by mass transport mechanisms, such as surface diffusion, which do not lead to shrinkage of the powder compact. The most common techniques for *in situ* monitoring of sintering powder compacts, dilatometric (density) measurements, are not sensitive to such mass transport mechanisms which may result in large changes in the particle contact morphology without shrinkage of the powder compact [10-12]. However, the most common techniques for measurement of specific surface area require the sample to be cooled to liquid nitrogen temperature, which precludes *in situ* measurement during sintering. Thus, there is a need for *in situ* measurements capable of providing quantitative information regarding the state of the sintering powder compact not available by conventional dilatometric techniques.

Ultrasonic velocity in powder compacts has been observed to increase significantly during the initial stage of sintering in which there is little or no change in the overall pore volume fraction [13-15]. This increase is attributable to the growth of interparticle necks, and the concurrent changes in pore morphology associated with sintering mechanisms, such as surface diffusion, which provide material transport without significant shrinkage. However, the quantitative interpretation of these data is not straightforward due, in part, to the complex pore/particle morphology present in real powder compact systems. The present work was designed to examine the correlation between ultrasonic velocity and surface area measurements performed post sintering. Evaluation of post sintering measurements creates a foundation upon which the interpretation of *in situ* ultrasonic data can be based.

Several series of zinc oxide and alumina samples were prepared by dry pressing of different initial powders. The samples were partially sintered to varying extents, and post sintering ultrasonic velocity measurements performed. The correlation between the longitudinal wave velocity and the surface area reduction during sintering was observed to be strongly affected by the particle size distribution in the initial powder. When the particle size distribution was narrow, the surface area varied linearly with the longitudinal wave velocity over most of the sintering cycle. When the particle size distribution was broad, the surface area-ultrasonic velocity relation was indicative of a percolated network of large particles with small particles situated in the interstices.

EXPERIMENTAL

Several series of green samples were prepared from different commercial zinc oxide and alumina powders. Due to differences in the particle size and degree of agglomeration, the preparation of the samples varied slightly among the different powders. The preparation and characterization techniques used in this study are described below. In all cases where uniaxial pressing was performed, a cylindrical mold with an inside diameter of approximately 31 mm was used. When cold isostatic pressing (CIP'ing) was performed after the uniaxial pressing step, the samples retained their cylindrical geometry.

The zinc oxide (ZnO) samples were prepared by uniaxial dry pressing of two high purity commercial ZnO powders (Kadox 930, Kadox 911, Zinc Corporation of America, Monaca, PA). The Kadox 930 was specified as 99.99% passing through a -325 mesh. Green samples were prepared by uniaxially pressing 20 g of powder without binder to 25 MPa. The average green sample thickness was 9 mm, and the average pore volume fraction was $P = 0.49$. The measured specific surface area was 3.54 m^2/g. The Kadox 911 was specified as 99.99% passing through a -325 mesh. The green samples were prepared by uniaxially pressing 10 g of powder, without binder, to 15 MPa. The resultant green samples had an average thickness of 5 mm and an average pore volume fraction of $P = 0.54$. The measured specific surface area was 8.08 m^2/g. Both ZnO powders were specified by the manufacturers as at least 99.6% pure.

Alumina (Al_2O_3) samples were prepared from three different commercial alpha alumina ($\alpha-Al_2O_3$) powders by uniaxial pressing 20 ± 1 g of powder, without binder, to15 MPa, followed by CIP'ing to 150 MPa. The CIP'ing was performed in order to minimize the elastic anisotropy associated with uniaxial pressing. The first powder (A-16SG, Alcoa Industrial Chemicals, Bauxite, AK) was specified as 99.9% passing through a -325 mesh screen, and 99.8% pure. It should be noted that this powder may have as much as 5% residual high temperature (γ) phase present. The measured specific surface area of this powder was 8.01 m^2/g. The average green sample specifications after CIP'ing were: diameter 30.5 mm, thickness approximately 12 mm, and pore volume fraction P=0.43. The second powder (SM-8, Baikowski International, Charlotte, NC) was specified by the manufacturer as >99.99% pure. The measured specific surface area of this powder was 10.14 m^2/g. The average green sample specifications after CIP'ing were: diameter 30.1 mm, thickness approximately

13.2 mm, and pore volume fraction P=0.47. The third powder (AKP-53, Sumitomo Chemical America, New York, NY) was specified by the manufacturer as >99.99% purity. The measured specific surface area was 13.19 m^2/g, with typical green sample specifications as follows: diameter 30.3 mm, thickness approximately 11.7 mm, and pore volume fraction P=0.42.

All green samples were dried at 105 °C for 20 hrs prior to sintering. Unless specified otherwise, sintering was performed by ramping from room temperature at a constant heating rate (Kadox 930: ~10 °C/min, Kadox 911: ~7 °C/min, all alumina samples: ~25 °C/min). For each ZnO sample the sintering run was interrupted at a predetermined temperature, and the sample was extracted from the furnace and cooled to room temperature in a refrigeration unit. Alumina samples were allowed to furnace cool. Bulk density was determined from dimensional measurement and weighing. Thus, pore volume fractions, as inferred from the bulk density measurements, relate to the total porosity in the sample. Pore volume fraction P was determined from

$$P = 1 - \frac{\rho}{\rho_0} \qquad (2)$$

where ρ represents the bulk density, as determined experimentally from mass and volume, and ρ_0 represents the theoretical (x-ray) density of ZnO, 5.67 g/cm^3, or of Al_2O_3, 3.99 g/cm^3.

Ultrasonic measurements were performed at room temperature using 2.25 and 3.5 MHz longitudinal wave transducers. Unless specified as 'radial', all measurements were performed along the axis of the cylindrical samples. The transducers were excited by timed pulses from an ultrasonic pulser-receiver. Time of flight measurements for the ultrasonic signals were performed by the pulse-echo overlap method using a digital oscilloscope. Sound wave velocity was determined by measuring the sample thickness with a micrometer, and dividing the pathlength (twice the measured thickness) by the time of flight. Considerable care was taken to avoid contamination of the porous samples with couplant. Each datum point presented in the ultrasonic data represents the numerical average of three independent measurements. Scatter in these measurements is typically less than 0.5%.

Specific surface area (m^2/g) of the porous samples was determined by analysis of gas adsorption data using the Brunnauer, Emmet and Teller (BET) equation. In the present work, 10-point analyses were performed using N_2 as the adsorbate (SA-3100, Coulter Corporation, Miami, FL). The accuracy of these measurements is typically better than ±5%. Particle size distributions were obtained using a sedimentation technique (Sedigraph 5100 , Micromeretics Instrument Corp., Norcross, GA).

RESULTS AND DISCUSSION

It has been reported that the reduction in specific surface area which takes place during the initial and intermediate stages of sintering in ZnO powder compacts occurs linearly with the increase in the longitudinal wave ultrasonic velocity [16].

This can be seen in Figure 1, where specific surface area is shown as a function of longitudinal wave ultrasonic velocity for two series of partially sintered ZnO powder compacts fabricated from different submicron ZnO powders. For each set of samples, the extent of sintering shown in Figure 1 ranges from completely unsintered, the lowest velocity/highest surface area samples, to pore fractions of approximately $P \sim 0.15$. Pore closure becomes significant at pore fractions $P \leq 0.15$, and the surface area measurements, which require gas infiltration of the pores, cease to be indicative of the total specific surface area in the powder compact. Linear regression analyses are shown for the two sets of data, and the respective correlation coefficients are $r^2=0.986$ for the Kadox 911 and $r^2=0.997$ for the Kadox 930. That the data range from green to the onset of pore closure indicates that the linear relation encompasses both the initial and intermediate stages of sintering [16]. *In situ* ultrasonic velocity measurements have been performed on sintering ZnO, and correlation with the specific surface area has been applied to evaluation of the kinetics of the initial stage of the sintering process [15]. It is of interest, therefore, to explore the generality of the surface area-ultrasonic velocity relation to other material systems, and to investigate the effects of such additional factors as particle size distribution.

In Figure 2, the specific surface area and the longitudinal wave ultrasonic velocity are plotted as a function of the fractional density, ρ/ρ_0, for samples prepared from the three alumina powders and sintered to different extents. Here ρ is the sample density and ρ_0 is the theoretical density of alumina. Consider first the ultrasonic velocity versus the fractional density. These data indicate, to different degrees, the expected behavior of initially rapidly increasing velocity with little increase in density, followed by a concurrent increase in both the velocity and the density. Note that the A-16 samples exhibit increasing velocity only when accompanied by significant densification. In contrast, the AKP-53 samples exhibit significant increases in the velocity with very little densification. The SM-8 samples exhibit an intermediate amount of constant-density increase in the ultrasonic velocity. This behavior is attributed to the relative breadth of the particle size distributions in the three powders, and will be discussed in more detail below. Note, also, that the initial density of the SM-8 samples is significantly lower than for the other two series of samples. This is attributable to the hard agglomerates present in the SM-8 powder. The specific surface area data for these three series of samples, also shown in Figure 2, indicate that in all cases notable surface area reduction occurs in the absence of significant densification.

The specific surface area is plotted as a function of the longitudinal wave ultrasonic velocity for the three series of alumina samples in Figure 3. The data include, for each powder, samples ranging from green to approximately 0.95 of the theoretical density. There are several features which are significant to the interpretation of these data. For the A-16 samples, the initial decrease in the surface area occurs much more rapidly than the increase in the ultrasonic velocity. At some point in the intermediate stage of sintering, above approximately 0.68 of the theoretical density, the data become linear. The nonlinearity exhibited by the SM-8 samples is much less pronounced, particularly when considered as a fraction of the initial specific surface area which was approximately 25% greater than the A-16 samples. Also, the point at

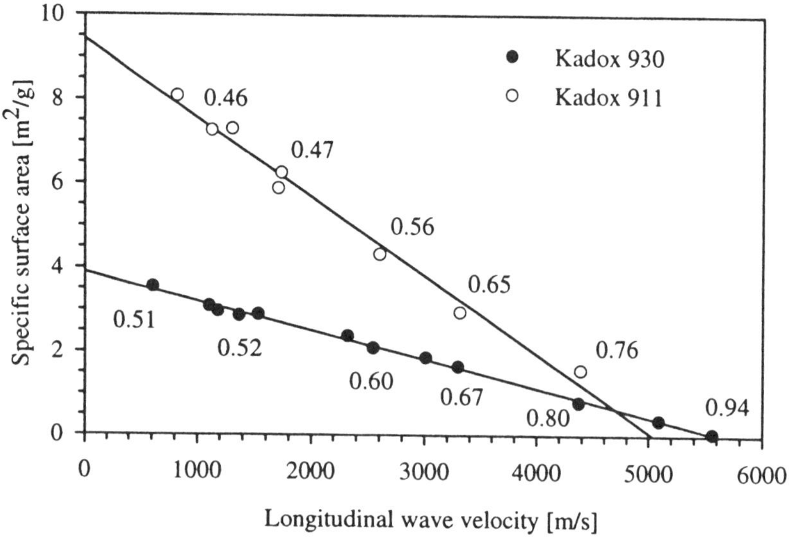

Figure 1: Specific surface area versus longitudinal ultrasonic wave velocity for two different ZnO powders partially sintered to varying densities. Labels indicate approximate relative density ρ/ρ_0.

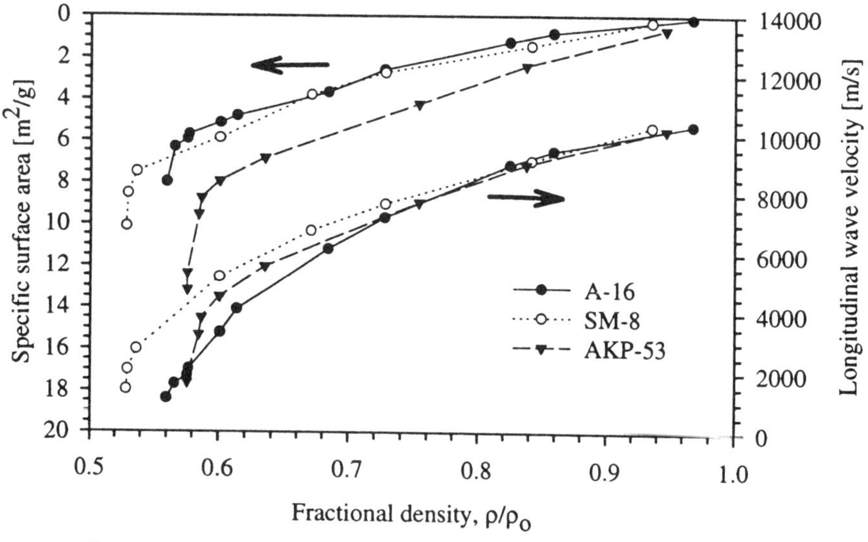

Figure 2: Specific surface area and longitudinal wave velocity versus fractional density for three series of alumina samples.

which the behavior approaches linear is approximately 0.60 of the theoretical density. The samples prepared from the AKP-53 powder exhibit a linear relation between the specific surface area and the longitudinal wave velocity over the entire range from 0.59 to 0.95 of the theoretical density. Recall that the green density for these samples was 0.58 of the theoretical. This linear relation can be expressed as

$$S=13.749-1.267\text{x}10^{-3}V_L \qquad [\text{m}^2/\text{g}] \qquad (3)$$

with a correlation coefficient of r^2=0.998. Here S represents the specific surface area and V_L represents the longitudinal wave velocity. It is apparent that two points in the early stages of sintering deviate significantly from the linear relation. This deviation appears to be different in nature from that exhibited by the A-16 and SM-8 powder samples in that the data fall above the extrapolation of the linear regime, rather than below. These data indicate the effects of low temperature (T=100-500 °C) heat treatment on the ultrasonic velocity, and therefore on the elastic moduli, in the samples prepared from the AKP-53 alumina powder. There is considerable experimental evidence that these effects are attributable to the removal of adsorbed surface species from the powder. Further investigation of these effects is currently being performed.

The linearity, or lack thereof, of the specific surface area versus longitudinal wave velocity relations shown in Figure 3 may be related to the particle size distributions in the three different alumina powders. Figure 4 shows the representative particle size distributions for samples of these three powders as determined by a sedimentation technique. The data are shown as cumulative mass percent smaller than the equivalent diameter spherical particle. The A-16 powder clearly exhibits the broadest size distribution, while the AKP-53 powder exhibits the narrowest. It should be noted that the sedimentation techniques used to gather the particle size data do not account for non-spherical particle shapes, and may not be sensitive to the presence of particles smaller than approximately 0.2 μm. The presence of sub-0.2 μm particles may result from the milling processes used to obtain the final powder. Preliminary electron microscopy indicates that both the particle size and shape are quite uniform in the AKP-53 powder, while both are highly irregular in the A-16 powder.

The role of the particle size distribution in determining the manner in which the surface area and elastic properties evolve during sintering may be described as follows. The mass transport occurring during the early stages of sintering is generally accepted to be strongly driven (thermodynamically) by the high curvature gradients associated with the particle contact regions [9]. These gradients increase rapidly with decreasing particle size, and thus the driving force for mass transport to the particle contact region (i.e. the interparticle neck) is much stronger for small particles. When the particle size distribution is broad, therefore, the driving force for sintering of the smaller particles is much stronger than for the larger particles. Also, due to their higher surface-to-volume ratio, the smaller particles must make a relatively greater contribution to the specific surface area than the larger particles. Thus, in the extreme case of a bimodal distribution with two very different sizes, the larger particles make an interconnecting, or percolated, network, and the smaller particles lie in the interstices between the large

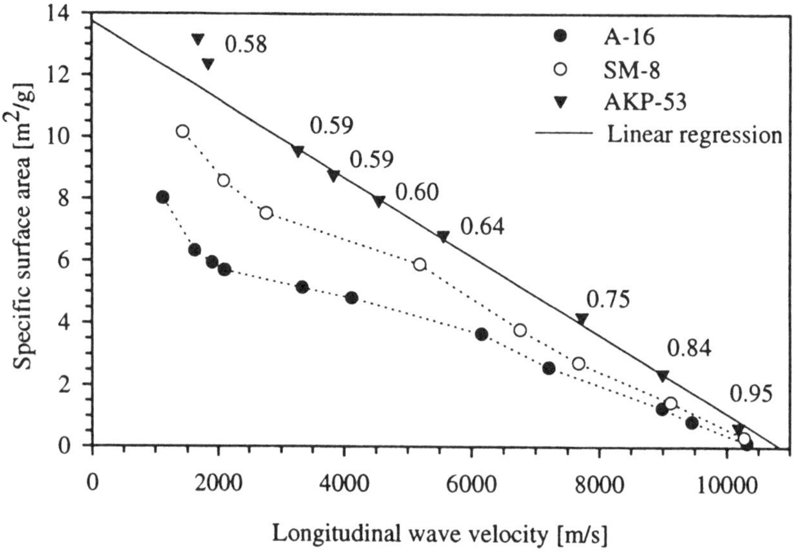

Figure 3: Specific surface area versus longitudinal wave velocity for three alumina powders sintered to different densities. Labels indicate relative density ρ/ρ_0 of the AKP-53 data.

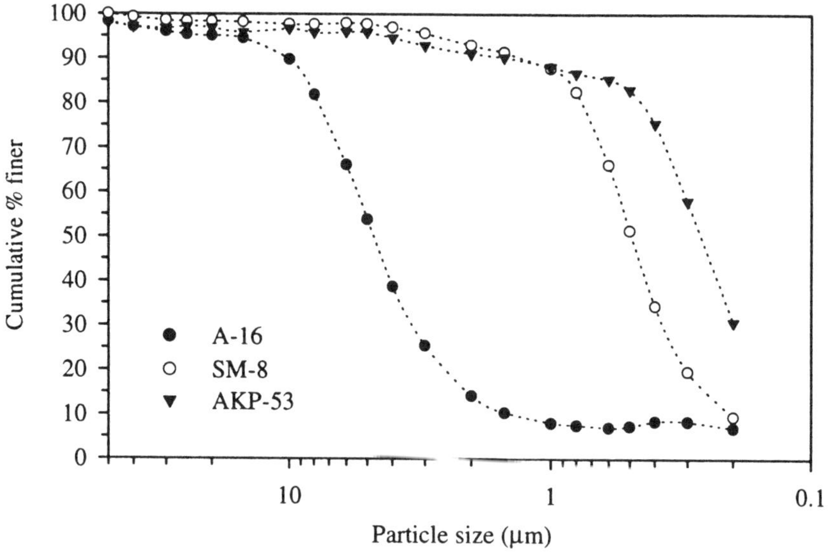

Figure 4: Representative particle size distribution for three Al_2O_3 powders.

particles. In this fashion, the small particles may sinter during the early stages without significantly affecting the interparticle neck regions between the large particles. This will serve to reduce the specific surface area without significantly affecting the elastic properties. Later in the sintering, the larger particles begin to sinter, and the increasing elastic moduli will be accompanied by a concurrent surface area reduction. This behavior is seen to some extent in the data for the A-16 samples shown in Figure 3. As the particle size distribution in the initial powder becomes narrower, for example in the AKP-53 powder, the difference in the driving force for sintering between the largest and smallest particles is reduced. In addition, the packing geometry described above is no longer feasible. Thus, the mass transport leading to surface area reduction must have a concurrent effect upon the elastic moduli.

It is interesting to note the very wide range of porosity over which the linear relation between surface area and longitudinal wave ultrasonic velocity is valid for the AKP-53 series. The porosity at the onset of this linearity is nearly equal to the green porosity (P=0.41 versus P=0.42 for the green samples). That the linear relation is valid, with a correlation coefficient of 0.998, to P~0.05 indicates that the correlation covers much of the initial and intermediate stages of sintering. The deviation of the samples shown in Figure 3 from the linear relation may be in part attributable to the presence of surface contamination residual from the powder processing. It is also possible that these data indicate the presence of very fine particles residual from the powder milling which was performed as part of the manufacturer processing. Such 'fines', ≤0.1 µm, will not be detected by the sedimentation techniques most commonly used to determine particle size distributions. Scanning electron microscopy is currently being performed in order to evaluate the particle size and shape distributions among the three alumina powders.

While the effects of irregular particle shape has not been discussed above, it may be proposed that the effect is roughly equivalent to that of a broad particle size distribution in the following sense. Irregularly shaped particles will possess regions of high curvature (i.e. sharp corners). These regions behave locally in much the same fashion as very small particles in that the driving force for mass transport is much higher when the radius of curvature is small. Thus, locally these regions will tend to sinter first, much as the small particles will sinter before the large ones.

SUMMARY AND CONCLUSIONS

The surface area reduction which occurs as a result of sintering has been correlated with the concurrent increase in the longitudinal wave ultrasonic velocity in several ZnO and Al_2O_3 systems. The observed correlation may enable surface area changes during sintering to be evaluated by ultrasonic velocity measurements which are relatively easy to perform *in situ*. This may enhance the present capability to monitor and interpret both sintering experiments and industrial sintering processes. The particle size distribution was observed to play a strong role in the relation between the surface area and the longitudinal wave velocity. Experiments performed using broad particle size distributions indicated that, during the early stages of sintering, the surface area reduction was dominated by the sintering of smaller particles. This process appears to have only a limited effect on the elastic properties, thus implying that the elastic properties are dominated by the larger particles in the powder compact. During the latter stages of sintering the elastic moduli increase is concurrent with the specific surface area reduction. It was also noted that the elastic properties in the green powder compact may be significantly affected by the presence of adsorbed species. This effect is outside the scope of the present work, and is currently being investigated. These observations contribute to the understanding of the dynamic behavior of sintering powder compacts in terms of the ultrasonic velocity, which may be determined non-intrusively during processing. This understanding facilitates the use of ultrasonic techniques for real-time monitoring and controlling of the sintering process in ceramic power compacts.

ACKNOWLEDGMENTS

The authors wish to express their appreciation for financial support from the U.S. Army Research Laboratory under award No. DAAL019620047, and from the. U.S. Department of Energy, Division of Advanced Energy Projects, under contract No. DE-FG02-94ER-12140

REFERENCES

1. J. P. Jernot, M. Coster and J. L. Chermant, "Model of the Variation in Specific Surface Area During Sintering," *Powder Tech.*, **30**, 21-29, (1981).
2. F. N. Rhines, R. T. DeHoff and R. A. Rummel, "Rate of Densification in the Sintering of Metal Powders," in *Agglomeration*, W. A. Knepper, ed. (Interscience Publishers, New York, NY, 1962) pp. 351-378.
3. R. M. German, "A Sintering Parameter for Submicron Powders," *Sci. of Sintering*, **10** [1], 11-25 (1978).
4. R. M. German and Z. A. Munir, "A Kinetic Model for the Reduction in Surface Area During Initial Stage Sintering," *Mat. Sci. Res.*, **10**, 249-257 (1975).
5. R. M. German, "Interpretation of Isochronal Temperature-dependent Alumina Sintering Data," *Powder Tech.,* **17**, 287-289 (1977).
6. S. H. Hillman and R. M. German, "Constant Heating Rate Analysis of Simultaneous Sintering Mechanisms in Alumina," *J. Mat. Sci.,* **27**, 2641-2648, (1992).
7. R. M. German and Z. A. Munir, "Morphology Relations During Surface-Transport Controlled Sintering," *Met. Trans. B*, **6B**, 289-294, (1975).
8. R. M. German and Z. A. Munir, "Morphology Relations During Bulk-Transport Controlled Sintering," *Met. Trans. A*, **6A**, 2229-2234, (1975).
9. G. C. Kuczynski, "Self-Diffusion in Sintering of Metallic Particles," *Trans. AIME*, **185**, 169-178 (1949).
10. W. S. Young and I. B. Cutler, "Initial Sintering with Constant Rates of Heating," *J. Am. Ceram. Soc.*, **53** [12], 659-663 (1970).
11. J. L. Woolfrey and M. J. Bannister, "Nonisothermal Techniques for Studying Initial-stage Sintering," *ibid.*, **55** [8], 390-394 (1970).

12. J. Wang and R. Raj, "Estimate of the Activation Energies for Boundary Diffusion from Rate controlled Sintering of Pure Alumina, and Alumina Doped with Zirconia or Titania," *ibid.*, **73** [5], 1172-1175 (1990).
13. K. L. Telschow, J. B. Walter and G. V. Garcia, "Laser Ultrasonic Monitoring of Ceramic Sintering," *J. Appl. Phys.*, **68** [12], 6077-6082 (1990).
14. J. H. Geiske and H. M. Frost III, "Technique for Measuring Ultrasonic Velocity and Attenuation Changes in Attenuative Materials at Temperatures such as During Sintering Processes," *Rev. Sci. Instrum.*, **62** [12], 3056-3060 (1991).
15. L. P. Martin and M. Rosen, "Analysis of Ultrasonic Velocity Measurements on Sintering Zinc Oxide," *Ultrasonics*, **35**, 65-73, (1997).
16. L. P. Martin and M. Rosen, "Correlation Between Surface Area Reduction and Ultrasonic Velocity in Sintered Zinc Oxide Powders," *J. Am. Ceram. Soc.*, **80** [4], 839-846, (1997).

ULTRASONIC NDE OF FINE CERAMICS WITH REFERENCE TEST PIECES

Yasushi Ikeda, Yasutoshi Mizuta, and Katsuhiro Onda
Japan Fine Ceramics Center
2-4-1, Mutsuno, Atsuta-ku, Nagoya, 456 Japan
81-528713500
and
Hitoshi Domon
Tokyo Engineering University
1404-1, Katakura-cho, Hachioji-city
Tokyo, 192 Japan
81-42637211

Takashi Udagawa
Nippon Steel Corp.
5-10-1, Fuchinobe, Sagaminhara
Kanagawa, 229 Japan
81-427-68-6037

ABSTRACT

We developed a highly sensitive ultrasonic testing with focusing type transducers using a water-immersion method. For speedy and reliable testing, some trial silicon nitride reference blocks(RBs) have been prepared and measured for their efficiency. Transducer characteristics, such as the effective frequencies, the focus sizes and the focal, were properly determined using the trial RBs. Some RBs with artificially induced surface flaws, such as slit-shaped defects and Vickers indentations were prepared. RBs with artificially induced inner flaws, such as spherical pores and flat-bottomed holes, were also prepared. The RBs were imaged with the ultrasonic c-scan method, and the echo-heights from these defects were measured with dB. For the RBs, quantitative relationships have been obtained between the sizes of defects and the echo-heights from them. It has been clarified that these RBs are very useful for the ultrasonic tests of ceramics.

INTRODUCTION

Small defects in fine ceramics, (e.g., smaller than 100 μm), can become the origins of fracture. It can be considered that the critical flaw size to guarantee quality is much smaller than 30 μm for surface flaws, and 300 μm for inner defects[1]. To detect such small flaw sizes in fine ceramics, ultrasonic testing with high frequencies is very effective[2-4].

The role of reference blocks is very important for ultrasonic tests[5,6,7]. Especially, for tests of fine ceramics, detection size can be very small, therefore, the selection of testing conditions is absolutely decisive for quality of testing. In order to obtain speedy and reliable testing conditions, it is important to compare known defects with unknown ones under same measuring conditions. When the calibration curve relating the defect size and the echo-heights from these defects has been obtained, the real size of an inner defect can be estimated from its echo-height when comparing it to the calibration data assuming it has a similar defect shape.

There are many ultrasonic reference blocks in the world, i.e., the ASTM-E127 and JIS Z 2345, which are most well known and the common ones in this area. However, these conventional standards are used mainly for the purposes of testing steel and welding. There are no authorized reference blocks for such small defects in fine ceramics and for tests involving higher frequencies greater than 20 MHz.

In the present paper, characteristic aspects of ultrasonic testing for surface and inner flaws of fine ceramics are described using some trial reference blocks (RBs) including artificially induced surface and inner defects. Such defects include the slit-shaped defects, Vickers indentation cracks, inner spherical pores and flat-bottomed holes.. These defects are very useful for speedy and reliable ultrasonic tests[7,8].

PREPARATION OF TRIAL REFERENCE BLOCKS AND ULTRASONIC TESTING

In order to detect and calibrate small defects in ceramics, the selection of the ultrasonic test conditions is very important. For this purpose, we prepared some trial reference blocks (TRBs). There are two types of defects in fine ceramics, ones which occur on or nearly on the surfaces of ceramic bodies (surface defects), and other ones that are included in the bodies (inner defects). In the former case, the stress concentration factor is large, and for latter case, the factor is usually not so large. On the bases of Griffith's theory of fracture mechanics, and many experimental data for the strength of silicon nitride, we have estimated the critical size of surface defects as 30 μm. On the other hand, for inner defects, the critical size depends largely on the inner stress distribution in the body. We made the rough approximation that sub-surface defects over 100 μm in silicon nitrides can affect the strength of the material.

In this line, we prepared some TRBs made of silicon nitride ceramics, including a series of slit-shaped TRB and Vickers indentation TRB, which are for surface defects. We have also prepared pore-induced TRB and flat-bottomed TRB, which are for inner defects. Table I shows the lists of these TRBs.

In order to detect and size small defects, the immersion methods of ultrasonic testing with high frequency focusing transducers have been employed. The high ultrasonic frequency used in the experiment ranges between 25MHz to 80MHz. For surface defects, normal beam methods with surface-focusing and with -defocusing conditions are tested. We also evaluated angle beam.

Table I. List of prepared trial reference Blocks

Name of TRB	Defect, size (μm) and position level	Location of defect
SNL-A	10 Slits (30/40 in width, 10-80 in depth)	Surface
Vickers ind.	Cracks (3 levels, 100-200 in length)	Surface
SNP-A Series	Spherical pores (50-500 in dia., 3 levels of thick.)	Inner
SNP-B	Spherical pores (50-500 in dia, 5 levels of depth)	Inner
SNH-4A	Flat-bottom holes (50-400 in dia., 1 mm in depth)	Inner

RESULTS AND DISCUSSION FOR C-SCAN TESTING

(1) Slit-shaped defects (SNL-A):

The slit-shaped TRBs (SNL-A) simulate a s et of 10 narrow surface flaws, of which the widths are 30 and 40 μm. The depths of the slits (30 μm) are ranging from 13 to 50 μm, and those of the slits (40 μm) are 10 to 80μm. The lengths of these slits are all 8 mm. The smallest depth of the slits is 10 μm, which is very close to the limit of detection with a usual ultrasonic test, and can be considered a "safe" defect size in this silicon nitride ceramic.

First, the angle beam technique using 50MHz has been used for detection of the slit-shaped defects on the SNL-A specimen. Figure 1 shows the C-scan image of SNL-A by the method of defocusing. In this case, quantitative relationships between the sizes of the slits and the echo-heights have been obtained. The normal beam technique under de-focusing conditions uses surface-leaky waves. In this case, the frequency was 50MHz, and the water path was shortened by 1 mm as compared to the condition of surface focusing. As shown in the figure, the dips of the echo-heights associated with each slit increase according to the slit. Fig.2 shows the echo-dips plotted against the slit depths. As shown in this figure, the relationship is not linear, but depends approximately on the logarithm of the depths. This calibration curve is very useful for the sizing of surface defects. The SNL-A specimen can be used as the comparator for surface cracks.

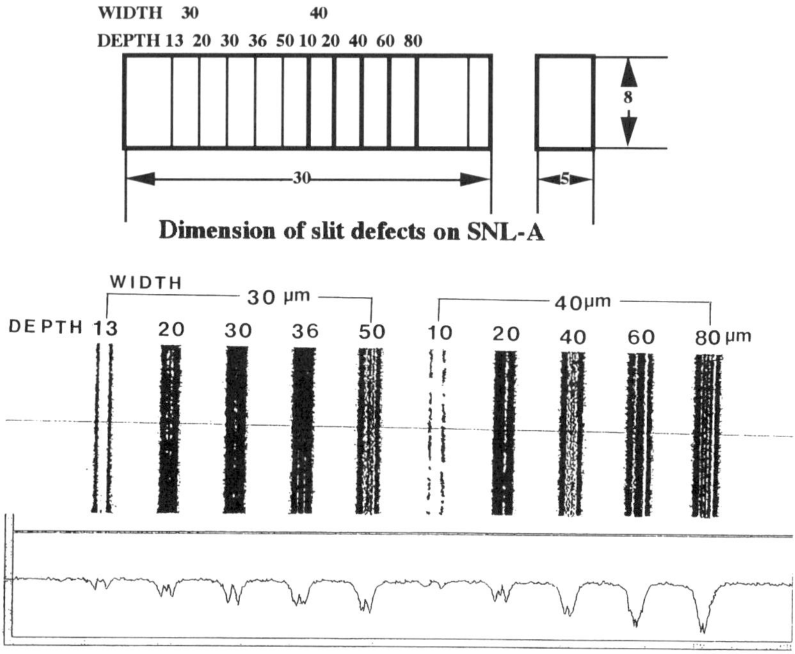

Fig.1 C-scan image of slits on SNL-A with normal beam technique (defocusing; surface, with 50MHz)

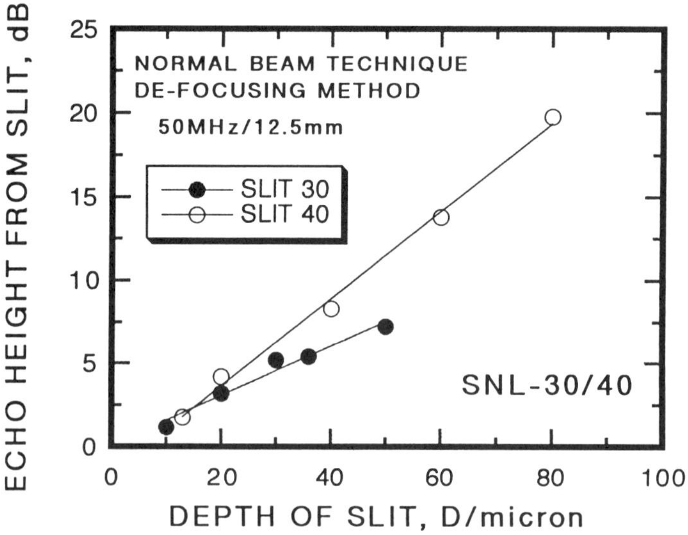

Fig. 2 Echo-heights versus the depth of slit

(2) Vickers indentation cracks (Si_3N_4 ceramics)

Vickers indentations with a load of 196, 294, and 490 N on silicon nitride ceramics were used to produce small surface flaws. In the Vickers indentation TRBs, the main indentations were eliminated with surface grinding, and 4 median cracks were left on the surface. The sizes of the indentations and the lengths of the cracks before grinding were measured. The widths of the cracks were very small, being a few μm or smaller. The depths of the cracks were not so clear and depended on the load. The crack lengths were about 90 μm for the load of 196N, and 200μm for the load of 490N. Fig.3 shows the C-scan images of the indentation cracks. The images of the two cracks were clearly separated with each other. Although the core of the indentation could not be found for both cases, the distance between the two highest points in the echo images was approximately equal to the size of the core of each indentation. Because the angle beam technique was used, the echo-images became fairly long. This result is very encouraging for ultrasonic testing in order to detect very small cracks and defects by the angle beam technique.

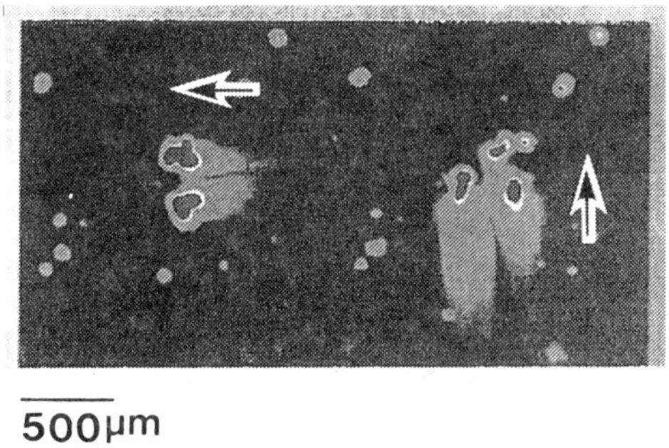

The indentation load is 490 N.
The arrows shows the direction of the C-scanning.

Fig.3 C-scan images of indentation cracks after removing the indentations

(3) Artificial pore-induced TRBs (SNP series)

Figure 4 shows the dimensions of the pore-induced TRB (SNP-A series). The pore-induced TRBs are specially fabricated in this series of works. These artificially induced pores simulate real pore-defects in ceramics. The pores are generated in the silicon nitride by placing plastic micro-beads into the ceramic body during the forming process. During sintering, the plastic beads pyrolize and leave the pores. As the sintering has been carried out at normal atmospheric pressure, usually the generation of natural pores in ceramics is unavoidable, but in this case, very careful techniques through the whole process from the preparation of powders to final sintering can successfully minimize the generation of natural pores except the desired artificial ones. The SNP-A series consisted of 3 silicon nitride plates with different thickness, but the same nominal sizes of 5 sets of spherical pores: 50, 100, 200, 300 and 500 μm, as shown in Fig.4.

Figure 5 shows the C-scan images of pores in the SNP-5A with a 50MHz transducer with 25 mm focal length. The average depth of the pores from the testing surface of the specimen was 2.5 mm. The attenuator gain was 30 dB, and the scanning pitch was 0.01 mm. All pores introduced from 50 to 500 μm were detected.

From these results, the echo height for each pore was measured. Figure 6 shows the relationship between the pore diameters (average) and the echo heights normalized with respect to those having nominal size of 500 μm. As can be seen for both cases the relative echo heights in dB are approximately linear with respect to the logarithm of pore diameters. The experimental result is very useful for quantitative determination of the sizes of defects which have similar reflecting characteristics as spherical pores.

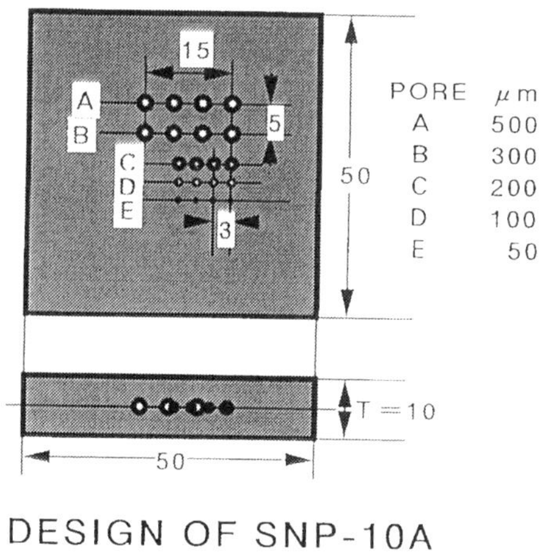

Fig.4 Dimensions of SNP-A series

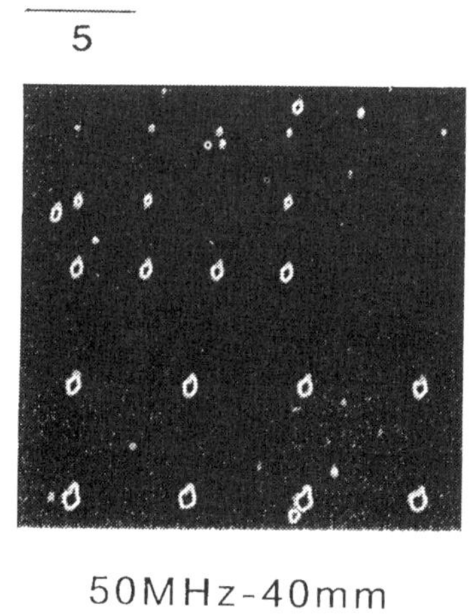

Fig.5 C-scan images of pore series in SNP-5A with 50 MHz

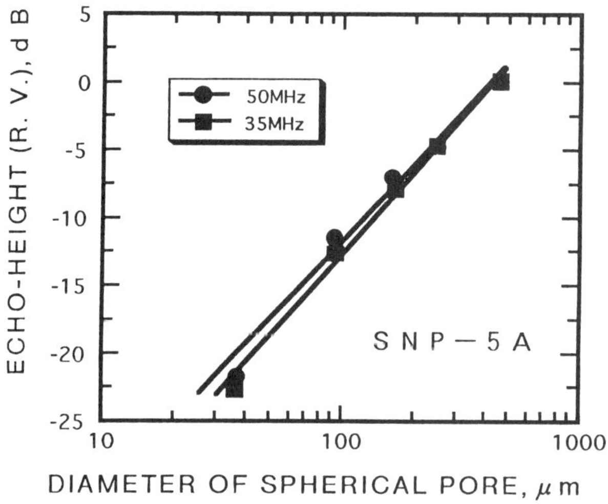

Fig.6 Echo-heights as a function of the diameters of pores for SNP-5A

Figure 7 shows the dimensions of another specimen (SNP-B), in which the depth of a given pore from the surface of the ceramic body is changed. The sensitivity of detection for such small pores in ceramics is significantly affected with the depth from the surface. Variations in the echo-heights as a function of the depth of each pore have been studied with SNP-B. The SNP-B test specimen has 5 sets of pores similar to the SNP-A series, each set having 4 pores of similar size. However, the depths of 4 pores are changed as 2.5, 3.2, 5.9 and 8.4 mm. Figure 8 shows one of the example of the C-scan image of the SNP-B (the transducer:frequency 35MHz, focal length 50mm; the water path:8mm). As shown in Fig.8, the pores at 5.9mm in depth are the clearest, and others are blurred or no detectable.

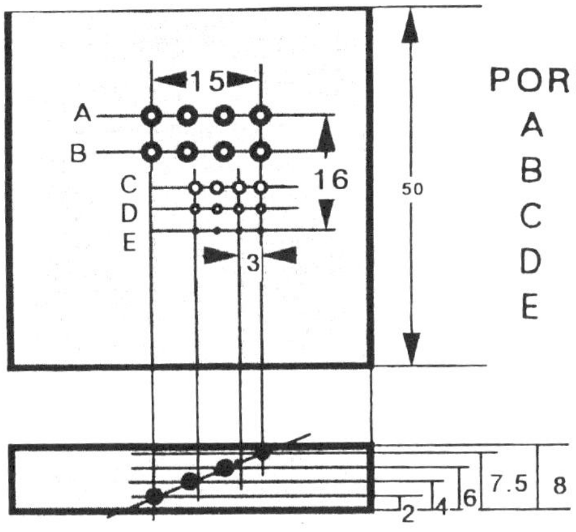

Fig.7 Dimensions of SNP-B

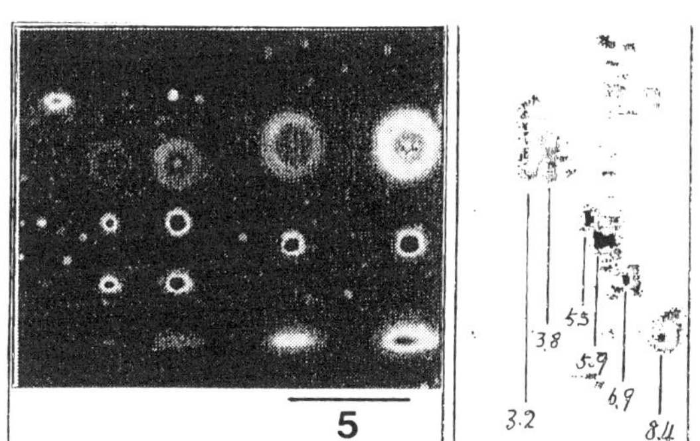

Fig.8 C-scan image of pores in SNP-B (frequency 35MHz, focal length 50mm)

Figure 9 shows the echo-heights as a function of the water path length for the ultrasonic tests. From the results, it is concluded that the transducer has rather narrow regions of detectable depths, which means that for ultrasonic testing of fine ceramics, repeated inspections are needed while changing the depth of focusing.

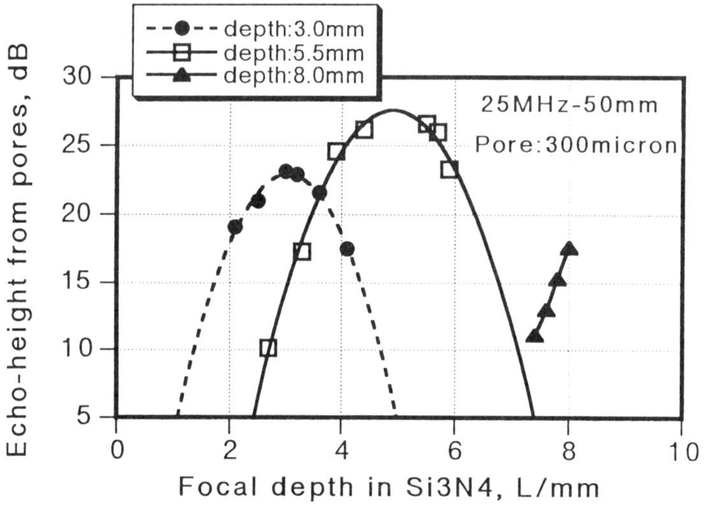

Fig.9 Variation of echo-heights from pores with the focal depth of beam in SNP-B

(d) Flat-bottomed hole TRB (SNH series)

In order to obtain the calibration curves for quantitative determination of defect sizes from their echo-heights, as like DGS curves for steel, a flat bottomed TRB has been prepared. The flat bottoms are prepared with drilling in forming process. After sintering, the drilled holes are designed to leave the flat bottoms. Figure 10 shows the illustration of the SNH-4A. The diameters of the holes range from 50 to 400 μm and the depth of them are about 1 mm. Figure 11 shows the relationship between the echo heights and the diameters of holes measured.

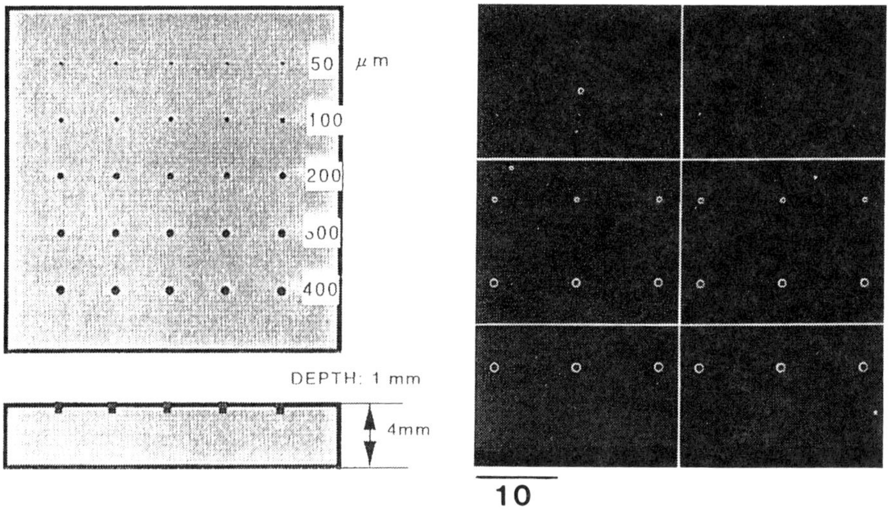

Fig.10 Dimension and C-scope image of flat-bottom holes in SNH-4A

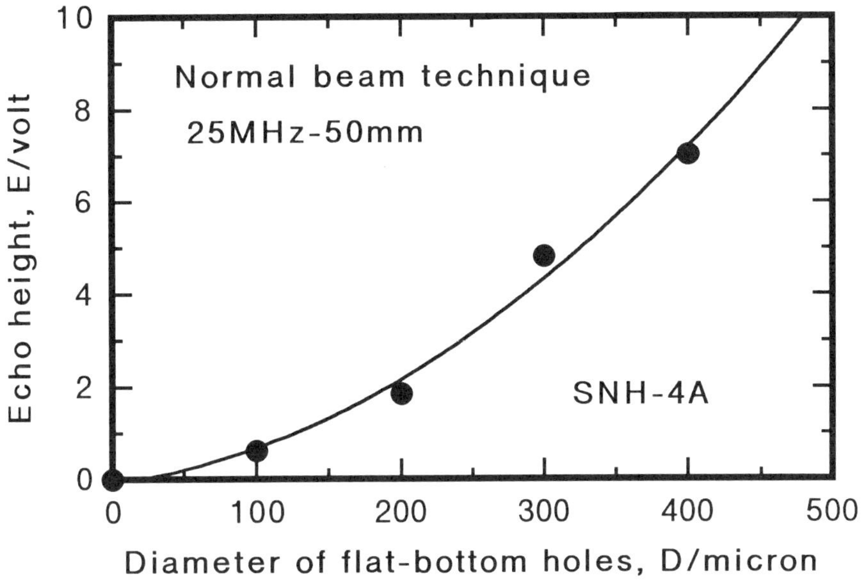

Fig.11 Variation of echo-heights with diameter of flat bottom holes

SUMMARY

Some trial reference blocks of silicon nitride for ultrasonic testing were fabricated for testing of fine ceramics. A quantitative relationship between the depth and the echo-height of slit-shaped defects has been obtained with the defocusing condition. The detectabilities for spherical pores and holes with flat bottoms have been studied. Quantitative relationships between the size and echo heights of the pores and holes have been obtained.

It is clarified that these trial reference blocks are very useful for setting the testing conditions and checking the detectability of the ultrasonic system with the testing condition.

ACKNOWLEDGMENT

This work was done at the activities of a sub-committee "NDE of New Materials" in an ad-hoc Committee "Survey and Research regarding Standardization of NDE" supported by the Ministry of International Trade and Industry of Japan.

The authors also greatly thank Dr.Sohji Sasaki, Mr.Hitoshi Kobayashi and Mr.Yutaka Abe for their cooperation in the sub-committee and also for the preparation of the data and samples.

REFERENCE

1) Y. Ikeda, T. Udagawa, S. Sasaki, H. Kobayashi and H. Domon; Proc. of 8th Asia-Pacific NDT, held in Taipei, p607 (1995).
2) P. S. Nicholson; Proc. of 5th Int. Sym. on ND Characterization of Mater. held in Karuisawa, p651(1991).
3) M.A. Mahmoud; Proc. of NDT of High-performance Ceramics, held in Boston, ASNT, p199(1987).
4) G. Y. Baaklini and P. B. Abel; Proc. of NDT of High-performance Ceramics, held in Boston, ASNT, p304(1987).
5) Y.Ikeda, T.Udagawa, S.Sasaki, H.Kobayashi and H.Domon,"Ultrasonic Flaw Detection for Structural Fine Ceramics" Proc. of 8th Asia-Pacific Conf. on NDT held in Taipei, pp607-614, 1995.
6) H.Domon, Y.Ikeda, Y.Abe, S.Sasaki, H.Kobayashi and K.Uemura, "Ultrasonic Standard blocks and Liquid penetrant Comparator for Fine Ceramics" Proc. of 1st US-Japan Symp. on Advances in NDT held in Hawaii, pp303-308, 1996.
7) Y.Ikeda, T.Udagawa, H.Kobayashi, S.Sasaki and H.Domon, "Ultrasonic Testing of Fine Ceramics (Surface Flaw)," Proc. 1st US-Japan Sym. on Advances in NDT held in Hawaii, pp344-338,1996.
8) Y.Ikeda, T.Udagawa, H.Kobayashi, S.Sasaki and H.Domon, "Ultrasonic Testing of Fine Ceramics (Inner Flaw)," Proc. 1st US-Japan Sym. on Advances in NDT held in Hawaii, pp339-344,1996.

Keywords; Silicon nitride ceramics, Ultrasonic testing, Reference blocks, Quantitative relationship, Echo-height, Size of defect

NONDESTRUCTIVE EVALUATION OF WOVEN FABRIC REINFORCED CERAMIC COMPOSITES

P.K. Liaw, J. Kim and N. Miriyala
Department of Materials Science and Engineering,
The University of Tennessee, Knoxville, TN 37996-2200
D.K. Hsu
Center for Nondestructive Evaluation, Iowa State University, Ames, IA 50011
N. Yu
Department of Mechanical and Aerospace Engineering and Engineering Science,
The University of Tennessee, Knoxville, TN 37996-2200
D.J. McGuire and W.A. Simpson, Jr.
Oak Ridge National Laboratory, Oak Ridge, TN 37831

ABSTRACT

Monolithic ceramics are highly sensitive to process and service-related flaws, making them inherently brittle. Due to their low toughness, these materials fail catastrophically. Continuous fiber reinforced ceramics (CFCCs) or whisker reinforced ceramic matrix composites (CMCs), however, can provide a significant amount of toughness as well as avoid catastrophic failure. Up to the present time, relatively little work has been performed on the nondestructive evaluation of continuous woven fabric reinforced CMCs. In this investigation, the ultrasonic scan and X-ray computed tomography were used for nondestructive evaluation of woven Nicalon™ fabric reinforced silicon carbide (SiC) composites, which were fabricated by the chemical vapor infiltration technique. Both the ultrasonic scan and X-ray computed tomography were used to characterize the type and the location of flaws in the composites. An ultrasonic C-scan was used to detect variations in porosity and to characterize the defect distribution in the composites. The relationship between the ultrasonic C-scan amplitude and porosity was quantitatively established through the result of X-ray computed tomography. Moreover, moduli of the Nicalon™/SiC composites were measured using an ultrasonic technique, and a theoretical model was formulated to predict the moduli of woven fabric composites.

INTRODUCTION

There has been considerable interest in the ceramic matrix composites (CMCs), especially for high-temperature structural applications, such as fusion reactors and aerospace industries due to their excellent stiffness to density and strength to weight ratios [1-5]. CMCs are lightweight structural materials that exhibit a much higher resistance to elevated temperatures and aggressive environments than other conventional engineering materials. Monolithic ceramics are highly sensitive to process and service-related flaws, making them inherently brittle. Due to their low toughness, these materials fail catastrophically. CMCs, however, can provide a significant amount of toughness as well as avoid catastrophic failure [6,7].

In order to stimulate wide applications of CMCs, the applications of nondestructive evaluation (NDE) methods on CMCs need to be explored. Various types of NDE techniques are available for CMCs [8]. In this investigation, ultrasonics (UT) and X-ray computed tomography (CT) were used for detecting internal flaws, determining the elastic moduli, and establishing the relationship between UT and CT in CMCs.

Up to the present time, only limited attention has been focused towards understanding NDE of woven fiber reinforced CMCs [4-7]. In the present study, both UT and CT were used to characterize the defect distribution in the composites, to develop a relationship between UT and CT, and to predict the elastic moduli of composites. Moreover, this paper provides a brief summary of our work on NDE of CMCs [3,4].

EXPERIMENTAL PROCEDURE

Materials

Two types of test samples were prepared from continuous Nicalon™ fiber reinforced SiC composites including plain-weave Nicalon™/SiC composites and cross-ply composites. For plain-weave composites, different fiber fabric lay-up sequences were used for this investigation, namely, [0/90], [30/120], and [60/150] for the moduli measurements and [0/90] for the ultrasonic C-scan. Note that the [0/90] cross-ply composites were used for both experiments of UT C-scan and CT. The Nicalon™ fibers were 10 to 15 μm in diameter, with a chemical composition of 59 wt % Si, 31 wt % C and 10 wt % O_2. Approximately 40 plies of fiber fabrics were present per centimeter thickness. The woven fibers were pyrolitically coated with 0.3 μm carbon by a chemical vapor deposition technique [7]. In this technique, the carbon coating is obtained by the decomposition of propylene at about 1100°C. The laminated fiber fabric preform was densified by the forced chemical vapor infiltration (FCVI) technique developed at the Oak Ridge National Laboratory [7]. Using the FCVI method, several disks of 7.6 cm in diameter and 1.25 cm in thickness were fabricated for the present investigation.

In Fig. 1, there are two major types of porosity in the Nicalon™/SiC composite, with a fiber fabric lay-up sequence of [0/90], [30/120], and [60/150], viz., porosity at the fiber tow intersection [Fig. 1(a)] and interlaminar porosity [Fig. 1(b)]. The extent of interlaminar porosity in the samples was found to be greater than the porosity at the fiber tow intersection. The Nicalon™/SiC composite is highly anisotropic, hence, it would be expected that the mechanical properties of the material will vary significantly along the in-plane (fiber fabric plane) and through-thickness directions.

The density (as measured from the specimen mass and dimensions) and the percentage porosity of each specimen are calculated by the following equation,

$$\% \text{ Porosity} = (1 - \text{measured density/theoretical density}) \times 100 \qquad (1)$$

The theoretical density of the composite is 2.96 g/ cm^3, which is based on a density of 2.6 g/cm^3 for the Nicalon™ fibers and 3.21 g/cm^3 for the SiC matrix, and a nominal fiber volume percentage of about 40. The density of the samples varied from 2.06 to 2.63 g/cm^3, and the porosity from 30.4 to 11.2 percent. The porosity content variation in the composite samples allowed us to evaluate the effect of porosity on the moduli of the material.

Modulus and Amplitude Measurements by Ultrasonics

Modulus measurements were performed using a dry-coupling ultrasonic technique. Specimens (25 mm × 25 mm × 5 mm) were machined from the Nicalon™ reinforced SiC composite disks for modulus determinations (Fig. 2). The coordinate system used for measuring moduli is shown in Fig. 2. Longitudinal moduli along the in-plane (fiber-fabric plane) direction (X_1 or X_2-axis) and through-thickness (X_3) direction were determined. Note that along the through-thickness direction, the specimen contained approximately 20 fiber fabric plies.

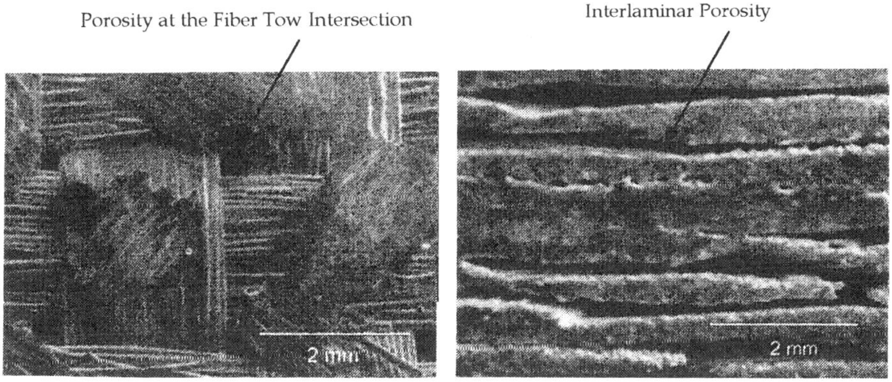

(a) Planar View (b) Cross-sectional View

Fig. 1. Fiber architecture of plain-weave Nicalon™/SiC composite with a [0/90], [30/120], and [60/150] fabric lay-up sequence (Optical microscopy).

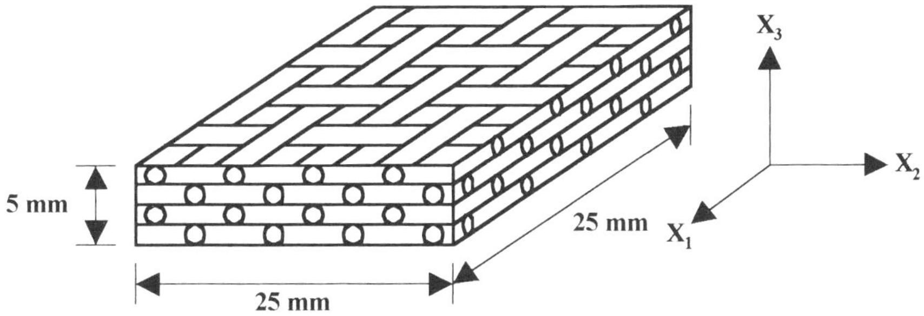

Fig. 2. Specimen geometry for modulus measurements of Nicalon™/SiC composites.

Longitudinal waves were used to make the ultrasonic measurements. The longitudinal modulus (C) is related to the longitudinal velocity by the equation,

$$C = \rho V_L^2 \qquad (2)$$

Since the samples used in this study were highly porous (Fig. 1), the use of liquid or gel couplant would change the value of the modulus and cause the contamination of the samples. To avoid these problems, ultrasonic measurements were performed using dry-coupling longitudinal transducers. Dry contact transducers had a thin elastomer face-sheet and coupled the ultrasonic propagation by pressure, without the aid of a liquid or gel couplant. The transducers had a crystal size of 19.1 mm and a center frequency of 0.5 MHz.

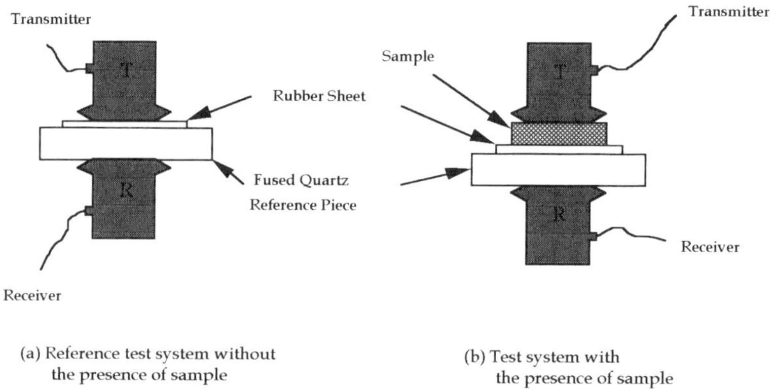

Fig. 3. Ultrasonic measurement setup for the modulus determination of Nicalon™/SiC composites.

A "pulse-echo overlap method" was used to measure the wave velocities. A schematic of the experimental setup is shown in Fig. 3. The reference signal was obtained by passing ultrasonic longitudinal waves through a reference test system composed of a rubber sheet and a fused quartz piece. The test sample was then placed between the transmitting transducer and the rubber sheet in the reference test system. Ultrasonic longitudinal waves were then passed through the setup. The signals were received by the receiving transducer and captured by an oscilloscope. The received radio frequency (RF) wave forms, with and without the sample in the acoustic path, were matched as closely as possible by shifting one to overlap with the other. Since the time shift required for the match is equal to the transmit time of the ultrasonic waves through the thickness of the sample, the ultrasonic longitudinal velocity is simply the sample thickness divided by the time shift. Because of the fact that the two signals are matched as closely as possible (i.e., overlapped), this technique is referred to as the pulse-echo overlap method. Both in-plane and through-thickness longitudinal ultrasonic velocities were measured in the present study, which yielded in-plane moduli (C_{11} and C_{22}) and through-thickness moduli (C_{33}), respectively, for the composite material.

Ultrasonic amplitude measurements were performed using a through-transmission C-scan mode at a frequency of 15 MHz, in an immersion tank (Fig. 4). Two different geometries of tensile specimens were used for amplitude measurements as shown in Fig. 5. For tensile specimens of CMCs, ultrasonic C-scans were conducted to investigate the material homogeneity. Figure 4 presents a schematic diagram of the through-transmission ultrasonic (TTU) setup using a pair of focused transducers. The scan is performed in an immersion tank. The transmitting transducer (pulser) and receiving transducer (receiver) are aligned with each other, and the beam path is always kept perpendicular to the test specimen during the scan. The two transducers are ganged together using a yoke arrangement that maintains the alignment of the focused beam.

The specimen is placed at the focal zone. In this work, the transducers have a diameter of 1.27 cm and a focus length of 5.08 cm.

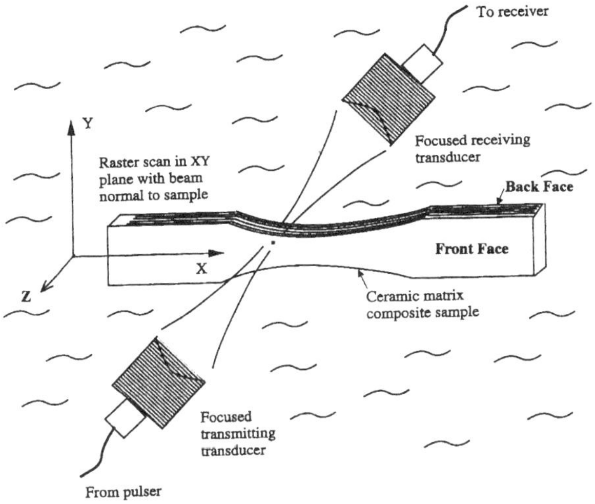

Fig. 4. Schematic diagram of a through-transmission mode of ultrasonic scanning in immersion.

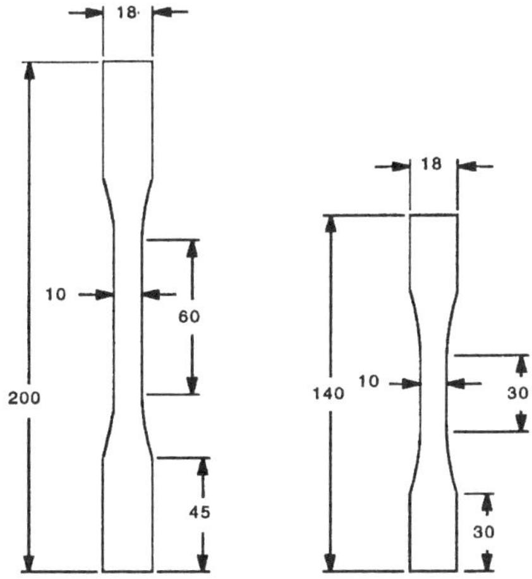

(a) Woven specimen (b) 0°/90° cross-ply specimen

Fig. 5. The geometry of tensile test specimens. All dimensions in mm.
Radius of curvature: 100 mm

In Fig. 4, the X and Y-axes are in the fiber fabric plane of the tensile sample, while the z-axis is the through-thickness direction of the sample. During the experiment, the front face of the tensile sample is defined as the surface encountering the pulser of the pair of the transducers and the back face meeting with the receivers (Fig. 4). The amplitude of the transmitted ultrasonic beam is measured by the receiving transducer, which is also focused and positioned symmetrically with respect to the transmitting transducer. The amplitude of the TTU signal is quite sensitive to the presence of internal defects, such as voids and delaminations, and to variations of the internal structure, such as the undulation of fiber tows. Any defects or internal material conditions that attenuate, scatter, or block the transmitting ultrasonic beam will result in a low TTU signal in the C-scan image.

X-ray Computed Tomography

Figure 6 illustrates a computed tomography setup for obtaining a CT scan image in the present investigation. The CT forms images of a cross-sectional slice parallel to a collimated X-ray beam. A thin slice of the specimen interacts with a thin beam of radiation, which is attenuated, as it passes through the specimen. The fraction of the X-ray beam that is attenuated is directly related to the density and thickness of the material through which the beam has traveled and to the composition of the material and the energy of the X-ray beam. The CT utilizes this information, from many different angles with each small rotation of the specimen, to determine cross-sectional configuration with the aid of a computerized reconstruction process. Note that in the present study, each incremental rotation angle equals 18°. The reconstruction process quantitatively determines the point-by-point mapping of the relative radiation attenuation coefficients from the set of one-dimensional radiation measurements. This CT reconstruction process yields a two-dimensional array of numbers corresponding to the cross section of the specimen in terms of pixels or picture elements. In this investigation, the CT is used to provide a cross-sectional view of each SiC composite sample.

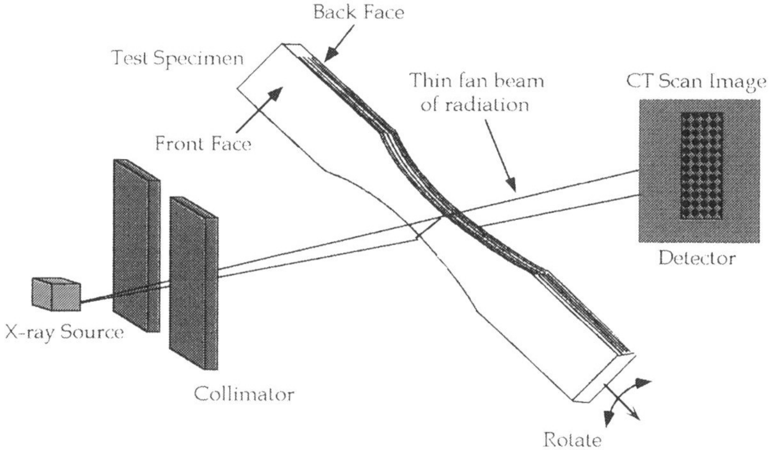

Fig. 6. Illustration of an experimental procedure for computed tomography setup to obtain a CT scan image.

RESULTS AND DISCUSSION

The ultrasonic velocities and longitudinal moduli values are plotted against percent porosity in Figures 7 and 8, respectively. The ultrasonic velocities (0.8 to 1.0 $\times$ 10^4 m/s) were greater in the in-plane direction than in the through-thickness direction (0.17 to 0.73 $\times$ 10^4 m/s).

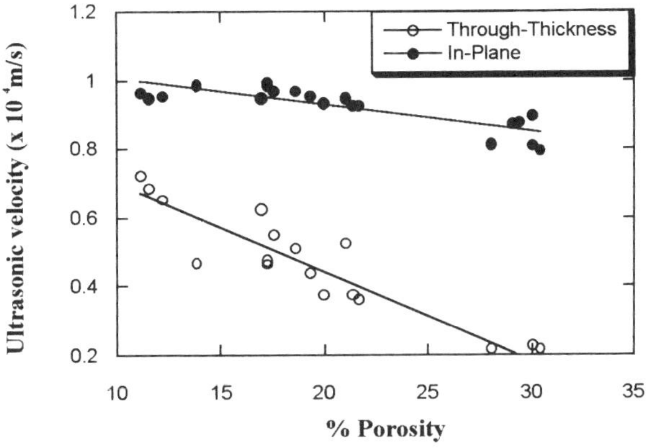

Fig. 7. Ultrasonic velocity vs. % porosity in Nicalon™/SiC composites.

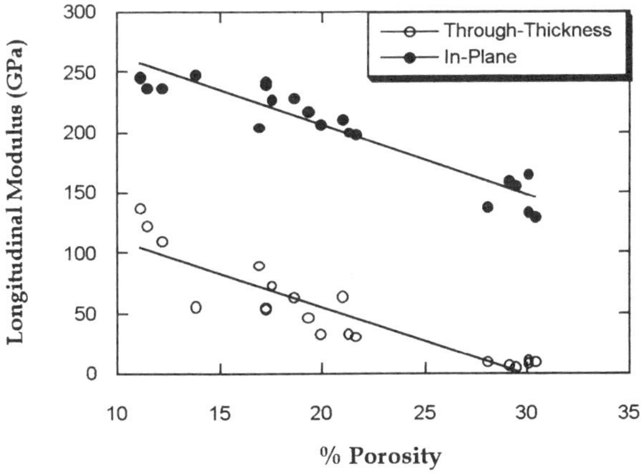

Fig. 8. Longitudinal moduli vs. % porosity in Nicalon™/SiC composites.

Consequently, the in-plane longitudinal moduli, C_{11}, (130 to 248 GPa) were greater than the through-thickness longitudinal moduli, C_{33}, (6 to 138 GPa), as is to be expected from the fabric lay-up. From Fig. 7, it can also be seen that porosity had a greater effect in reducing the ultrasonic velocity in the through-thickness direction than in the in-plane direction. These results derive from the greater interlaminar porosity compared to the porosity at the fiber tow intersection (Fig. 1). Furthermore, increased porosity significantly decreased both in-plane and through-thickness moduli, as shown in Fig. 8. It should be noted that the ultrasonic measurements yielded comparable C_{11} and C_{22} values along the fiber fabric plane (Fig. 2).

Figures 9 and 10 illustrate the results of the ultrasonic C-scan versus the X-ray computed tomography of the plain-weave and the cross-ply specimens, respectively. As shown in Figs. 9 and 10, the color scheme is such that blue and green colors (dark areas in the black and white photograph) correspond to high amplitudes of transmitted ultrasonic signals, which are less attenuated by the anomalies and derive from the overall "good" or "dense" region in the composites. The red and yellow colors (bright areas in the black and white photograph) correspond to low amplitudes, which are greatly attenuated by the defects and result from generally "bad" or "porous" areas. This color scheme is based on the relative amplitude range of transmitted ultrasonic waves. The range of the relative amplitude is from '0' to '1' with '0' corresponding to the smallest relative amplitude and '1' the largest amplitude. Defects distributions have been designated as a C-scan in light of the UT transmitted amplitudes (Figs. 9 and 10). In other words, for an area containing a greater amount of defects, more UT waves will be deflected or scattered by the defects, resulting in lower amplitudes of transmitted UT waves.

The purpose of using CT was to demonstrate the correlation between the transmitted ultrasonic amplitude of the C-scan and the extent of porosity and/or internal flaws revealed by CT. The defects of a specimen seem to be distributed through the whole specimen, as shown in Figs. 9 and 10 for both woven and cross-ply composites, respectively. The CT scans were made through three different cross sections of each specimen, indicated by the letters A, B, and C (Figs. 9 and 10). These sections were carefully chosen in specific areas to correlate the UT C-scan data with the CT scan results; that is, the 3A section in Fig. 9 was composed of a series of relatively higher values of UT transmitted amplitudes along the cross-sectional direction of the plain-weave specimen, which is perpendicular to the length direction of the sample. The 3B section consisted of a cross section generally containing the points of the lowest UT transmitted amplitude values. The intermediate amplitude distributions between the 3A and the 3B sections were chosen as the 3C section. A similar selection scheme was used on the cross-ply specimen, sample 6 (Fig. 10); i.e., the 6B section has the lower values of the UT transmitted amplitude than 6A or 6C.

A visual examination of the C-scan image versus CT-scan shows a reasonably good correlation between the amount of porosity and transmitted amplitudes. Note that the dark areas in the CT-scans (Figs. 9 and 10) indicate the presence of porosity and/or defects in the composites. In the case of the woven specimen (Fig. 9), the 3A section includes less defects or greater UT wave amplitudes than any other section, such as 3B or 3C, from the C-scan data. The results of CT-scans for these sections displayed a good correlation with the C-scan data. Specifically, the CT-scan of the 3A section has seemingly less porosity than that of the 3B or 3C section, which is in agreement with those of the UT C-scans. Similarly, in the case of the cross-ply composite, the 6B has greater porosity than 6A or 6C, which corresponds to lower UT transmitted amplitudes in 6B than 6A or 6C.

THEORETICAL MODELING

Modulus Measurements

In the present work, the composites are considered to be infinite solids containing periodically distributed inhomogeneities (particles, fibers, etc.). The periodic model was also used to estimate the effect of porosity content on the elastic properties of woven fiber composites [4]. As shown in Fig. 11, a representative unit cell consists of eight woven fiber tows in the shape of elliptic cylinders and one parallelepiped of voids that is used to model the interlaminar porosity, as indicated in Fig. 1(b). The distribution of the inhomogeneities is such that the composite is composed of a collection of unit cells that are repeated in all directions. In view of

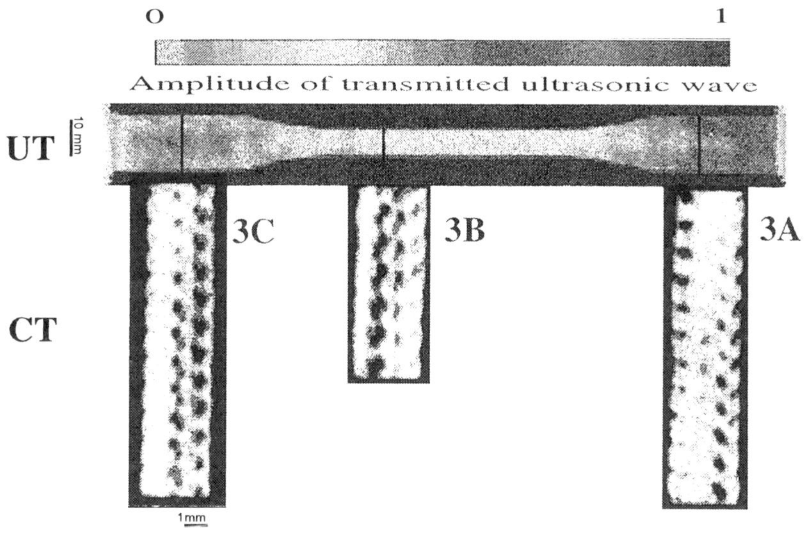

Fig. 9. Ultrasonic C-scan versus CT-scan for woven Nicalon fiber reinforced composite (sample 3).

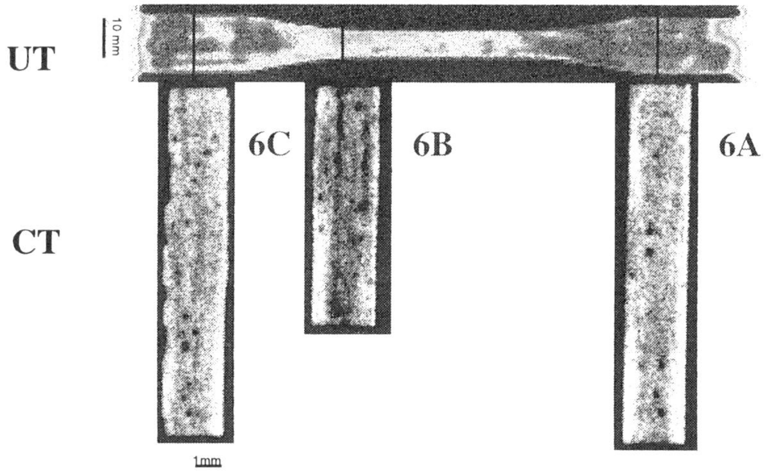

Fig. 10. Ultrasonic C-scan versus CT-scan for 0°/90° cross-ply Nicalon fiber reinforced composite (sample 6).

the periodicity, the response of the unit cell is representative of that of the infinite composite, under uniform applied loads.

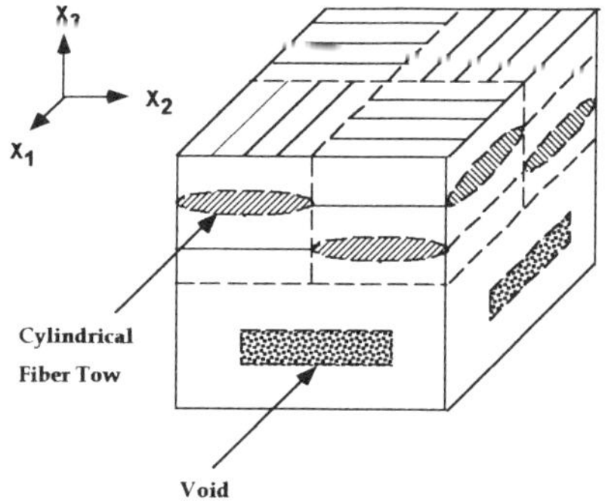

Fig. 11. A model based on the periodic microstructure of Nicalon™/SiC composites.

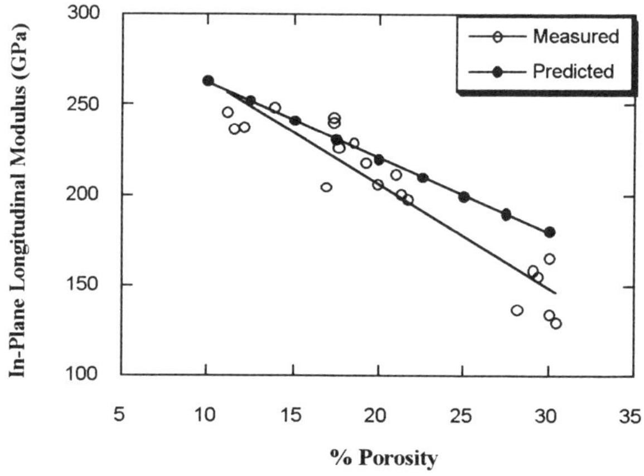

Fig. 12. Comparison of predicted and measured in-plane longitudinal moduli in Nicalon™/SiC composites.

When the composite with periodic microstructure is subjected to homogeneous strains, ε^0 (or stresses, σ^0), the resulting strains and stresses are $\varepsilon^0 + \varepsilon^P (X_1, X_2, X_3)$ and $\sigma^0 + \sigma^P (X_1, X_2, X_3)$, respectively, due to the existence of periodically distributed inhomogeneities. The homogeneous (average) strains and stresses are ε^0 and σ^0, and ε^P and σ^P are periodic disturbance strains and stresses, which are functions of positions and can be expressed in Fourier series [9]. The elastic stiffness constant matrix, $[\overline{C}]$, of the composite is defined by

$$\{\sigma^0\} = [\overline{C}]\{\varepsilon^0\} \qquad (3)$$

On the basis of Eshelby's concept of stress-free strains [10], one can replace the heterogeneous unit cell by a homogeneous one that is composed of matrix material only, and prescribe stress-free strains, ε^* (X_1, X_2, X_3), within each of the N inhomogeneities (Ω_i, i = 1, 2,..., N), such that the average stress and average strain remain the same everywhere in the unit cell before and after the homogenization. The equivalent inclusion method was extended in terms of eigenstrain by Mura [11] and Nemat-Nasser and Hori [12]. For the average stresses to remain the same before and after homogenization, the following consistency conditions must hold in Ω_i:

$$\left[C^{\Omega_i}\right]\left\{\varepsilon^0+\left\langle\varepsilon^P\right\rangle_{\Omega_i}\right\}=\left[C^M\right]\left\{\varepsilon^0+\left\langle\varepsilon^P\right\rangle_{\Omega_i}-\varepsilon^{*i}\right\} \quad \text{for i = 1, 2,..., N} \qquad (4)$$

where the angle brackets with the subscript, Ω_i, represent the volume average over Ω_i; $C^{\Omega i}$ and C^M are the stiffnesses of the i-th inhomogeneity and the matrix, respectively; ε^{*i} are the stress-free homogenization strains prescribed in Ω_i and are assumed to be constant. The average periodic disturbance strains can be related to the stress-free strains by [13]

$$\left\{\left\langle\varepsilon^P\right\rangle_{\Omega_i}\right\}=\sum_{j=1}^{N} f_j \sum_{(n_1,n_2,n_3)\neq(0,0,0)}^{\pm\infty} g_j g_i \left[\overline{\Gamma}\right]\left[C^M\right]\left\{\varepsilon^{*j}\right\} \qquad (5)$$

where the term (n_1, n_2, n_3) = (0, 0, 0) is excluded in the summation; f_j is the volume fraction of Ω_j, and the components of the six-by-six matrix, $\left[\overline{\Gamma}\right]$, are given by

$$\overline{\Gamma}_{\alpha\alpha}=\frac{1}{G^M}\left(\overline{\zeta}_\alpha^2-\frac{1}{2\left(1-\nu^M\right)}\overline{\zeta}_\alpha^4\right) \quad \text{for } \alpha = 1, 2, 3$$

$$\overline{\Gamma}_{\alpha\beta}=\overline{\Gamma}_{\beta\alpha}=\frac{-1}{2G^M\left(1-\nu^M\right)}\overline{\zeta}_\alpha^2\overline{\zeta}_\beta^2 \quad \text{for } (\alpha, \beta) = (1,2), (2,3), (3,1)$$

$$\overline{\Gamma}_{\alpha\alpha}=\frac{1}{G^M}\left(\frac{1}{4}\left(\overline{\zeta}_\beta^2+\overline{\zeta}_\gamma^2\right)-\frac{-1}{2\left(1-\nu^M\right)}\overline{\zeta}_\beta^2\overline{\zeta}_\gamma^2\right)$$

for (α, β, γ) = (4,2,3), (5,3,1), (6,1,2)

$$\overline{\Gamma}_{\alpha\beta}=0 \quad \text{otherwise,}$$

with

$$\overline{\zeta}_i=\frac{n_i}{a_i\sqrt{\left(\frac{n_1}{a_1}\right)^2+\left(\frac{n_2}{a_2}\right)^2+\left(\frac{n_3}{a_3}\right)^2}} \quad \text{for i = 1, 2, 3.}$$

G^M and ν^M are the shear modulus and Poisson's ratio of the matrix; a_i (i = 1, 2, 3) are the dimensions of the unit cell; and n_i are integers. The shape and size of the inhomogeneities are accounted for in the g-parameter, which is defined by

$$g_i = \frac{1}{V_{\Omega i}} \int_{\Omega i} \cos\left(\frac{\pi n_1 x_1}{a_1}\right) \cos\left(\frac{\pi n_2 x_2}{a_2}\right) \cos\left(\frac{\pi n_3 x_3}{a_3}\right) dV$$

After solving the consistency conditions for suitable ε^{*i} which preserve the same stress and strain fields before and after homogenization, the elastic stiffness constant matrix, $[\bar{C}]$, of the composite are, thus, determined by

$$\{\sigma^0\} = [\bar{C}]\{\varepsilon^0\} = [C^M]\{\varepsilon^0\} - \sum_{j=1}^{N} f_j [C^M]\{\varepsilon^{*j}\} \qquad (6)$$

The following input data were used in the calculation of the composite moduli [14]: the shear modulus and the Poisson's ratio of the Nicalon™ fiber are 80 GPa and 0.12, respectively, and those of the matrix are 146 GPa and 0.2. Figure 12 presents the predicted in-plane moduli, C_{11} (= C_{22}), of the Nicalon™/SiC composites versus percent porosity. Increased porosity decreased the stiffness, consistent with what was observed experimentally (Fig. 8). Moreover, there is a reasonably good agreement between the predicted and the measured moduli, as can be seen from Fig. 12. Note that a detailed description of the calculation of the moduli can be found in Ref. 4.

The Relationship between UT and CT

A wave propagation in an attenuating medium shows a decrease in the amplitude as a function of the distance X following the equation.

$$A = A_0 \exp(-\alpha x) \qquad (7)$$

where A_0 and A are the incident and transmitted amplitudes, respectively, of a section of material having a thickness, X, and an attenuation coefficient, α.

Gubernatis and Domany [15] and Adler, Rose and Mobley [16] have obtained exact expressions for estimating the attenuation in porous media. The general solution of these formal expressions is intractable. They introduced the approximation that correlations between scatterers are unimportant. They showed that the following equation is yet a further approximation to this result.

$$\alpha = \frac{1}{2} n \gamma(k) \qquad (8)$$

where n is the number density of pores, and γ (k) is the total scattering cross section which is related to scattering by each pore. Also, γ (k) is a function of the geometric factors, such as the pore radius, the volume fraction of pores. Equation (8) has previously been given by Evans et al. [17]. Note that in Eq. (8), Adler, Rose and Mobley have emphasized a small volume fraction of pores less than 6% in structural materials to simplify the scattering and attenuation effect between ultrasonic waves and porosities [16].

Now we can obtain the relationship between the relative ultrasonic transmitted amplitude and the attenuation in following equation,

$$\frac{A}{A_0} = \exp(-\alpha x) \qquad (9)$$

where the ratio, A/A_0, is the relative ultrasonic transmitted amplitude.

In this investigation, since the pore scattering effect by the ultrasonic transmitted amplitude has been ignored, we have obtained a relationship between the relative ultrasonic transmitted amplitude and the number density of porosity of the X-ray computed tomography by the combination of Eqs. (8) and (9); that is, now, we have the relation between the relative ultrasonic transmitted amplitude and porosity, as shown in the following equation,

$$\frac{A}{A_0} = \exp(-nf) \qquad (10)$$

where n is the number density of porosity and f is a value which is related to the geometric factor ($f = \frac{1}{2}\gamma(k)x$). In Eq. (10), we can directly expect that the ratio A/A_0 is an exponential function of n.

In this derivation, we should note that most theories treating the ultrasonic attenuation are for simple systems, which are homogeneous and isotropic. Moreover, voids are usually assumed to be small and uniform in size and shape, and the volume percent of voids is low. This is especially not true for the present porous CMCs. Pore scattering, which plays a major role in affecting the ultrasonic transmitted amplitude, has been considered in the term, γ (k), in Eqs. (8)-(10). It is the main reason for the signal amplitude reduction. In a phenomenological sense, the attenuation coefficient α in Eq. (9) contains all mechanisms for signal reduction, mainly scattering. Also, we have to note that Eq. (10) follows from Eq. (8), but Eq. (8) may be too simplified to be valid for CMCs. Further theoretical work needs to be developed for modeling the ultrasonic response of CMCs. Equation (10) only provides a guide for the present CMCs.

To quantify the CT results, the relative porosity calculation was performed using a National Institutes of Health (NIH) Image analyzer. On the NIH Image analyzer, the region containing porosity and/or defects has been assumed to be of black color on the CT-scan, whereas the white region regarded as a defect-free area. A relative color distinction between white and black made possible the calculation of the amount of relative porosity and/or defects on the CT results.

The correlation between the UT transmitted amplitudes and the amount of relative porosity is presented in Fig. 13. We can note that the UT transmitted amplitude decreases when porosity increases, which is represented as an exponential function for both woven and cross-ply composites (Fig. 13). This trend results from the fact that the greater the amount of porosity, the less the transmitted ultrasonic amplitude is. The exponential relationship between the UT transmitted amplitude and the amount of relative porosity is shown in Fig. 13 and is also theoretically substantiated in Eq. (10). Currently, more experimental work on CT is in progress to further substantiate this exponential relationship between the UT amplitude and relative porosity. In summary, a reasonable correlation is observed between the ultrasonic C-scan amplitude and the amount of relative porosity determined from computed tomography. Thus, both UT and CT can be used to detect variations in porosity as well as characterize the defect distributions in ceramic matrix composites.

CONCLUSIONS

For woven Nicalon™ fiber reinforced silicon carbide composites, both through-thickness and in-plane moduli were measured using dry-coupling ultrasonic methods. The in-plane moduli

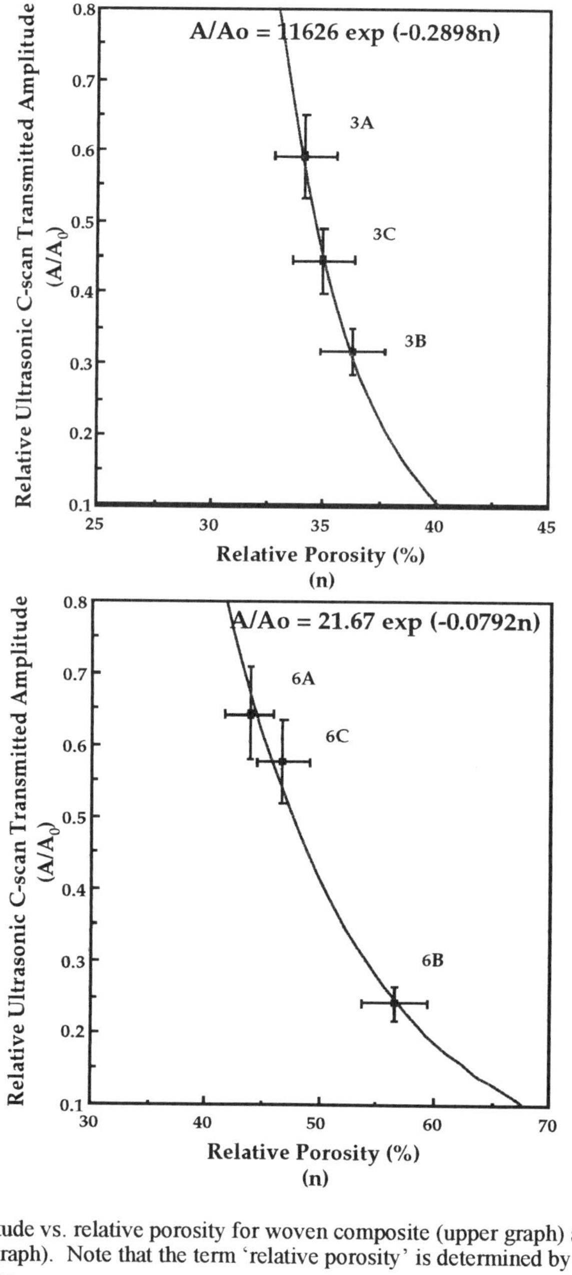

Fig. 13. UT amplitude vs. relative porosity for woven composite (upper graph) and for cross-ply composite (lower graph). Note that the term 'relative porosity' is determined by the image analysis of CT-scans

were found to be greater than the through-thickness moduli. Increased porosity in the composite resulted in a significant decrease in both through-thickness and in-plane moduli.

A theoretical model, using a homogenization technique, was developed to predict the effect of porosity on the elastic moduli of woven fiber reinforced ceramic matrix composites. The predicted elastic moduli are in good agreement with the experimental results for woven fabric composites investigated in the present study.

Computed tomography images can be related to the ultrasonic scan results, and the relative transmitted ultrasonic amplitude decreases, as the amount of relative porosity increases. Furthermore, an exponential relationship between the ultrasonic transmitted amplitude and the extent of porosity seems to be present for Nicalon™/SiC composites, which is in a general agreement with theoretical predictions.

ACKNOWLEDGMENTS

This work is supported by the National Science Foundation, under contract No. EEC-9527527 with Mrs. Mary Poats as a contract monitor, and under contract 11X-ST270V with the Oak Ridge National Laboratory, Martin Marietta Energy Systems, Inc.

REFERENCES

[1]B. Marshall and B. N. Cox, Acta Metall., vol. 35, 1987, p. 2607.

[2]R.L. Lehman, S.K. El-Rahaiby, and J.B. Wachtman, Handbook on Continuous Fiber-Reinforced Ceramic Matrix Composites, The Am. Ceram. Soc, 1995, p. 495.

[3]J. Kim, P.K. Liaw, D.K. Hsu, D.J. McGuire, R.A. Lowden, and T. Moudden, Symposium on "Nondestructive Evaluation and Materials Properties (III)," P.K. Liaw, O. Buck, R.J. Arsenault and R.E. Green, Jr., eds., TMS Fall Meeting, Cincinnati, OH, October 6-10, 1996, p. 55.

[4]P.K. Liaw, D.K. Hsu, N. Yu, N. Miriyala, V. Saini, and H. Jeong, Acta Metall., vol. 44, n. 5, 1996, p. 2101.

[5]E. Vagaggini, J.-M Domergue, and A. G. Evans, J. Am. Ceram. Soc., vol. 78, n. 10, 1995, p. 2721.

[6]K.K. Chawla, Ceramic-matrix Composites, Chapman & Hall, London, 1993, p. 4.

[7]T.M. Besmann, B.W. Sheldon, R.A. Lowden and D.P. Stinton, Science, 253, 1991, p. 1104.

[8]Y. Bar-Cohen, A.K. Mal, ASM Handbook, vol. 17, Nondestructive Evaluation and Quality Control, ASM International, 1992, p. 231-277.

[9]S. Nemat-Nasser and M. Taya, Q. Appl. Math., 39, 1, 1981, p. 43; also, 43, 2, 1985, p. 187.

[10]J.D. Eshelby, Proc. R. Soc. London, A241, 1957, p. 376.

[11]T. Mura, Micromechanics of Defects in Solids, Second Edition, M. Nijhoff Pub, 1987.

[12]S. Nemat-Nasser and M. Hori, Micromechanics: Overall Properties of Heterogeneous Materials, North-Holland, 1993.

[13]S. Nemat-Nasser, N. Yu, and M. Hori, Mech. Mater., 15, 1993, p. 163.

[14]J.L. Bohet, Ph. D Thesis, Bordeaux, France, 1993.

[15]J.E. Gubernatis and E. Domany, in Review of Progress in Quantitative Nondestructive Evaluation, edited by D.O. Thompson and D.E. Chimenti, Plenum, New York, 1983, vol. 2A, p. 833; Wave Motion 6, p. 579 (1984)

[16]L. Adler, J.H. Rose, and C. Mobley, J. Appl. Phys., 59 (2), 1986, p. 336.

[17]A.G. Evans, B.R. Tittamnn, L. Ahlberg, B.T. Khuri-Yakub, and G.S. Kino, J. Appl. Phys. 49, 1978, p. 2669.

Table I. Applied heat treatment to SiC disk pairs

Specimen	$T_{dwell,1}$ (°C)	$t_{dwell,1}$ (h)	$T_{dwell,1}$ (°C)	$t_{dwell,1}$ (h)	No. of Cycles
1	200	1	1000	2	3
2	400	5	1100	6	1

To determine how the joint was affected by temperature transients, both specimens were subjected to a high frequency thermal cycle test. The thermal cycle test was conducted between 500°C and 1000°C with a cycle time of approximately 3 minutes, an average cooling rate of 20°C sec^{-1}, and an average heating rate of 3.5°C sec^{-1}. A total of 10,000 such cycles were performed, after which the specimens were re-examined by ultrasonic microscopy.

Upon completion of the thermal cycling test, specimen 1 was fractured and characterized using optical microscopy, electron backscatter, and high frequency ultrasound. A 150 MHz (2 mm focal length, 6.35 mm diameter) spherically focused transducer was used to characterize the specimen fracture surface. Adequate sampling was ensured by using a 250 MHz bandwidth receiver and a 1 GHz analog-to-digital converter (no equivalent time sampling was required). The images were then statistically analyzed and correlated to the nondestructive test both pre- and post-thermal cycling. Post thermal cycle ultrasonic test data on specimen 2 has been obtained but not compared to destructive evaluation.

RESULTS

Figure 1 shows a typical A- and B–scan trace of an ultrasonic waveform through the total thickness of specimen 2 that was joined by the microwave process. As shown in Figure 1, two interfacial signals and the specimen backsurface are evident. Similar time traces were used to calculate the speed of sound (11.76 mm/µsec) from reference, unjoined SiC disks of equal thickness and composition. Figure 2 shows the C–scan (top view) of the first signal reflected from the interface (back surface of first disk) in specimens 1 and 2 before and after high frequency thermal cycling. Intuitively, a higher quality bond would allow the sound energy to pass through the interface without being reflected, thus the higher amplitude regions indicate relatively less bonding compared to the lower amplitude regions. Finally, Figure 3 shows optical, acoustic and electron backscatter images of the fracture surface as well as the C-scan for specimen 1 before thermal cycling.

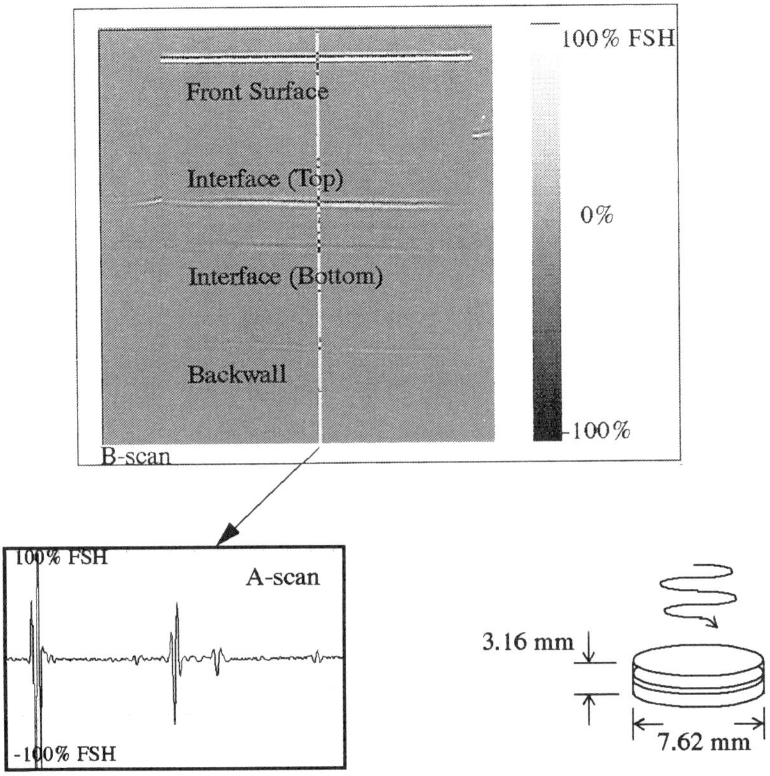

Figure 1. Ultrasonic B- and A- scans of a microwaved SiC-SiC joint.

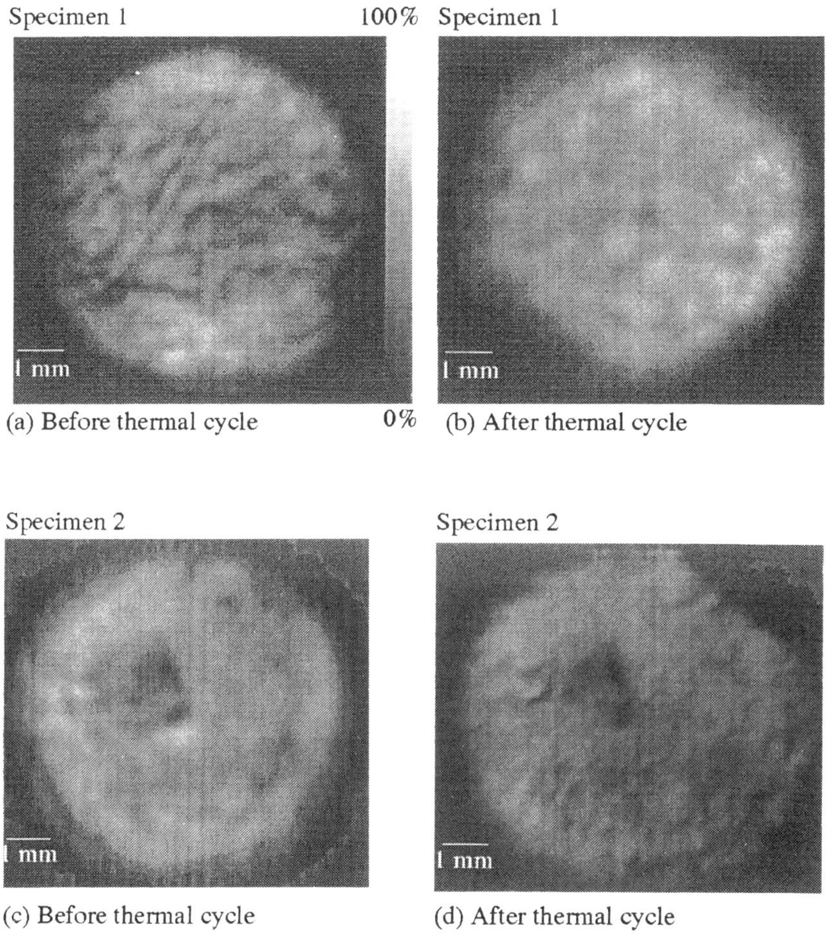

Figure 2. Ultrasonic C-scan images (taken at 50 MHz) of specimens 1 and 2 before and after high frequency thermal cycle testing.

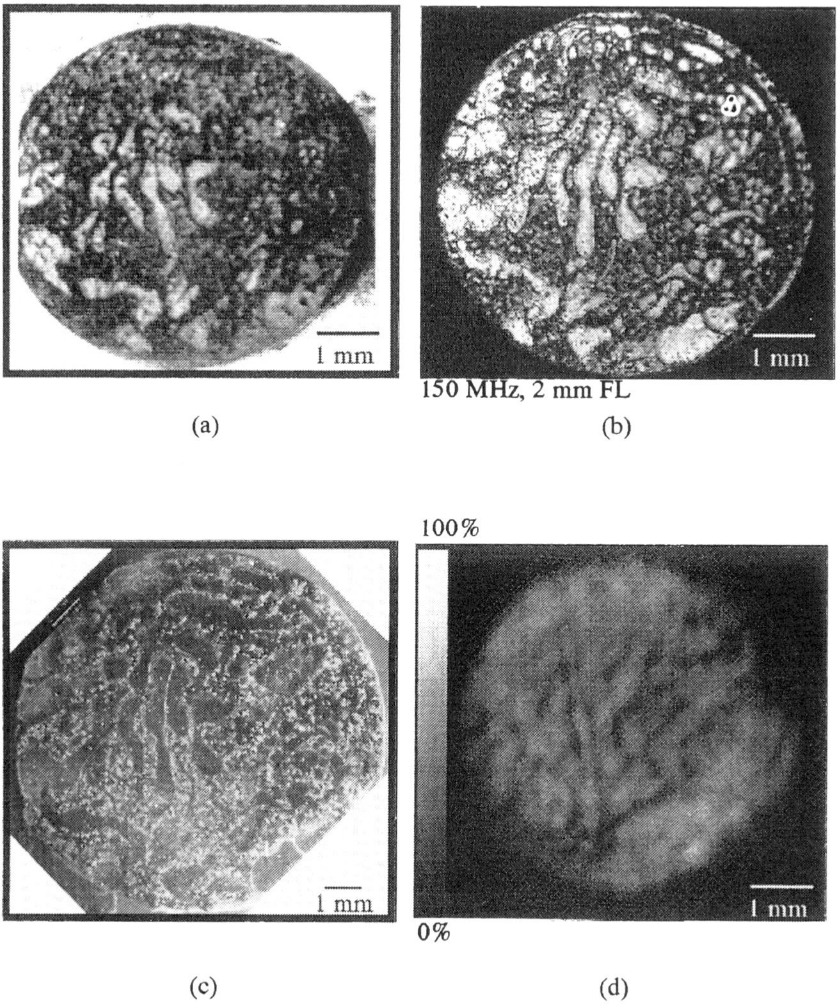

Figure 3. (a) Optical fractograph, (b) 150 MHz acoustic surface scan of fracture surface, (c) electron backscatter images of fracture surface and (d) C-scan of joint interface prior to high frequency thermal cycle testing. All images are of specimen 1.

DISCUSSION

The two microwave joined samples showed non–uniform reflectance indicating variable quality bonding with regions of lower bond quality (light regions of Fig. 2) near the specimen edges. The information contained in the C-scan images is confirmed by the acoustic and optical fractographs. Electron backscatter microscopy further confirmed regions that were successfully microwave bonded from other non-bonded regions based on magnified microstructural features. While the ultrasonic test could not estimate the degree or strength of the bond, it accurately imaged those regions that were bonded from those regions that were unbonded.

Frequency domain information was calculated to investigate pre- and post-thermal cycle test conditions. A frequency downshift of the 50 MHz waveform was calculated (including the 95% confidence limits) point-by-point from the radio frequency waveform data over the entire scanned surface and is given in Table II for specimens 1 and 2.

Table II. Frequency analysis of reflected interface signals

Specimen Condition	Interface Signal	Reflected Peak Frequency	
		Specimen 1	Specimen 2
Before Thermal Cycle	Top	36.3±5.1 MHz	33.9±5.78 MHz
	Bottom	27.9±5.5 MHz	26.5±4.46 MHz
After Thermal Cycle	Top	36.7±4.3 MHz	28.7±7.72 MHz
	Bottom	25.7±4.7 MHz	26.0±26.2 MHz

A downshift of 2.2 MHz in mean frequency from the bottom interface signal and a 0.4 MHz upshift in mean frequency from the top interface were observed in specimen 1 after thermal cycling. The selective attenuation of higher frequency ultrasound at the bottom interface surface suggests (not conclusively given the scatter in the data) the deterioration of the bond quality during thermal cycling. Such a condition was not seen in specimen 2 due to large scatter in the peak frequency data upon thermal cycling. The scatter itself suggests the presence of damage.

CONCLUSIONS

The ultrasonic test was successful in indicating regions that were well bonded from regions that were unbonded and provided information on the capability of high frequency ultrasonic imaging for future material characterization efforts. The test was sensitive to damage induced in the SiC-SiC joint by thermal cycling, as demonstrated by observable changes in the frequency spectrum data. The ultrasonic tests were confirmed by acoustic, optical, and electron backscatter characterization of the fracture surfaces. Such concurrent engineering of nondestructive tests and material fabrication processing offers to aid in the development of future production inspections while also improving bond formation processes.

REFERENCES

1. T.T. Meek and R.D. Blake, U.S. Patent Nos. 4,529,856 and 4,529,857 (July 16, 1985).
2. D. Palaith, R. Silberglitt, C.C.M. Wu, R. Kleiner, and E.L. Libelo, in *Microwave Processing of Materials*, edited by W.H. Sutton, M.H. Brooks, and I.J. Chabinsky (Mater. Res. Soc. Symp. Proc. 124, Pittsburg, PA, 1988) p. 255.
3. H. Fukushima, T. Yamanaka, and M. Matsui, in *Microwave Processing of Materials*, edited by W.H. Sutton, M.H. Brooks, and I.J. Chabinsky (Mater. Res. Soc. Symp. Proc. 124, Pittsburg, PA, 1988) p. 267.
4. I. Ahmad, W.M. Black, and R. Silberglitt, *Ceram. Eng. Sci. Proc.* 13 (7,8) (1992) p. 250.
5. L. Kolaya, G Borges and N. Lewis, "The influence of carbon on the structure of pre-ceramic polymer derived SiC" 21st Annual Coca Beach Conference and Exposition, Jan. 12-16, 1997.
6. B.T. Khuri-Yakub et al., "Nondestructive evaluation of ceramics" Review of Progress in QNDE, v 1 New York, NY: Plenum Press (1982).
7. N. Kulkarni et al., "Ultrasonic characterization of green and sintered ceramics: ii frequency domain" *American Ceramic Socity Bulletin* v 73 n 7 July 1994, p. 83.
8. D. Palaith et al., "Microwave joining of ceramics" in *Microwave Processing of Materials*, edited by W.H. Sutton, M.H. Brooks, and I.J. Chabinsky (Mater. Res. Soc. Symp. Proc. 124, Pittsburg, PA, 1988) p. 255.

NONDESTRUCTIVE EVALUATION OF BRITTLE-MATRIX COMPOSITES DURING LOADING BY ELECTRICAL RESISTANCE MEASUREMENT

Shoukai Wang, Xuli Fu and D.D.L. Chung
Composite Materials Research Laboratory
State University of New York at Buffalo
Buffalo, NY 14260-4400

ABSTRACT

Brittle-matrix composites with matrices that were not totally electrically insulating were found to change electrical resistance upon loading such that reversible strain caused the resistance to change reversibly and damage caused the resistance to change irreversibly. This electromechanical behavior allows dynamic strain and fatigue damage to be monitored simultaneously without the use of embedded or attached sensors, as the composites are themselves sensors. This new technique has been demonstrated for carbon-matrix and cement-matrix composites. The carbon-matrix composite with continuous carbon fibers was particularly effective for damage sensing; even damage in the first cycle of tensile loading at a stress amplitude of just 20% of the fracture stress was detected. The cement-matrix composite with short carbon fibers (0.2-0.5 vol.%) was particularly effective for strain sensing; the gage factor (fractional change in resistance per unit strain) was as high as 500, compared to 2 for resistive strain gages.

INTRODUCTION

Although crack deflection by fibers in a brittle-matrix composite makes the composite less brittle than the matrix itself, the composite remains brittle from the point of view of a design engineer, who demands durability and reliability. One way to alleviate this problem is to continuously or periodically monitor (i.e., nondestructively evaluate) the structural health of the composite structure while the structure is in use. In addition to monitoring the damage, the dynamic strain can be monitored in real time for the purpose of structural control, which is important for smart structures. The monitoring of both damage and dynamic strain has the additional function of providing information on which part of which load cycle (not necessarily periodic) damage occurs.

The monitoring mentioned above is conventionally achieved by embedding into the composite or attaching on the composite one or more sensors, such as piezoelectric sensors, magnetostrictive sensors, piezoresistive sensors, optical fiber sensors, etc., or by acoustic emission detection. However, the sensors suffer from high cost, poor durability, limited sensing volume and, in the case of embedded sensors, degradation of the mechanical properties of the composite. Acoustic emission is not sensitive to slight damage or dynamic strain. Instead of using embedded or attached sensors, this work uses the composite itself as the sensor. In other words, the composite is both the structural material and the sensor. Such self-monitoring structural materials do not suffer from the problems described above for the embedded or attached sensors.

This paper addresses two kinds of self-monitoring brittle-matrix composites. They are a carbon-matrix composite with continuous carbon fibers and concrete with short carbon fibers. In both cases, the matrix is not totally electrically insulating and the fibers are electrically conducting. As a result, the electrical resistance of the composite changes upon loading. The scientific origin of the electromechanical effect differs between the two kinds of composite.

CARBON-CARBON COMPOSITE

Carbon-carbon composites with continuous carbon fibers and a carbon matrix are used for high-temperature aerospace structures and biomedical implants, due to the high-temperature resistance and biocompatibility of carbon. The carbon matrix, though much more high-temperature resistant than a polymer matrix, is much more brittle than a polymer matrix. This brittleness makes carbon-carbon composites prone to matrix cracking. As a result, there is a need for monitoring the condition or health of a carbon-carbon composite structure while the structure is in use, in order to minimise hazards. This monitoring is conventionally conducted by acoustic emission, which is capable of detecting substantial cracking, but not slight cracking. Exceptional sensitivity to even slight damage can be obtained by using the carbon-carbon composite itself as the damage sensor to monitor the composite's own damage (i.e., to self-monitor the damage), using the increase in electrical resistivity as a measure of the irreversible composite damage [1]. The high sensitivity to damage is due to the high conductivity of the carbon matrix, compared to the cement and polymer matrices, and the importance of matrix cracking in the mechanism for damage in a carbon-carbon composite.

Figure 1 shows the stress (curve (a)) and the fractional DC resistance increase ($\Delta R/R_0$) (curve (b)) obtained during static tension up to failure for a carbon-carbon composite having two-dimensionally (90°) woven fibers and a heat treatment temperature of 2000°C, with the resistance and stress in the direction of one of the two perpendicular sets of fibers. $\Delta R/R_0$ increased monotonically with strain, such that the increase was gradual (only slightly above the increase in $\Delta R/R_0$ due to the changes in dimensions, curve (c) in Fig. 1) at low strains and abrupt at high strains.

Figure 2 shows $\Delta R/R_0$ obtained during cyclic tension to a stress

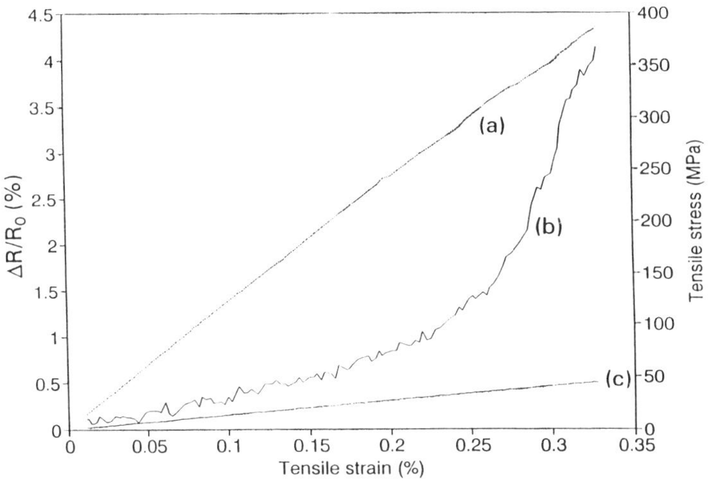

Fig. 1 Plots of (a) tensile stress vs. strain, and (b) $\Delta R/R_0$ vs. strain, obtained simultaneously during static tension up to failure for a carbon-carbon composite. Curve (c) is the calculated $\Delta R/R_0$ based on dimensional changes.

amplitude (360 MPa) equal to 94% of the breaking stress. The tensile strain was almost totally reversible. The irreversible strain was 0.040% at the end of the first cycle, and increased very slightly with increasing cycle number. $\Delta R/R_0$ increased upon loading in every cycle, such that it irreversibly increased slightly after every cycle and the irreversible increase in $\Delta R/R_0$ was particularly large for the first cycle, as shown in Fig. 2. At fatigue failure, $\Delta R/R_0$ abruptly increased, such that $\Delta R/R_0$ did not more rapidly increase irreversibly near the end of fatigue life. Fig. 3 shows the peak $\Delta R/R_0$ values in a cycle as a function of cycle number throughout the fatigue life up to failure. The peak $\Delta R/R_0$ increased with cycle number significantly during the first 500 cycles and gradually during all subsequent cycles up to failure. The small step increases in the peak $\Delta R/R_0$, for example at ~ 1350 cycles, are not experimental artifacts but are attributed to damage occurring at those cycle numbers, similar to the step increases observed for a continuous carbon fiber polymer-matrix composite under similar cyclic loading [2].

The reversible part of $\Delta R/R_0$ is mainly due to reversible dimensional changes and correlates with reversible strain. The irreversible part of $\Delta R/R_0$ is due to damage. Although the increase in irreversible strain and decrease in Young's modulus also indicate damage, the changes in these parameters are very small compared to the change in the irreversible part of $\Delta R/R_0$. The great sensitivity of the irreversible part of $\Delta R/R_0$ to damage is also shown by the significant non-zero value of the irreversible part of $\Delta R/R_0$ after merely the first cycle, even at a stress amplitude of just 20% of the fracture stress (Fig. 4).

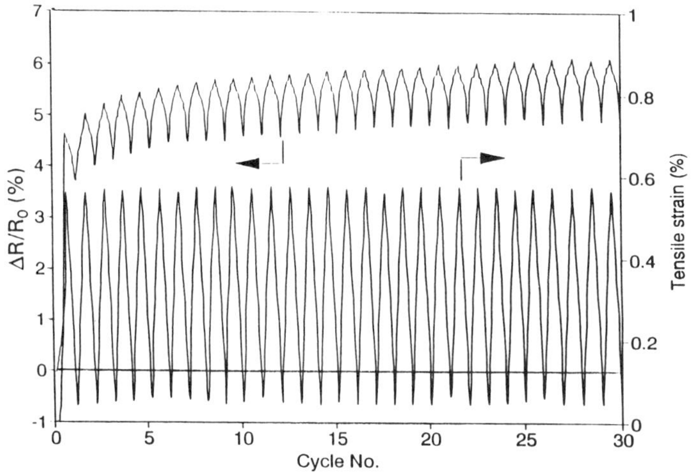

Fig. 2 Plots of $\Delta R/R_0$ vs. cycle No. and of tensile strain vs. cycle No., obtained simultaneously during first cycle tension at a stress amplitude of 94% of the fracture stress for a carbon-carbon composite.

However, the incremental rise in irreversible $\Delta R/R_0$ beyond ~ 500 cycles was small. The composite damage probably involved fiber-matrix interface weakening, matrix cracking and fiber breakage; these origins of damage could not be distinguished through the experimental technique used. Nevertheless, the increase of the irreversible part of $\Delta R/R_0$ as cycling progressed provided a continuous indication of the extent of damage. That the reversible part of $\Delta R/R_0$ also increased with cycling and that an abrupt increase of the irreversible part of $\Delta R/R_0$ is associated with an abrupt increase in the reversible part of $\Delta R/R_0$ suggest that the reversible part of $\Delta R/R_0$ is partly associated with a phenomenon which intensifies as damage increases, although it is mostly associated with dimensional changes. This phenomenon may be reversible crack opening during tension, as cracks are expected to increase in size and/or density as cycling progresses. This interpretation is consistent with the observation that an abrupt increase in the reversible part of $\Delta R/R_0$ is associated with an abrupt increase in reversible strain and that the abrupt increase in reversible strain occurs at stress amplitudes beyond the range in which the reversible strain is linear in relation to the stress amplitude. As a result of this phenomenon, the gage factor (reversible $\Delta R/R_0$ per unit reversible strain) increases slightly with cycle number. The dependence of the gage factor on the cycle number complicates the practical use of the carbon-carbon composite as a strain sensor.

CONCRETE

Carbon fiber reinforced concrete is attractive for civil structures due to

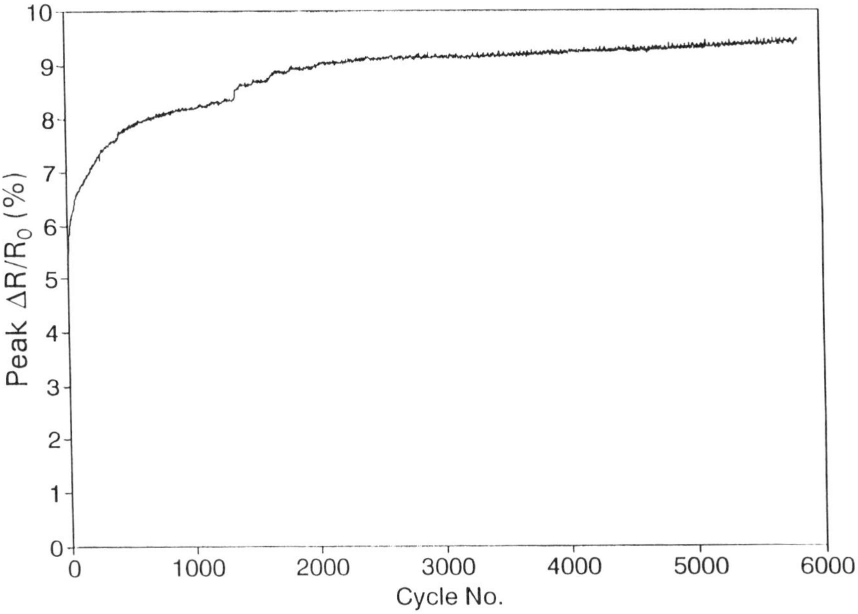

Fig. 3 Variation of the peak $\Delta R/R_0$ with cycle No. throughout the entire fatigue life at a stress amplitude of 94% of the fracture stress for a carbon-carbon composite.

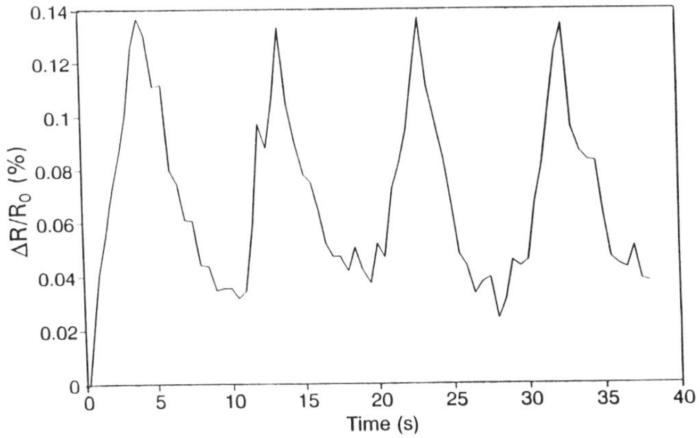

Fig. 4 Plot of $\Delta R/R_0$ vs. time during the first four cycles at a stress amplitude of 20% of the fracture stress for a carbon-carbon composite.

its high tensile and flexural strengths, high tensile ductility, high flexural toughness, low drying shrinkage [3-13], strain sensing ability [14-22] and low cost. This section focuses on its strain sensing ability. Damage sensing ability exists in this material, but it is not as powerful as the strain sensing ability.

Self-monitoring in concrete was achieved with concrete containing short, electrically conducting microfibers, preferably carbon fibers, in an amount as low as 0.2 vol.%. The DC volume electrical resistivity of the concrete changes reversibly upon reversible strain, such that the fractional change in electrical resistance per unit strain (i.e., the strain sensitivity or the gage factor) reaches 500 (extraordinarily high compared to conventional strain gages). The effect has been observed in cement paste, mortar as well as concrete, under tension, compression and flexure, though the effect is more pronounced in cement paste and mortar than in concrete. The reversible electromechanical effect stems from the reversible increase in the contact electrical resistivity between fiber and matrix upon slight opening (< 1 μm) of the crack which the fiber bridges (i.e., slight fiber pull-out), and the consequent reversible increase in the volume electrical resistivity of the cement-matrix composite [14-21].

The evidence in support of the above-mentioned origin of the self-monitoring ability of intrinsically smart concrete is summarized below.

1. The sensing ability was present when the fibers were conducting (i.e., carbon or steel) and absent when the fibers were non-conducting (i.e., polyethylene) [18].
2. The sensing ability was absent when fibers were absent [14,15,18,20].
3. The sensing ability occurred even at low carbon fiber volume fractions, which were associated with little effect of the fiber addition on the concrete's volume electrical resistivity [18,23,24].
4. There was no maximum volume electrical resistivity required in order for the sensing ability to be present [18].
5. The sensing ability was present when the carbon fiber volume fraction was as low as 0.2% (below the percolation threshold) [18,20].
6. Fracture surface examination showed that the fibers were separate from one another [18].
7. The fractional increase in electrical resistance ($\Delta R/R_0$) upon straining essentially did not increase with increasing carbon fiber volume fraction, even though the increase in fiber volume fraction caused large decrease (by orders of magnitude) in the volume electrical resistivity [18].
8. The electrical resistance increased upon tension (fiber pull-out) and decreased upon compression (fiber push-in), except for the first compressive strain cycle at 7 days of curing. The resistance change during the first strain cycle is consistent with the need to weaken the fiber-matrix interface prior to fiber pull-out in case the interface was relatively strong to start with [18,19,21].
9. The presence of carbon fibers caused the crack height to decrease from

~ 100 μm, as observed after deformation to 70% of the compressive strength [20].

10. The presence of carbon fibers caused the flexural toughness and tensile ductility of the composite to greatly increase [24,25].
11. The stress required for fiber pull-out in the short fiber composite was consistent with the shear bond strength between carbon fiber and cement paste, as obtained by single fiber pull-out testing [17].
12. The contact electrical resistivity between carbon fiber and cement paste increased during debonding [17].

Due to the small volume fraction of carbon fibers, the decrease of the concrete's volume electrical resistivity due to the fiber addition is small [18,23]. In order to help the fibers disperse, admixtures such as silica fume, methylcellulose and latex are used along with the fibers [23,24]. The fibers are the most inexpensive type of carbon fibers, namely those based on isotropic pitch.

Due to the high electrical contact resistivity between steel rebar and concrete [26] and the low ability of an electrical current applied on one surface of a concrete structure to penetrate the concrete to depths more than 1 mm, the presence of steel rebars in concrete does not interfere with the performance of the self-sensing concrete around it unless the electrical contacts are around the whole cross-section of the concrete (rather than on one side) and unless the concrete is inside a cage made of the rebars.

Figure 5 gives the fractional DC resistance increase ($\Delta R/R_0$) during first cyclic compressive loading of mortar with 0.24 vol.% carbon fibers at a stress amplitude of 16 MPa, or a strain amplitude of 8 x 10^{-4}, which was within the elastic regime at 28 days of curing. (The compressive strength was 45 MPa). The resistance was in the stress direction, as measured by the four-probe method (i.e., the outer two probes for passing current and the inner two probes for measuring the voltage), with silver paint for the electrical contacts. Both stress and strain returned to zero at the end of each cycle. The $\Delta R/R_0$ decreased during compressive loading in each cycle and increased during unloading in each cycle. This is due to fiber push-in during loading and fiber pull-out during unloading. At the end of the first cycle, $\Delta R/R_0$ was positive rather than zero. This resistance increase is attributed to damage of the fiber-cement interface due to the fiber push-in and pull-out. As cycling progressed, both the maximum $\Delta R/R_0$ and minimum $\Delta R/R_0$ in a cycle decreased. This is attributed to damage of the cement matrix separating adjacent fibers at their junction; this damage increased the chance for adjacent fibers to touch each other, thereby decreasing the resistivity. This decrease from cycle to cycle persisted for the first ~ 150 cycles, after which the maximum and minimum $\Delta R/R_0$ did not change with cycling.

At 28 days of curing (Fig. 5), since the amplitude of resistance variation in a cycle is 0.4 and the strain amplitude is 8 x 10^{-4}, the gage factor is 0.4

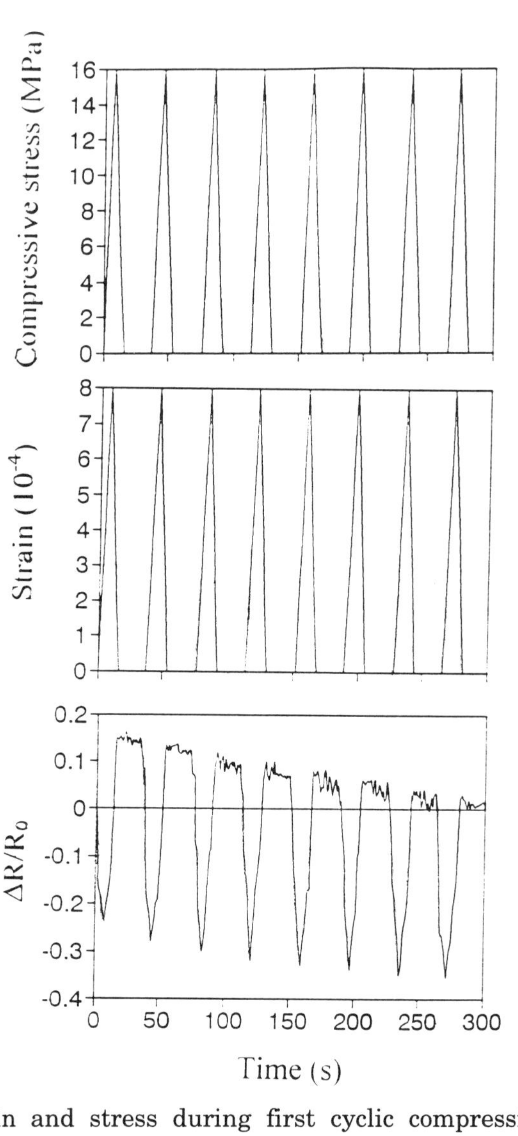

Fig. 5 $\Delta R/R_0$, strain and stress during first cyclic compressive loading of mortar at 28 days of curing.

divided by 8 x 10^{-4}, or 500. At 7 days of curing (Fig. 6), the gage factor is 375. These exceptionally large values of the gage factor means that the mortar is an extremely sensitive strain gage. The gage factor of a conventional resistive strain gage is around 2.

Figure 7 shows $\Delta R/R_0$ versus strain and stress versus strain during first

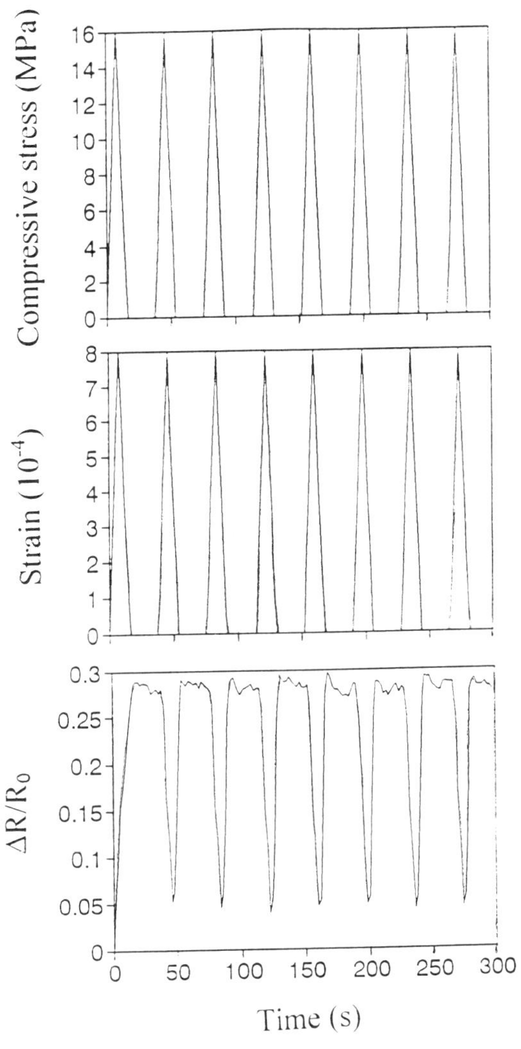

Fig. 6 $\Delta R/R_o$, strain and stress during first cyclic compressive loading of mortar at 7 days of curing.

static compression up to failure of mortar with 0.24 vol.% carbon fibers at 28 days of curing. The resistance decreased monotonically upon loading up to failure.

Results similar to Figs. 5 and 7 were obtained at 14 days of curing, except that the decrease in maximum $\Delta R/R_o$ and minimum $\Delta R/R_o$ from cycle to cycle was less pronounced. However, at 7 days of curing, the resistance

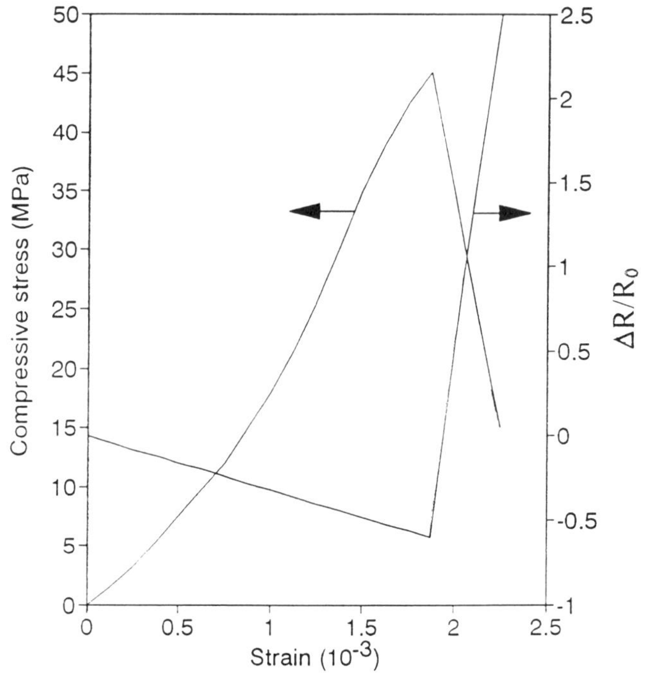

Fig. 7 $\Delta R/R_0$ versus strain and stress versus strain during first static compression up to failure of mortar at 28 days of curing.

increased monotonically upon first static compression up to failure (Fig. 8). Consistent with the static compression result of Fig. 8 is the cyclic compression result of Fig. 6. During the first cycle, the resistance increased upon loading, increased further upon subsequent unloading (due to fiber pull-out), decreased upon loading in the second cycle (due to fiber push-in), and increased upon unloading in the second cycle (due to fiber pull-out); the behavior was similar in second and subsequent cycles, but was different in the first cycle. The behavior at 7 days of curing is attributed to the relatively strong fiber-matrix bonding at 7 days and the consequent need to weaken the bond prior to fiber pull-out. Bond weakening is accompanied by irreversible increase in the contact electrical resistivity [17], which results in the irreversible resistance increase observed in the first cycle [18]. This irreversible resistance increase was larger in magnitude when the stress amplitude increased [18]. The fiber-cement bond strength decreased with increasing curing time from 7 to 14 days, while the contact resistivity increased, as shown for stainless steel fiber [19]. At 14 or 28 days of curing, the bond strength was weak to start with, so bond weakening was not necessary prior to fiber pull-out. The monotonic resistance increase up to failure at 7 days of curing is due to the bond weakening at least

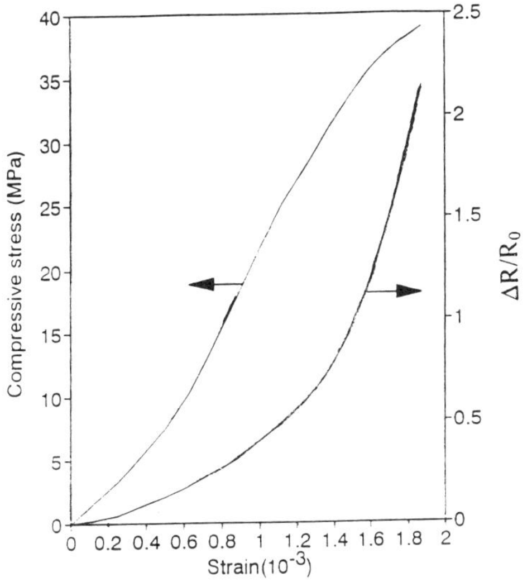

Fig. 8 $\Delta R/R_0$ versus strain and stress versus strain during first static compression up to failure of mortar at 7 days of curing.

in the low stress regime; in the high stress regime, it is probably due to damage. Note from Fig. 6 that the maximum $\Delta R/R_0$ and minimum $\Delta R/R_0$ did not change from cycle to cycle, in contrast to the decrease in these quantities at 14 and 28 days of curing. This effect of the curing age is attributed to the decrease in ductility with increasing curing age and the resulting increased tendency for repeated fiber pull-out and push-in during cyclic loading to cause damage to the cement matrix separating adjacent fibers. The effect of the curing age is not due to the effect of the moisture content, as the humidity during curing had essentially no effect on the electromechanical behavior [18] and the volume electrical resistivity of the mortar increased negligibly with curing age [27].

During the few cycles before fatigue failure at 28 days of curing, the resistance did not increase irreversibly and significantly, indicating that the mortar is poor in its ability to monitor damage. This poor damage sensitivity is also reflected by the absence of a resistance increase prior to static failure at 28 days (Fig. 7).

CONCLUSION

Nondestructive evaluation of carbon-matrix and cement-matrix composites during loading was achieved in real time in terms of both fatigue damage and dynamic strain by measurement of the electrical resistance. The carbon-matrix composite was highly effective for damage sensing (i.e., sensitivity even to the damage after the first cycle of tensile loading within the elastic regime), because the matrix was conducting and its fracture caused the resistivity to increase irreversibly. The carbon-carbon composite was also a strain sensor, due to the dimensional changes during tension and the resulting reversible increase in the resistance. Concrete containing short carbon fibers was highly effective for strain sensing (gage factor as high as 500), due to slight fiber pull-out during strain and the resulting reversible increase in resistivity. The concrete was also a damage sensor, due to fiber and matrix fracture causing the resistivity to increase irreversibly.

REFERENCES

1. S. Wang and D.D.L. Chung, Carbon, in press.
2. X. Wang and D.D.L. Chung, Smart Mater. Struct., in press.
3. P. Chen and D.D.L. Chung, Composites 24, 33-52 (1993).
4. P. Chen and D.D.L. Chung, Composites: Part B 27B, 269 (1996).
5. L.A. Kuhlmann, Transportation Research Record 1204, 52 (1988).
6. X. Yang and D.D.L. Chung, Composites 23(6), 453 (1992).
7. N. Banthia, ACI SP-142, Fiber Reinforced Concrete, J.I. Daniel and S.P. Shah, eds., ACI, Detroit, Mich., 1994, p. 91-120.
8. N. Banthia, A. Moncef, K. Chokri and J. Sheng, Can. J. Civ. Eng. 21, 999 (1994).
9. V.C. Li and K.H. Obla, Composites Engineering 4(9), 947 (1994).
10. S.B. Park and B.I. Lee, Cem. Concr. Composites 15(3), 153 (1993).
11. P. Soroushian, M. Nagi and J. Hsu, ACI Materials J. 89(3), 267 (1992).
12. H.A. Toutanji, T. El-Korchi and R.N. Katz, Cem. Concr. Composites 16, 15 (1994).
13. K. Zayat and Z. Bayasi, ACI Materials J. 93, 178 (1996).
14. P. Chen and D.D.L. Chung, Smart Mater. Struct. 2, 22-30 (1993).
15. P. Chen and D.D.L. Chung, J. Am. Ceram. Soc. 78(3), 816-818 (1995).
16. D.D.L. Chung, Smart Mater. Struct. 4, 59-61 (1995).
17. X. Fu and D.D.L. Chung, Cem. Concr. Res. 25(7), 1391-1396 (1995).
18. P. Chen and D.D.L. Chung, Composites: Part B 27B, 11-23 (1996).
19. X. Fu and D.D.L. Chung, Cem. Concr. Res. 26(1), 15-20 (1996).
20. P. Chen and D.D.L. Chung, ACI Materials J. 93(4), 341-350 (1996).
21. X. Fu and D.D.L. Chung, Cem. Concr. Res. 27 (1997).
22. X. Fu, Erming Ma, D.D.L. Chung and W.A. Anderson, Cem. Concr. Res. 27 (1997).
23. P. Chen and D.D.L. Chung, J. Electron. Mater. 24(1), 47-51 (1995).
24. P. Chen, X. Fu and D.D.L. Chung, ACI Mater. J., in press.
25. P. Chen, X. Fu and D.D.L. Chung, ACI Mater. J. 93(2), 129-133 (1996).
26. X. Fu and D.D.L. Chung, Cem. Concr. Res. 25(7), 1397-1402 (1995).
27. X. Fu and D.D.L. Chung, Cem. Concr. Res. 25(4), 689-694 (1995).

Surfaces and Films

QUANTITATIVE NONDESTRUCTIVE EVALUATION ON SURFACE ROUGHNESS USING ULTRASONIC REFLECTIVITY AND SURFACE WAVE VELOCITY

Tatsuhiko Aizawa
Department of Metallurgy, University of Tokyo,
7-3-1 Hongo, Bunkyo-ku, Tokyo 113, Japan

Ikuo Ihara
Department of Mechanical Engineering, Nagaoka Institute of Technology and Science,
kamitomioka 1603-1, Nagaoka 940-21, Niigata, Japan

ABSTRACT

Acoustic spectro-microscopy has been developed and successfully applied to quantitative nondestructive evaluation on the elastic characterization of monolithic ceramics, metals and ceramic-coated materials. In those studies, the surface of material is assumed to be completely flat, and no wrought layers nor residual stresses are present at the vicinity of the surface. In the actual situation, however, those irregularities exist in materials to be working as a disturbance to the measured acoustic structure. Hence, precise understanding on the effect of those geometric and structural irregularities on the acoustic structure enables us to evaluate the actual surface roughness, the wrought layers and the residual stress distributions in materials from the measured acoustic data. In the present study, samples with seven model surface roughness profiles are prepared and used to investigate the effect of surface roughness on the acoustic reflectivity. This close relationship between the surface roughness and the measured reflectivity enables us to make nondestructive detection of the surface roughness profile by linear scanning. Furthermore, the apparent dispersion of the surface wave velocity is found to be a base to evaluate the geometric morphology of surface roughness. Finally, some comments on the effect of wrought layer and residual stresses on the surface wave velocity are made to stimulate the importance of quantitative nondestructive measurement of stresses by using the acoustic spectro-microscopy.

INTRODUCTION

Acoustic spectro-microscopy (ASM) 1,2,3,4,5) has been developed and successfully applied to quantitative nondestructive evaluation on the elastic characterization of monolithic ceramics, metals and ceramic-coated materials. In those studies, the surface of material is assumed to be completely flat, so that the ultrasonic lens system with transducers is spatially accommodated for working with the focus on the surface, moving in the circumferential direction with the right angle to the surface and in the rotational direction parallel to the surface. Furthermore, no wrought layers nor residual stresses are thought to be present at the vicinity of the surface. In actual situation, most of material surfaces have their intrinsic roughness on it, and various wrought layers and residual stresses are often introduced during machining processes and heat treatment. The surface roughness has the following two influences on the acoustic structure: 1] Effect on the reflectivity in measurement and 2] Effect on the surface wave propagating on the surface. As depicted in Fig. 1, intensity of coherent waves in reflection and transmission significantly reduces and enhancement of incoherent field is measured with increasing the surface roughness (6,7), resulting in the significant decrease of the signal to noise ratio in the reflectivity measurement. This decrease of S/N ratio is further enhanced with increasing the frequency (8). On the other hand, qualitative discussions (8,9,10) have been performed on the effect of surface roughness on the surface wave dispersion.

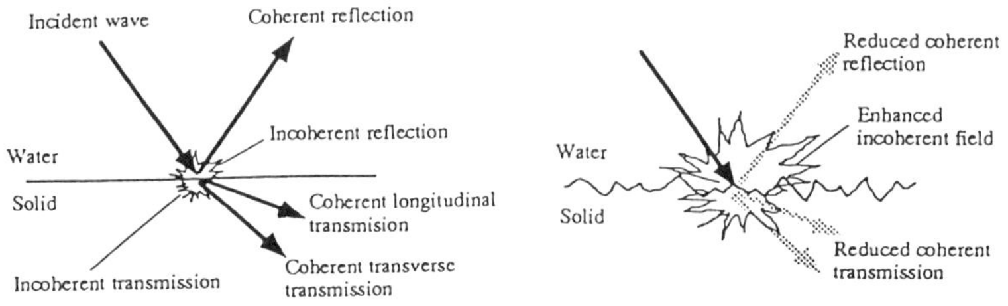

Fig. 1 Effect of surface roughness on the acoustic waves in reflection and transmission.

When the materials are subjected to a stress field, both acoustic velocity and attenuation are changed by the acousto-elastic effect (11,12). In particular, at the presence of the wrought layers and the residual stresses at the vicinity of the surface, the acoustic structure is expected to be different from that at the absence of those irregularities and stresses.

In the present study, the effect of the surface roughness, the wrought layers and the residual stresses on the acoustic structure is discussed through the systematic experiments. First of all, seven samples with different surface roughness are prepared by buffing, filing and using shaper. Direct comparison of the measured phase shift and power amplitude in the function of incident angle and frequency enables us to understand the large influence of the surface roughness on the acoustic structure. The average surface roughness (Ra) is one of the parameters affecting on the surface wave velocity dispersion apparently measured by ASM. Since other parameters of surface roughness profile also have influence on this apparent dispersion behavior, the actual geometric modulation on surface can be described by precise analysis of these close relationship between surface roughness and surface wave dispersions. Finally, some comments are made on the effect of wrought layers and residual stresses on the surface wave velocity.

ACOUSTIC SPECRO-MICROSCOPY

Our developing acoustic spectro-microscopy is briefly introduced here to stress the importance on the reflectivity measurement and its linear scanning, and the surface wave velocity measurement. Different from the conventional ultrasonic microscopy, the spherical-planar-pair lens (SPP lens) is used and controlled to move in five directions (in the parallel plane (x-y plane) to the surface of a sample, in the perpendicular direction (z-axis) to the surface, in the circumferential direction (θ - axis) and in the rotational direction (ϕ -axis)). The focus of this lens system is fixed at the specified point on the surface of a sample within the spatial mismatch less than 100 nm by using x-, y-, and z-axis driving units. The incident angle can be directly controlled by using the θ - axis driving unit; three different lens systems enable us to make measurement for $10° < \theta < 50°$. ϕ -axis control is necessary to make evaluation on the anisotropic properties of materials. To be noted, since a ultrasonic pulse is injected into a sample through a coupler material (usually, water), data acquisition tools including the Fast-Fourier-Transform are available to make spectroscopic analysis of the reflected signals. From directly received signal, both the phase shift and the power amplitude distributions are obtained at every point of measurement in the function of the incident angle and the frequency. Linear scanning of SPP lens system provides us the variation of these phase shift and power amplitude distributions along the specified direction.

Through the data acquisition by varying the incident angle, the critical angle, where the surface wave is activated, can be detected as a significant amount of phase shift and a noticeable intensity reduction. From this measured critical angle (θ_c), the surface velocity (Vsaw) can be determined by

$$Vsaw = Vw / \sin(\theta_c) \quad (1)$$

where Vw is the acoustic velocity of a coupling material (usually, water). Since Vw is functional in temperature, the difference in temperature of the coupler is controlled to be less than 0.1 K during measurement. This Vsaw is a constant value indifferent to the ultrasonic frequency for the homogeneous material with a completely smooth surface. As have been reported in Refs. (13, 14), the multi-layered material system has a surface wave velocity dispersion, which is intrinsic to each system. The measured dispersion curve often becomes a basic information for elastic characterization and defect detection in the quantitative nondestructive evaluation of the structured materials.

EFFECT OF SURFACE ROUGHNESS ON ACOUSTIC STRUCTURE

S45C (steel with 0.45 wt% C) specimen after annealing at 973 K was used as a model specimen to investigate the effect of surface roughness on the acoustic structure. Buffing, filing and shaper were used to control the surface roughness to yield seven specimens with different surface profile. As shown in Fig. 2, P-direction denotes for the working direction, and N-direction is perpendicular to the working lines. The surface-profilometer (Surfcom-470A) was employed to determine the surface profile of surface along the N-direction in Fig. 2. Little or no significant roughness can be seen for samples prepared by buffing and #1500 filing, roughness was enhanced with reduction of mesh number in filing from #600 to #120, and the surface was pitched by shaper. Table 1 lists the working method to be used and the measured average surface roughness (Ra).

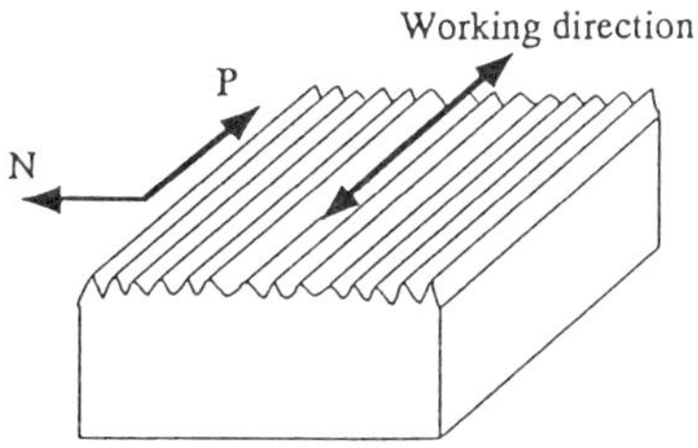

Fig. 2 Geometric coordinates of the wrought material surface.

Table 1 Working methods to be used and the measured average surface roughness for seven samples.

Working method		Surface roughness Ra, μm
Buffing	1μm alumina particle	0.019
Filing	#1500 emery paper	0.07
	#600 emery paper	0.23
	#320 emery paper	0.49
	#120 emery paper	0.88
shaper	50μm pitch	2.20
	100μm pitch	4.10

First of all, in order to investigate the effect of surface roughness on the signal-to-noise ratio, let us make comparison of reflected wave signals for various incident angles θ of ultrasonic wave. Figure 4 shows the measured ultrasonic signals for $\theta = 28^\circ$, 30° (near the critical angle θ_c to activate the surface wave) and 32°. Little or nearly no effects of surface roughness on the S/N ratio can be seen for Ra < 0.88 μ m. When using shaper, the wave form of the activated surface wave at $\theta = \theta_c$ was severely deteriorated so that accuracy in measurement of reflectivity should be worse. In the present acoustic spectro-microscopy, the measured acoustic structure can be characterized by both the power amplitude and the phase shift of reflectivity in the

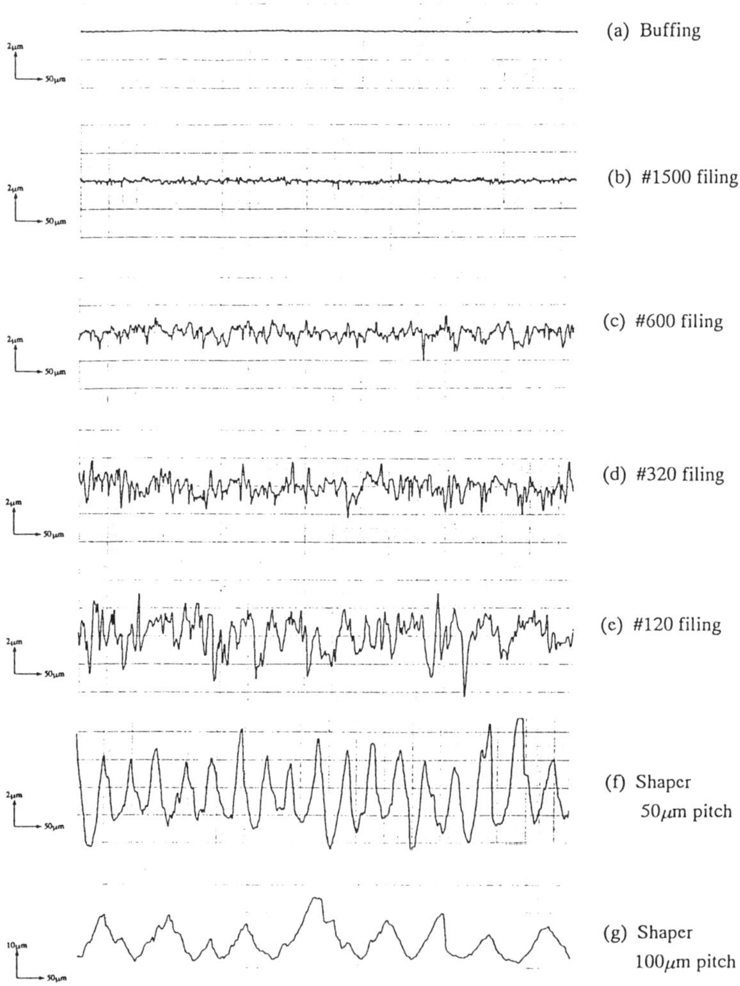

Fig. 3 Typical surface profile for seven samples with different average surface roughness.

function of the incident angel and the frequency. Figures 5 and 6 show the measured distribution of the phase shift and the power amplitude. Although no differences can be distinguished for buffed and #600-filed surfaces, the phase shift has an apparent frequency dependency with increasing Ra. In particular, significant deterioration of phase shift can be seen in the higher frequency range for the rough surface with 50 μ m pitch. This is because the apparent attenuation increases by scattering on the rough planes. Further increasing Ra, when the surface has a periodic irregularities of pitches by shaping, no significant phase shift can be obtained for f > 30 MHz. To be noticed, this disturbance in the phase shift for f > 30 MHz is not attributed to scattering from rough surface but due to mirror reflection from the skewed surface by working. In other words, geometrical irregularity of each asperity reflects on the phase shift distribution when Ra exceeds 1.0 μ m. In fact, for the rough surface with 100 μ m pitch by shaping, although no significant phase shifts are also measured in this case, abrupt change of phase shift with increasing the frequency must be attributed to the phase delay of reflective wave from the skewed plane of asperity. Looking into the change of intensity distribution with increasing Ra, reduction of intensity at the vicinity of the critical angles is enhanced with increase of Ra. Just as the phase shift change observed in Fig. 5, significant disturbance of intensity distribution can be seen for rough surfaces with 50 to 100 μ m.

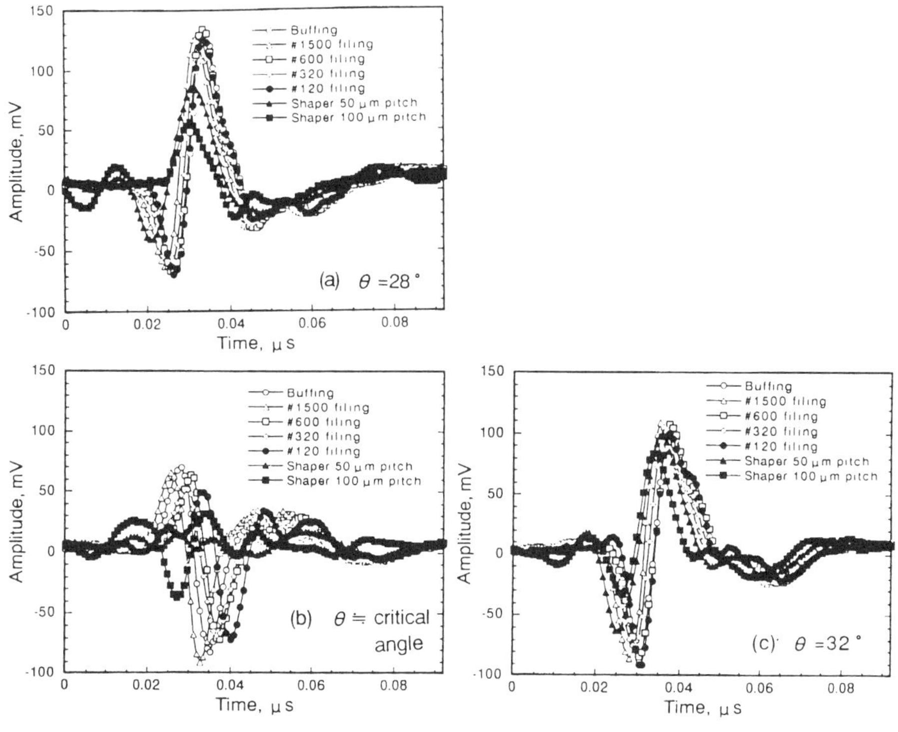

Fig. 4 The measured ultrasonic signals at the incident angle θ = 28°, 30° and 32° : (a) θ = 28°, (b) θ = 30° and (c) θ = 32°

STM (Scanning Tunnel Microscopy) and AFM (Atomic Force Microscopy) are effective to measure small irregularity of asperity in the localized area. Considering the effect of the surface roughness on the acoustic structure in the above, it might be possible to make nondestructive inspection of the surface roughness by the present approach. As also stated in Ref. (13), the linear scanning reflectivity measurement by the acoustic spectro-microscopy was applied here to detect rough surface. To be noticed, the sensitivity of reflectivity at the vicinity of critical angle to the surface roughness is too much to continue the linaer scanning; then, the incident angle of ultrasonic pulse must be shifted from this critical angle ($\theta = 30^\circ$). In the following experiments, the incident angle is kept constant by $\theta = 32^\circ$ during the linear scanning in order that the ultrasonic wave should be absolutely reflected from the surface. The linear scanning range is assumed to be 500 μm, and the ultrasonic lens is controlled to move at the interval of 10 μm, which is selected to be smaller than the pitch of surface irregularity. Figure 7 lists the variation of the measured

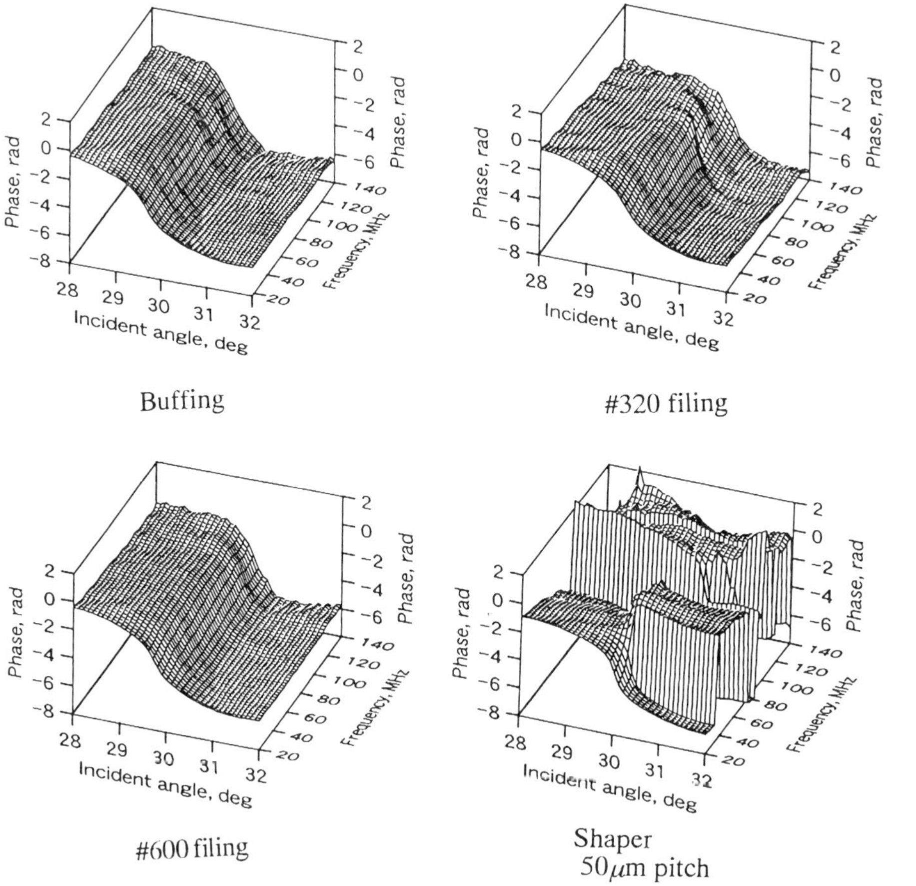

Fig. 5 The measured phase shift distribution in the function of the incident angle and the frequency for seven specimens with different surface roughness.

phase shift and intensity distribution in the function of the relative displacement on the surface and the frequency with increasing Ra. As shown in (a), the frequency domain without disturbance corresponds to the range where the accurate reflectivity can be measured without any disturbances from surface roughness. On the other hand, the observed disturbance in the linear scanning is corresponding to actual surface roughness in geometry. With increasing Ra, frequency domain with disturbance is enlarged.

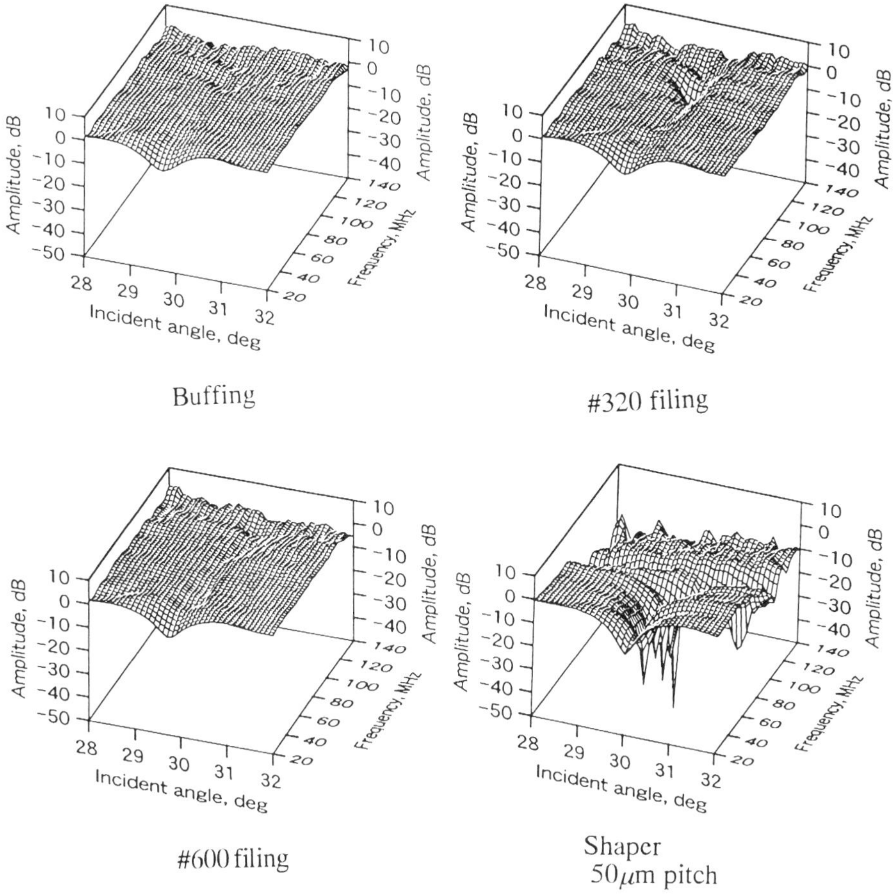

Fig. 6 The measured intensity distribution in the function of the incident angle and the frequency for seven specimens with different surface roughness.

FINE CHARACTERIZATION OF SURFACE ROUGHNESS

Although the surface roughness has strong influence on the measurement of reflectivity, the surface wave velocity can be accurately determined from the measured phase shift at the specified frequency. These obtained surface velocities are thought to have close relationship not only with the average surface roughness but also the roughness profile in

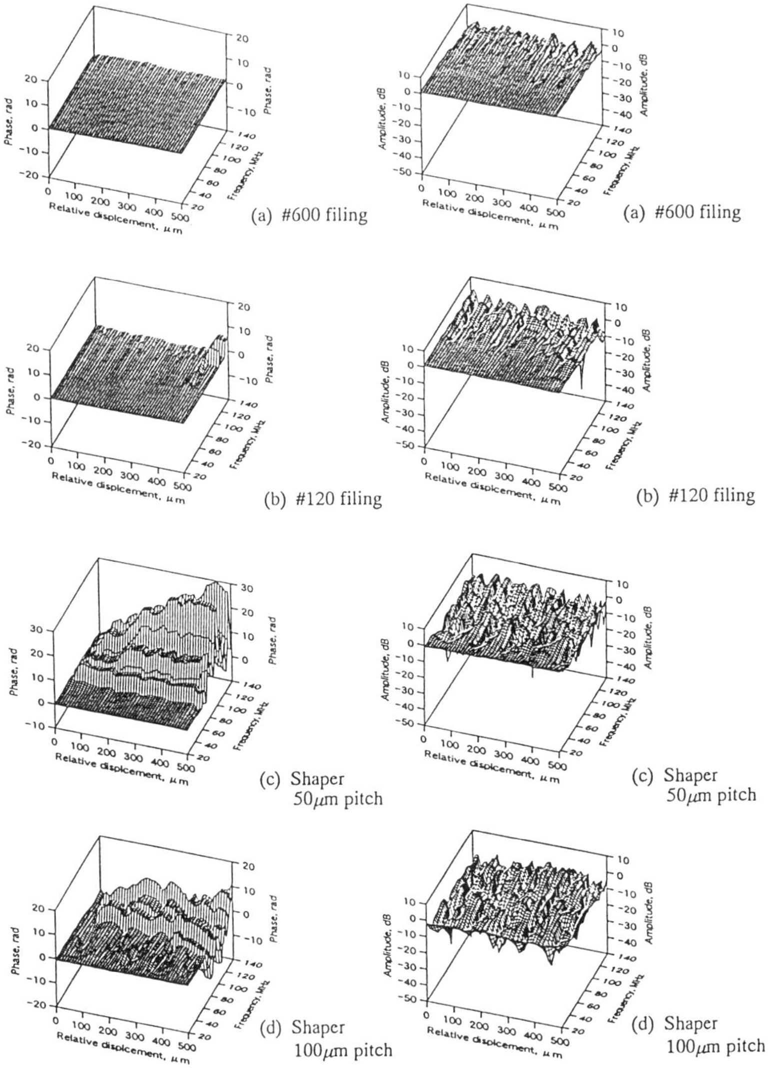

Fig. 7 Variation of the measured phase shift and intensity distributions by the linear scanning reflectivity measurement with increasing Ra.

geometry. In the following experiments, the surface wave velocities were measured for each specimen both in the N- and P- directions in Fig. 2. Although no distinct differences in the measured reflectivity can be recognized even by changing the location of measurement in the N-direction, the averaged surface wave velocity from ten measured data was used in the following discussion.

Figure 8 summarizes the measured dispersion of the surface wave velocity Vsaw in the frequency for seven samples. Considering that any dispersion never appears in the surface wave velocity for any material without coatings nor wrought layers but with completely flat surface, the measured constant surface wave velocity indifferent to the frequency for the buffed sample must be a true surface wave velocity V'saw for a material with completely flat surface. On the other hand, apparent dispersion of the surface wave velocity can be found in Fig. 8. This is because the measured acoustic structure should be corresponding to a system with the surface layer including roughness on the true matrix material. As noticed here, larger dispersion can be seen for measurement in the N-direction than in the P-direction. Let us evaluate the dependence of the surface wave velocity with Ra. Figure 9 compares the relationship between the surface wave velocity and the averaged surface roughness for different frequencies and scanning directions. The measurement by using higher frequency is sensitive to this surface structure change with increasing Ra. As has been reported in Refs. (9,10), Vsaw decreases linearly with Ra. Different from that reported results, however, the decreasing gradient in this linear relationship becomes larger in the N-direction than P-direction for any ultrasonic frequencies to be used. This distinct difference of sensitivity of Vsaw to the scanning direction must be attributed to the elastic anisotropy for the two-layered system where the soft layer including an air among asperities (or height profiles) in the surface roughness is joined with the matrix material. Since this surface layer becomes soft in elasticity, larger dispersion of Vsaw in frequency is measured in the N-direction.

Sinclair (15) proposed a model where Vsaw should be decreasing with the height of asperity by assuming the asperity is approximated by an added mass to the matrix material. Owing to his model, Vsaw is related to the geometry of asperity by

$$\Delta \mathrm{Vsaw} = (\mathrm{Vol}) \cdot f / 2B \qquad (2)$$

where Δ Vsaw = V'saw - Vsaw, Vol is a volume fraction of asperity in the unit cross-section, f the frequency, and B the constant dependent on the transmission mode. Assuming that asperity is geometrically approximated by an equilateral triangle, V = 2 Ra. Hence, the equation (2) is rewritten by

$$\Delta \mathrm{Vsaw} = C \cdot \mathrm{Ra} \cdot f \quad \text{for } C = 1/B. \qquad (3)$$

The whole measured data are summarized again in the diagram of Δ Vsaw - Ra · f. Figure 10 shows the linear relationship between Δ Vsaw and Ra · f for the samples with the filed and the shaping-machined surfaces. Although this C value is theoretically constant by C = 2.5 in Ref. (15), the measured C value is strongly dependent on Ra, and C varies from 0.37 to 1.64 in (a) and from 0.16 to 0.87 in (b). This discrepancy between theoretical and experimental values for C was also stated in Ref. (10); this might be because no coupling materials were taken into account in the theoretical calculation. From this linear relationship, Ra value can be roughly estimated from the measured Δ Vsaw for the prescribed frequency.

To be noticed here, this C is strongly dependent on the working procedure to make a surface profile and the direction of measurement. If Δ Vsaw were only dependent on Ra or Vol, variation of C value would be much smaller than observed in Fig. 10. In other words, the measured surface wave velocity distribution has close relationship with the effective elastic anisotropy for a soft-layered material. The stiffness of this soft layer varies with the geometric asperity profile results in its intrinsic surface wave velocity dispersion. Hence, these measured apparent dispersions can be utilized as a key to quantitatively describe the surface roughness profile even by the nondestructive evaluation.

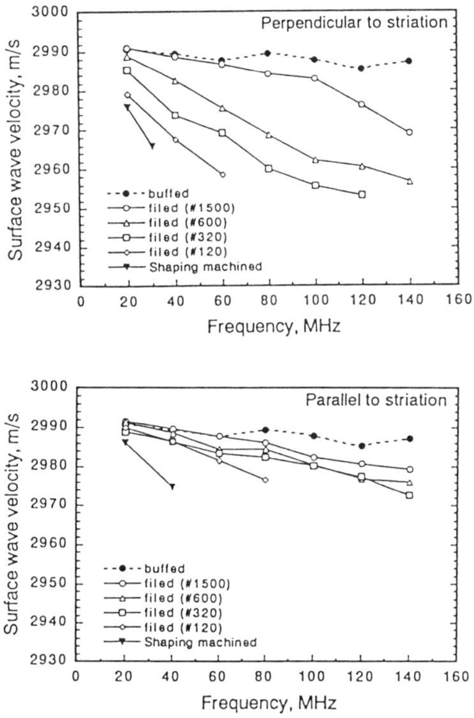

Fig. 8 Surface wave velocity dispersion for seven samples both in the N- and P-directions.

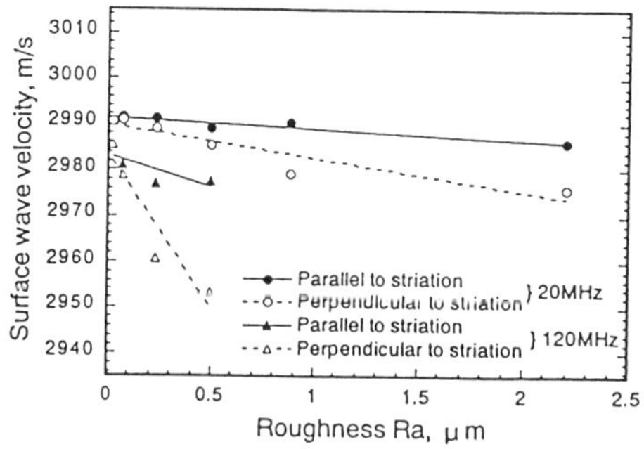

Fig. 9 Relationship of the surface wave velocity with the average surface roughness (Ra).

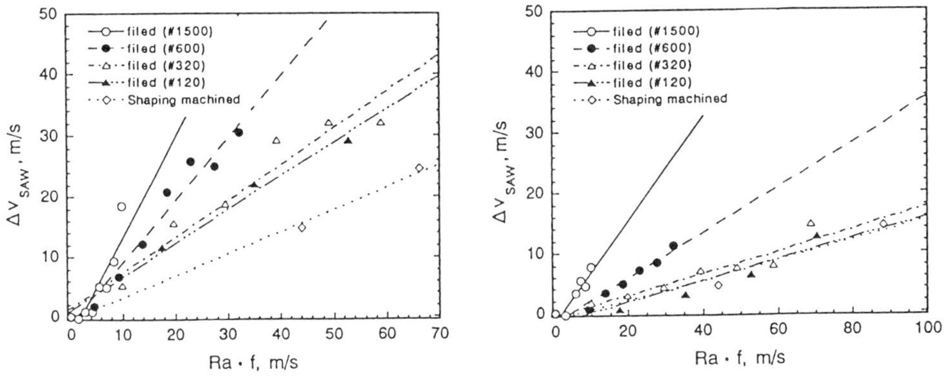

Fig. 10 Linear relationship between Δ Vsaw and Ra · f for the samples with the filed and the shaping-machined surfaces: (a) In the N-direction and (b) In the P-direction.

CONCLUSION

The acoustic spectro-microscopy is effective to make quantitative non-destructive evaluation on the surface roughness. First, both phase shift and intensity distributions in the incident angle and the frequency can be available even to make detection on the geometric change of asperity; the asperity with Ra from sub-micron to 10 μ m can be directly detected and its geometry can be distinguished. In the present method, the ultrasonic pulse works for nondestructive sensing the geometric modulation at the specified local area just as a microprobe used in STM, AFM and optical near-field microscope. Furthermore, measurement of apparent anisotropic dispersion behavior due to the surface roughness becomes a key to understand the influence of geometric modulation in the surface roughness on the effective stiffness of the surface layer including the surface roughness. In other words, since one-layer model corresponds to the measured apparent dispersion curve, elastic anisotropy can be estimated via the inverse analysis; the roughness profile can be described from this apparent elastic stiffness in anisotropy in terms of micromechanics.

As precisely discussed in Refs. (5, 13), although the residual stresses have significant influence on the surface wave velocity, the acousto-elastic theory suggested that the relative variation of the surface wave velocity should become at most 0.0288 % even when the maximum residual stress of 550 MPa is present in materials: absolute difference of Vsaw is less than or nearly equal to 1 m/s. However, in evaluation of ion-plated TiN coatings in practice, noticeable difference is often measured by ASM in the surface wave dispersion. In order to experimentally demonstrate the validity of residual stress analysis by ASM, a systematic works are still necessary where the sensitivity of Vsaw dispersions to the applied and the residual stresses. The related study will be reported in near future.

REFERENCES

1) T. Aizawa, et al.: Nondestructive evaluation of PVD/CVD TiN coated WC/Co by acoustic spectro-microscopy. Acosutics-Optics and Acoustic Microscopy. ASME AMD-140 (1992) 55-70.
2) T. Aizawa, et al.: Nondestructive quality assurance of PVD/CVD ceramic coated WC/Co ceramics. ASME MD-44 (1993) 111-132.
3) I. Ihara, T. Aizawa and J. Kihara: Determination of ultrasonic attenuation in surface layer of materials by ultrasonic reflectivity measurement. Electronics Letters. 29-22 (1993) 1999-2003.
4) T. Aizawa and J. Kihara: Nondestructive evaluation of Co-agglomeration and WC growth in WC/Co superhard alloys. Nondestructive Characterization of Materials. VI (1994) 575-583.
5) T. Aizawa, et al.: Acoustic diagnosis for nondestructive evaluation of ceramic coatings on the steel substrate. ASME. PVP-302 (1995) 257-270.
6) J.A. Ojilvy: Theory of wave scattering from random surfaces. (1991) Adam Higer, Bristol.
7) P.B. Nagy and J.H. Rose: Surface roughness and the ultrasonic detection of subsurface scatters. J. Appl. Phys. 73-2 (1993) 566-580.
8) P.B. Nagy, L. Adler and J.H. Rose: Effects of acoustic scattering at rough surfaces on the sensitivity of ultrasonic inspection. Review of Progress in Quantitative Nondestructive Evaluation. 12 (1993) 1775-1782.
9) T. Nakayama, et al.: Observation of surface roughness by using the surface waves. J. J. Appl. 42-10 (1973) 994-998.
10) M. Obata, et al.: Effect of surface roughness on quantitative measurement by ultrasonic microscopy. JNDI. 35-12 (1987) 888-893.
11) M. Hayes and R.S. Rivlin: Surface waves in deformed elastic materials. Archive of Rational Mechanics and Analysis. 8 (1961) 358-380.
12) M. Hirano, et al.: Acoustoelastic effect of Rayleigh surface wave in isotropic material. J. Appl. Mech. 48 (1981) 119-124.
13) I. Ihara: Surface wave spectroscopy by reflectivity measurement for quantitative nondestructive material evaluation. PhD Thesis of University of Tokyo (1996).
14) T. Aizawa and J. Kihara: Mechanical characterization of PVD/CVD TiN coated WC/Co by acoustic spectro microscopy. J. Faculty of Engineering, University of Tokyo. XLII (1993) 155-168.
15) R. Sinclair: Velocity dispersion of Rayleigh waves propagating along rough surfaces. J. Acoust. Soc. Am. 50-3 (1971) 841-845.

MICROSCOPIC CHARACTERIZATION OF MACHINED HIGH-PERFORMANCE CERAMICS USING ACOUSTICAL, OPTICAL AND MECHANICAL NDT TECHNIQUES

V. Scherer, S.U. Fassbender, G. Weides, and W. Arnold
Fraunhofer Institute for Nondestructive Testing (IZFP)
University, Bldg. 37, D - 66123 Saarbruecken, Germany
vscherer@izfp.fhg.de

ABSTRACT

Ground surfaces of alumina (Al_2O_3), silicon infiltrated silicon carbide (SiSiC) and pressureless sintered silicon carbide (SSiC) have been investigated nondestructively by employing Scanning Acoustic Microscopy (SAM), Atomic Force Microscopy (AFM) and Confocal Scanning Laser Microscopy (CSLM). Subsurface microcrack densities, evaluated by SAM images, were correlated with grinding parameters such as abrasive grain size and grinding speed. Furthermore, microcrack densities were found to be decreased after tribological tests. The surface topography of machined surfaces was measured on a sub-micrometer scale with the AFM. The AFM images distinguished between different phases in multiphase ceramics. CSLM images of Al_2O_3 samples provided information about the sample microstructure and material removal mechanism by grinding.

1. INTRODUCTION

High-performance ceramics meet the requirements for various applications due to their temperature and chemical resistance. However, due to their low fracture toughness (K_{IC}) brittle fracture tends to occur with high performance ceramics. When residual microscopic imperfections such as voids and microcracks are loaded, unstable crack growth occurs, leading to failure of the ceramic component. Since in most applications, the load is predominantly applied to the surfaces, surface and near-surface imperfections are more sensitive than bulk imperfections of the same size. Thus, surface integrity plays a key role for quality assessment of ceramic components. The surface integrity is determined by the microstructure as well as by the machining process which in turn can substantially change the microstructural properties in the subsurface zone. To simulate the complex relationship between surface quality and machining parameters, knowledge of the surface integrity is a crucial factor. For surface characterization at a high lateral resolution, optical microscopy and Scanning Electron Microscopy have been successfully employed. These methods allow surface imaging and evaluation of surface cracks down to the micrometer scale. They are, however, either not capable of imaging cracks on a finer scale and/or do not provide quantitative information about crack depth unless applied to cross sectional cut samples.

Both CSLM and AFM are nondestructive techniques that allow topography mapping, combined with a lateral resolution on the order of 1 μm for CSLM and 0.1 nm for AFM. Initially, due to its capability of obtaining 3D-images, particularly on transparent specimens, CSLM was predominantly used in biological applications[1]. Semiconductor metrology and film thickness measurements followed, even though CSLM has not yet drawn much attention from materials scientists and engineers. The results of the present work showed that CSLM images can provide useful information on engineering surfaces. 2D and 3D images can be obtained with a lateral resolution of about 0.5 μm and a depth sensitivity on the order of a few hundred nanometers. One of the fastest growing microscopy techniques is undoubtedly AFM[2]. It can be a contacting, intermittently contacting or a noncontacting technique in which a miniaturized profiler is used to determine the topography of various surfaces, conducting or insulating. The result is a 3D topography image providing a lateral resolution less than 0.1 μm and a vertical resolution on the order of 0.1 nm. AFM is becoming a useful tool for imaging the topography as well as for investigating physical and mechanical properties of engineering surfaces[3].

SAM operates with a focused acoustic beam scanned across the surface under inspection[4]. The reflected beam can be used to generate elasticity maps based on the distribution of the surface wave velocity or local acoustic impedance. Imaging surface and near-surface defects such as microcracks with a lateral resolution of up to a few μmicrometers can be done as well[5]. To obtain this resolution, ultrasonic frequencies on the order of 1 GHz must be used, limiting the penetration depth to approximately ten microns on engineering ceramics. From the SAM images, one can count the number of microcracks per unit area, even on surfaces that reveal relatively high topographical perturbation. A high surface roughness makes it difficult to count cracks from optical microscopy and electron microscopy images.

2. High-Performance Ceramics for tribomechanical applications

Machining of advanced materials is subjected to steady development and improvement. In the present study, engineering ceramics used for tribomechanical applications such as sealing rings and sliding bearings were investigated. High hardness, thermal conductivity and chemical resistance makes high-performance ceramics, such as Al_2O_3 and SiC, preferable for those applications. Both SSiC and SiSiC are now commonly used as materials for sealing rings and sliding bearings. SSiC is pressureless sintered from monodisperse SiC primary powder. With 2-4% boron and amorphous high-reactive carbon and organic binder added, the submicron size powder is sintered at 2100° C in an Ar atmosphere. A linear shrinkage of 20% can be observed during the sintering process. A remaining porosity is desirable to reduce friction and abrasive wear in tribological applications. Thus, the sintering process is stopped when a porosity of 2-4% is reached. The microstructure is an α-SiC with an average grain diameter of a few microns. SiSiC, on the other hand, is pressed from α-SiC primary powder with carbon and organic binder. The green bodies are Si infiltrated and sintered under pressure at a temperature of

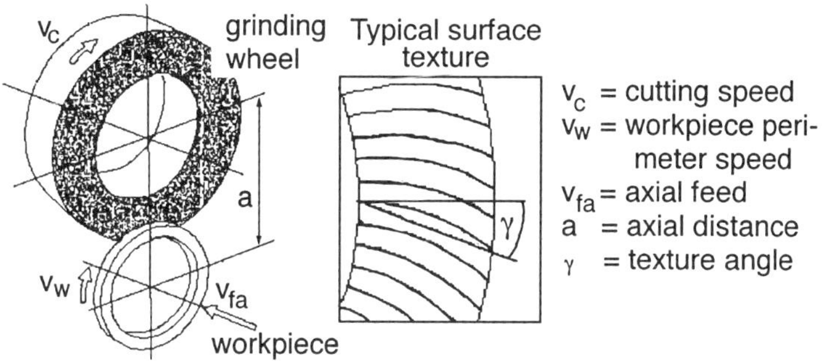

Fig. 1: Setup for longitudinal-face cross-grinding with rotating workpiece[7]

1700-1900° C. The liquid Si reacts with C to β-SiC which then segregates at the α-SiC grain boundaries. Besides α-SiC and β-SiC, the microstructure reveals remaining amorphous Si and C. Both of the additional phases influence the mechanical properties and determine the machinability and tribological properties of the machined surfaces. Finally, Al_2O_3, with or without sintering additives, with or without particle reinforcement, is a material of common use for tribomechanical applications, e.g. sealing rings, artificial joints and more. The material under inspection for this work was a pure Al_2O_3, sintered from primary powder, as well as a SiC fiber-reinforced Al_2O_3 with no further specifications.

3. Machining of High-Performance Ceramics

The mechanical and chemical resistance of high-performance ceramics makes them difficult to machine - for shaping as well as for surface finishing. Machining hard materials usually requires high forces which in turn affects the surface integrity of the machined product. Advanced grinding processes are desired that will reduce time and cost with an optimum surface quality and surface textures suited for tribomechanical components[6]. To evaluate and improve the grinding process, it is necessary to optimize process parameters. Surface characterization data can be a useful feedback parameter after machining both prior to and after tribological tests. High-performance ceramics are usually shaped in a costly fine-machining process with loose abrasive diamond particles i.e., polishing and lapping. Fine grinding with bonded abrasive grains offers ecological and economical advantages. A new method, longitudinal-face cross-grinding with rotational workpiece motion, was introduced to manufacture ceramic components with a rotational symmetry, leading to a surface texture favorable for tribological applications[7].

3.1 Sample preparation - Al_2O_3

Al_2O_3 samples were prepared by creep feed grinding and alternating grinding. In creep feed grinding the abrasion process only takes place while the grinding wheel moves in one direction with a depth feed in subsequent strokes. When alternating grinding is employed, the workpiece is subjected to abrasion while the grinding wheel moves back and forth with a given depth feed. Machining parameters and roughness values are shown in Table 1. Generally, creep-feed grinding is the more time consuming

Table I: Machining parameters and roughness of the Al_2O_3 samples for optimum surface quality

Machining process	In-feed mm	Feed rate mm/s	Cutting speed m/s	Material removal rate $mm^3/mm \cdot s$	Roughness R_a µm
Alternating grinding	0.03	167	25	5	0.7
Creep-feed grinding	0.5	10	25	5	0.5

and costly grinding technology, but it tends to be more mild than alternating grinding at the same material removal rate. X-ray measurements reveal higher subsurface residual compression stress for alternating grinding as compared to creep feed grinding[8].

3.2 Sample Preparation - SiSiC and SSiC

SiSiC and SSiC were finished by longitudinal-face cross-grinding with rotational workpiece motion as indicated in Fig. 1. The grinding wheel, with diamond grains as abrasive elements, is rotated against the rotating workpiece with an axial distance a. The machining parameters for the surfaces generated for this study are given in Table 2. As the workpiece is fed towards the grinding wheel, grinding marks are produced as drawn in Fig. 1. Their texture can be determined by the process parameters. This surface pattern is desirable for tribological applications, because transport of lubrication medium within the interface of sealing ring pairs is facilitated. All machining was performed using a mineral oil as a coolant. The ground SiSiC and SSiC surfaces both revealed a typical surface roughness of $R_a = 0.2$ µm. Another set of SiSiC samples was prepared by alternating and creep feed grinding. The machining data are given in Table 3. Roughness data were not taken for this set, but are comparable to those given above.

Table II: Machining parameters for longitudinal face cross-grinding of SiSiC and SSiC samples

Sample	Tool	material removal rate per unit time Q'_W $mm^3/mm \cdot s$	material removal rate V'_W mm^3/mm	Cutting speed v_C m/s	workpiece perimeter speed V_W m/s	axial depth feed v_{fa} mm/min.	axial distance mm
S1	K+4821 C100	1.8	220	40	5.8	0.5	177.5
S2	K+8821 C100	1.8	220	30	5.8	0.5	197.5

Table III: Machining parameters for alternating and creep-feed grinding of SiSiC samples

Sample	Machining process	In-feed mm	Feed rate mm/s	Cutting speed m/s	Material removal rate $mm^3/mm \cdot s$
A1	Alternating grinding	0.03	167	25	5
A2	Alternating grinding	0.03	84	25	2.5
C	Creep-feed grinding	0.5	10	25	5

3.3 Tribological Tests

In order to characterize the effect of tribomechanical load on the surface quality of the SiSiC and SSiC samples, sealing rings were tested with a pin-on-disk tester. The rings were loaded with a polished SiSiC pin at a normal load of 10 N. Sliding velocity and sliding distance were 1.83 m/s and 1.65 km. After the test, samples were again inspected with nondestructive techniques.

4. NDT Techniques

4.1 Confocal Scanning Laser Microscope

Unlike conventional optical microscopy, the CSLM only uses the part of light that is reflected from a selected focal plane of the object, see Fig. 2. Light beams passing the confocal pinholes i.e., originating from the focal plane only, can reach the photodetector and contribute to the image formation. As the object is scanned underneath the incident light beam, the detector records the intensity of light reflection originating from the focal plane across the scanned area. The result is an area scan with the reflected light intensity displayed as a greyscale. By moving the object towards the lens, the focal plane is chosen to be in a certain depth of the sample. Commercial instruments allow recording a set of images at different focal planes along the vertical axis, called an "extended-focus image". From this set of data, appropriate software can calculate the surface relief and roughness data. Additional information can be obtained by scanning the object in the x-z plane. The x-z image reveals information about the reflectivity in the subsurface, e.g. from imperfections or second phases. In ultrasonic NDT, this kind of image is commonly called B-scan whereas the equivalent for extended-focus images is a C-Scan.

We used a Leica TCS NT (Leica, Bensheim, Germany), operating with an Ar laser light source. The lateral resolution of the CSLM is given by the wavelength (λ_{Ar}: 488 and 514 nm). The depth of focus i.e., vertical resolution, is limited by the numerical aperture of the setup. In our case it is limited to 200 to 400 nm. The maximum size of x-y scans depends on the instrument type, in our case $1 \times 1\ \mu m^2$.

Confocal microscopy is only capable of imaging materials with a low reflectivity, metals cannot be examined. Ceramic surfaces allow the light to penetrate the surface. The coefficient of reflection can even be diminished by using immersion liquid between lens and object. By using immersion oil, the maximum depth of reflections to be observed in an Al_2O_3 can be increased to the order of 1 μm.

4.2 Atomic Force Microscopy

The AFM works similar to a record player, but with other geometrical dimensions: A microfabricated tip of 4 μm length with a tip radius of less than 50 nm, is attached to a 1 μm thick cantilever. As it is scanned across the sample, the average distance between tip and sample is maintained approximately constant by attractive and repulsive interatomic forces - we say the tip is in contact with the surface. Thus, the cantilever is deflected according to the surface topography of the sample. The beam deflections are detected by an optical lever method where a laser beam is focused on the back side of the tilted cantilever and reflected onto a position sensitive photodiode, see Fig. 3. The detection scheme has a vertical resolution of 0.1 nm. The vertical signal (VS) on the

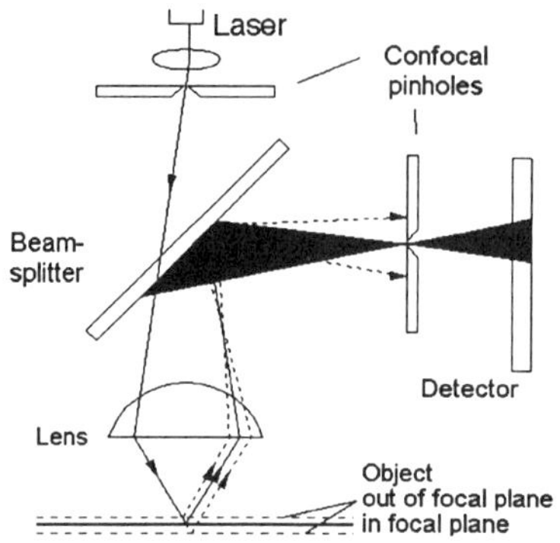

Fig. 2: Principle of Confocal Scanning Laser Microscope

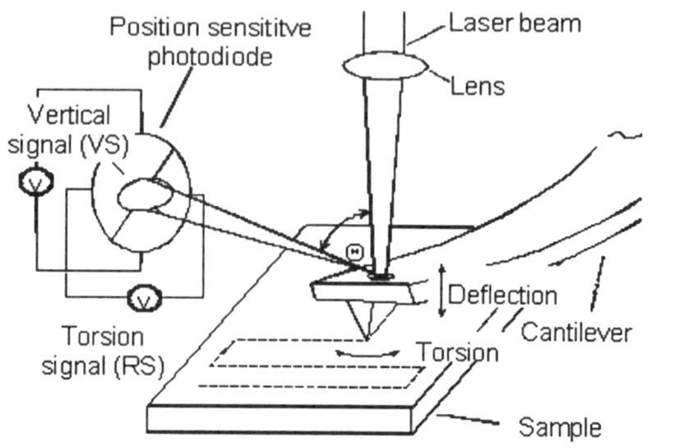

Fig. 3: Schematics of an Atomic Force Microscopy in contact mode

photodiode represents the cantilever deflection and is read out while the sample is scanned underneath the probe tip. The result of an AFM scan is a topography map of the scanned surface with the height data displayed as a color value. We used a commercial available instrument, a NanoScope III AFM/LFM of Digital Instr., Santa Barbara, CA, with a maximum scan size of $100 \times 100\ \mu m^2$ and a maximum vertical stroke of $4\ \mu m$. The AFM software displays images, cross-sections and processes roughness data.

4.3 Scanning Acoustic Microscopy

An ELSAM, Fa. Leica, Bensheim, Germany, was used to characterize the machined ceramics. An ultrasonic beam, focused by an acoustic sapphire lens, and coupled by water onto the sample, is scanned across the surface. The aperture angle of the lens is large enough to allow the excitation of Surface Acoustic Waves (SAW) in the sample as indicated in Fig. 4. The SAW travels along the sample surface but part of leaks back through the water back into the lens and finally to the transmitter/receiver transducer. The amplitude of the reflected beam is recorded and displayed as a greyvalue in the acoustic image. Any change in local elasticity at the sample surface can be observed at the receiver, thus contributing to the contrast formation of acoustic images. By display-

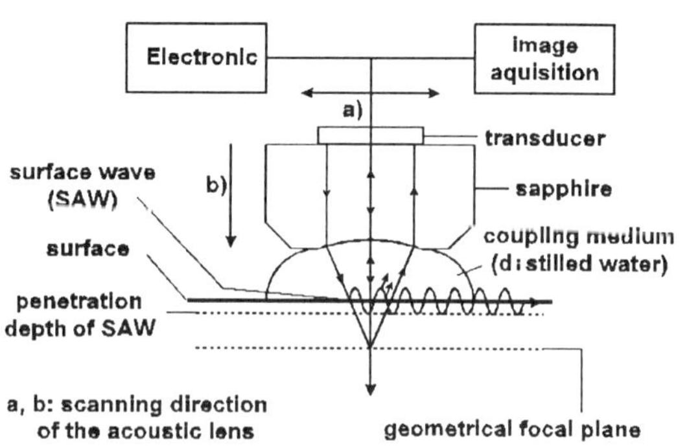

Fig. 4: Principle of SAM

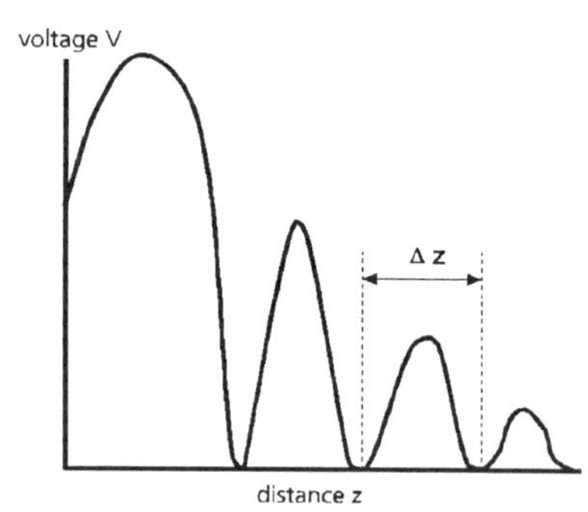

Fig. 5: V(z)-curve on a ceramic surface

ing the contrast as datapoints of a 2D scanned surface, we can image objects with a lateral resolution corresponding to the wavelength in use. For an ultrasonic frequency of 1 GHz, the SAW wavelength λ is roughly 1-5 μm on ceramics. However, cracks whose crack opening displacement is narrower than the wavelength can still be detected as there is no traction between the crack walls; the surface waves perform both longitudinal and shear elastic deformation. For surface breaking cracks perpendicular to the surface, shear deformations are hardly transmitted between the crack flanks[4]. Thus, even on cracks as narrow as a few tens of nanometers, SAW are perturbed and scattered, resulting in a formation of contrast in the scanned image. On engineering surfaces, however, this contrast may be weak; due to the surface roughness, acoustic images tend to be noisy.

Whereas SAM images allow counting of defects that can be imaged directly, so-called V(z) curves give qualitative information about the defect density, including cracks narrower than the spatial resolution of the technique. With the V(z) technique, the acoustic lens is moved towards the specimen (z-axis) at a single location on the sample, the rasterscan is disabled. A reflected signal amplitude-vs.-defocus curve can be observed as shown in Fig. 5. The theory of the oscillatory behavior observed is complex but, nevertheless, known and described[4]. Employing a complex beam theory that accounts for all participating rays, the SAW velocity can be derived from the distances Δz of the minima in the V(z)-curve[9]. Since the surface waves experience dispersion due to scattering at the narrow defects, the SAW velocity is a measure of the defect density on a given material[10].

5. Experimental Results

5.1 CSLM Measurements on Al_2O_3

To illustrate the performance of the CSLM on machined surfaces, a creep-feed ground surface was examined. Fig. 6 shows an extended focus image of 50×30 μm^2 scan size. The surface texture from grinding can be seen as diagonal lines. In addition, we observe a granular texture, interrupted by relatively smooth areas. Presumably, brittle material removal has taken place leaving a fracture-like surface. On the smoother areas, material removal was accompanied by higher compressive stress, indicating that the workpiece material was deformed plastically.

Fig. 7 shows an x-z image on a similar surface. The ground surface, however, was postpolished to avoid light scatter from surface roughness, thus increasing the pene-

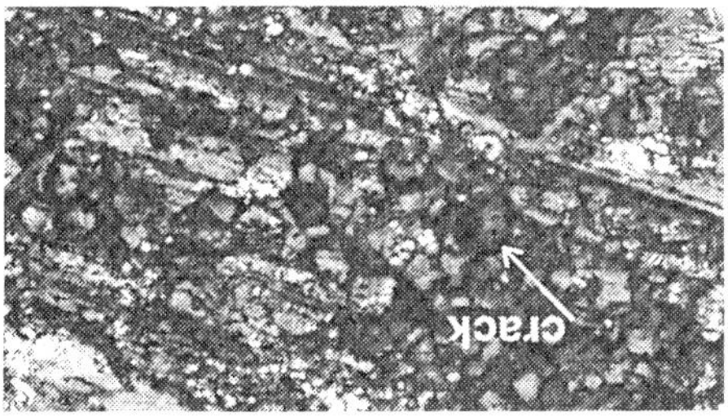

Fig. 6: Extended Focus Image of a ground Al_2O_3 surface, 50×30 μm^2 area, crack (arrow)

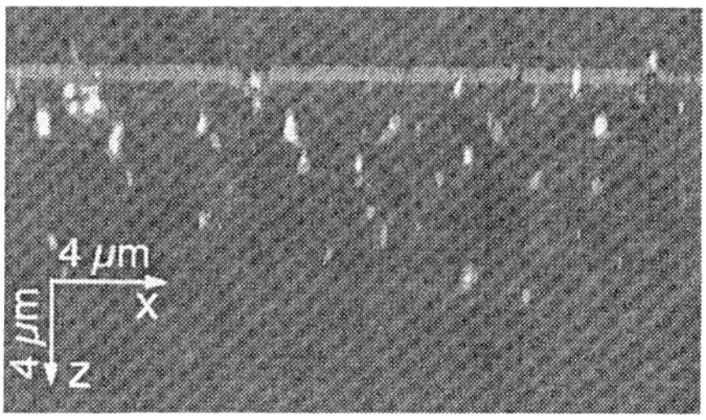

Fig. 7: x-z image on ground Al_2O_3, bright spots indicate second phases

tration depth of the light. The bright line is the surface reflection, which is unavoidable despite using immersion oil as a coupling medium between lens and object. The part of the light that is not reflected passes the surface and is either absorbed from the material or reflected from interfaces within the material. The interfaces reflect part of the light back to the surface where again a part is reflected and the other part transmitted into the immersion oil following the optical path to the detector. Thus, interfacial reflections can be seen as bright spots within a dark alumina matrix. Interfaces can result from the presence of either voids or second phases. In this sample, metallography and electron microanalysis revealed a second phase material. The grain diameter and distribution of the second phases correspond reasonably well with the data obtained from CSLM images.

5.2 AFM Measurements on SiSiC

Ground surfaces of SiSiC were imaged with the AFM, an example is shown in Fig. 8. The $100 \times 100\ \mu m^2$ topography map on the left shows the height distribution on a vertical scale of roughly 1 μm from dark to bright. It clearly reveals the grinding marks. The surface was partially worn in a brittle mode whereas other areas were deformed in a ductile mode providing a smoother surface. From the distribution of brittle/ductile deformed areas, we determine the SiC-phase as the ductile and the amorphous Si matrix as the brittle deformed areas. Due to the difference in fracture toughness K_{IC} (1 MPa·√m for bulk Si and 5 MPa·√m for sintered SiC[11]), brittle deformation is more likely to occur with Si than with SiC. The profiles in Fig. 8 are cross sectional profiles along an extrusion and along a SiC phase. It can be seen that the surface profile on the plastically deformed SiC phase (upper graph) is smoother (Peak-to-Valley roughly 0.1 μm) than the

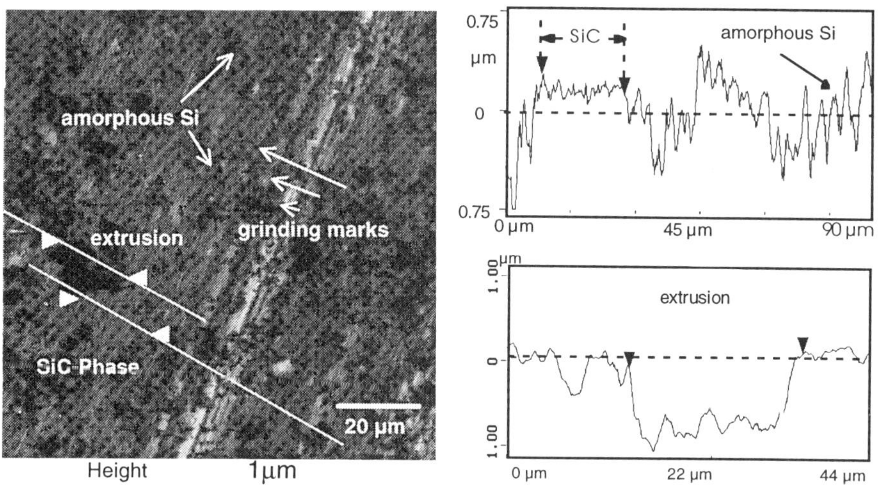

Fig. 8: Topography map (left) of a SiSiC, ground surface. Surface profiles along the white lines are shown on the right. The SiC phase is smoother than the amorphous Si (upper right). Depth of extrusion (lower right) is roughly 1 μm.

surrounding Si matrix (P-V ≈ 0.5 µm). Furthermore, the surface reveals extrusions (lower profile) of about 10 µm diameter and 1 µm depth. Presumably, these large holes were already introduced by the ceramics processing, i.e. porosity, rather than by grinding the surface. Extrusions originating from brittle material removal by machining cannot be distinguished from porosity of the bulk ceramics.

5.3 Microcrack Imaging on SSiC and SiSiC with the SAM

The ceramics were machined by creep feed grinding and alternating grinding as described in section 2. Typical SAM images taken on the SiSiC and SSiC samples are shown in Fig. 9. The α-SiC phase can be recognized by a difference in contrast on the SiSiC material (left), resulting from both a smoother surface and a change in Young's Modulus. Some crack like indications are indicated by arrows. Dark spots on the SSiC surface (right) result from a surface relief caused by the porosity of the material. Crack-like indications can be seen here, too. The number and size of crack-like indications were determined from the SAM images and the density of those indications correlated with machining parameters. The results are given in the sections below.

5.3.1 Abrasive Grain Size: The influence on the size of the diamond grains used for longitudinal-face cross-grinding on the surface quality is shown in Fig. 10. The specimens were machined by longitudinal cross-feed grinding as given in Table 2, S1. The average number of crack-like indications, hereafter referred as to crack density, is shown for abrasive grain sizes of 46, 76, 91 and 126 µm. SSiC samples machined with 91 and 126 µm grain size were unavailable for measurements. There is no significant difference in crack density between the two materials. A tendency to lower crack density with larger grain size is observed, however any clear dependance of crack density on grain size is obscurred by errors of counting cracks.

5.3.2. Cutting Speed: Fig. 11 and 12 shows the data for crack density and crack length for different cutting speeds (light columns) on SiSiC, samples S1, machined according Tab. 2. A polished sample was measured as a reference. We observe a low

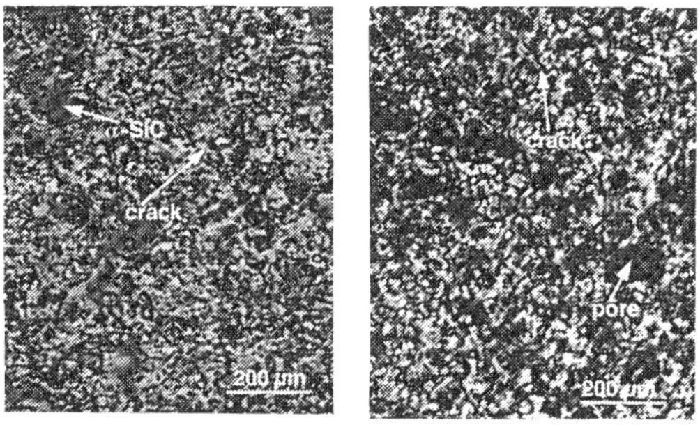

Fig. 9: SAM image of SiSiC (left) and SSiC (right). Crack-like indications, α-SiC phases and pores are indicated by arrows

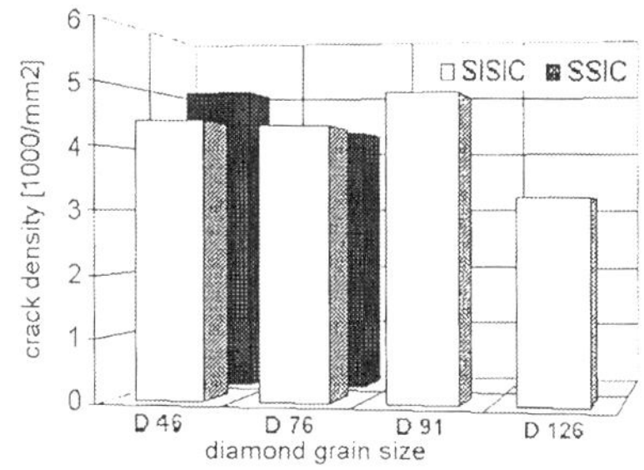

Fig. 10: Density of crack-like indications on surfaces ground with different diamond grain size, determined from SAM images

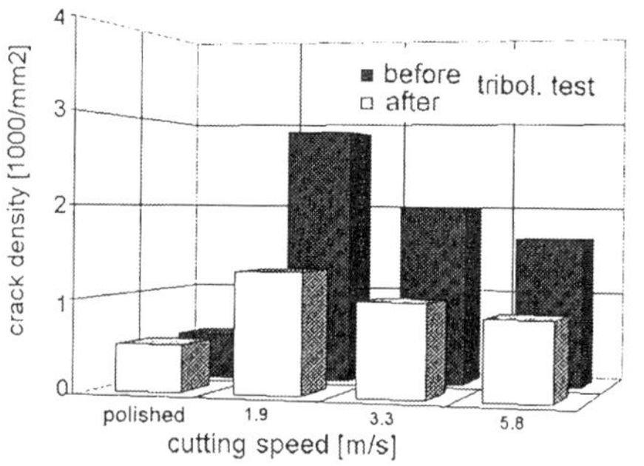

Fig. 11: Crack density obtained from SAM images as a function of cutting speed before (dark) and after (light) tribological tests.

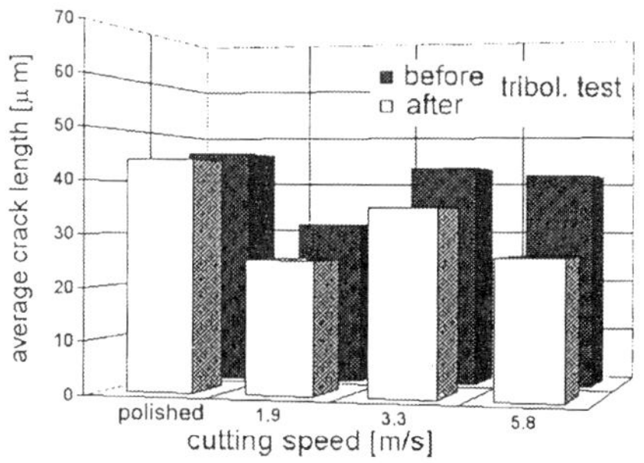

Fig. 12: Average crack length obtained from SAM images as a function of cutting speed before (dark) and after (light) tribological tests

number of relatively large cracks on the polished sample as compared to the ground surfaces. The cutting speed has a strong influence on the crack density, but less influence on the crack length.

5.3.3 Tribological Tests: The white columns in Fig. 11 and 12 represent crack density and length for the same surfaces after the tribological tests. It can be noticed that on the polished reference surfaces neither length nor density of the cracks are altered by the tests. On the ground surfaces, however, number and length of the cracks is decreased by up to a factor of 2 after tribological tests. This observation can be explained by the fact that plastic deformation occurs in the subsurface zone during tribological load and cracks are filled and closed by deformed material.

5.3.4 Alternating and Creep Feed Grinding of SiSiC: SiSiC samples were prepared by alternating and creep feed grinding with the parameters according to section 3.2 such that three grooves were ground in the same sample, see Fig. 13. SAM images were taken both in the groove center and near the edges. The crack density (squares) for the measured areas is shown with the values for the fracture toughness (circles) for creep feed and alternating ground samples, taken from the 3-point bending test.

Alternating grinding with a high material removal rate (A1) causes the highest crack density, a decrease of depth-feed resulting in a lower material removal rate (A2) lowers the subsurface damage substantially. Creep-feed grinding (C), even with a high material removal rate, shows the least crack density; it is the milder of both grinding technologies. From the data we conclude that there is more damage near the groove edges, where the number of cracks can exceed twice that of the groove center. Fracture toughness data obtained from bending tests correlate reasonably well with the subsurface damage measured with the acoustic microscope.

5.4 Depth of Subsurface Damage Measured by the V(z)-Technique

So far, acoustic microscopy has been employed to image surface breaking cracks with dimensions larger or comparable to the resolution of the microscope of about 1μm. As stated in section 4.3, the V(z) technique can measure the surface damage including

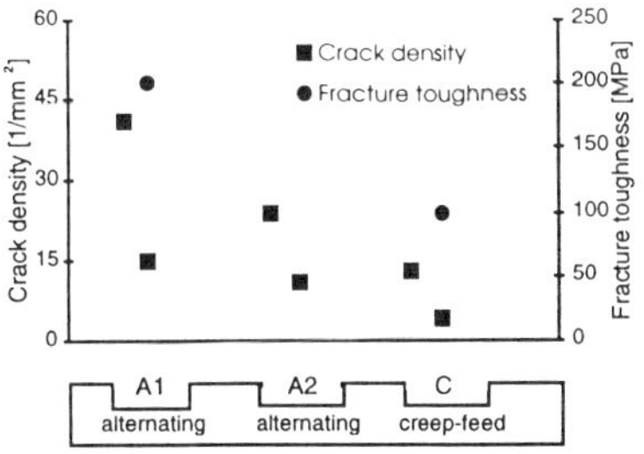

Fig. 13: Crack density (squares) for both groove center and edge on SiSiC. Fracture toughness (circles) for creep- feed and alternating ground samples obtained from 3-point bending tests

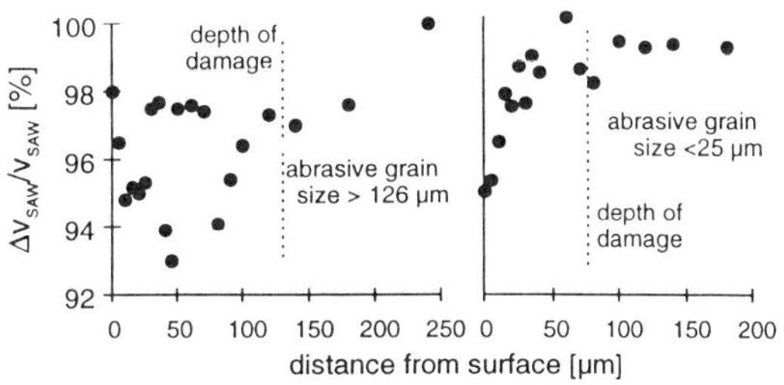

Fig. 14: Relative decrease of SAW velocity on Al_2O_3 (normalized on a damage free surface) , obtained from V(z) data indicates damage. Depth of damage is larger for high abrasive grain sizes, ca. 140 µm, as compared to ca. 80 µm for smaller abrasive grains

imperfections much smaller than the spatial resolution. To obtain such information as a function of the depth into the sample, cross sections of ground SiC fiber-reinforced Al_2O_3 specimens were cut and polished. The SAW velocities were taken with the V(z) technique on the polished cross section on different spots with a given distance from the surface. Two surfaces, machined by alternating grinding with two different abrasive grain sizes, >126 µm (left) and <25 µm (right), were prepared and measured. The relative SAW velocity, normalized on a damage-free polished reference surface is shown as a function of the distance from the sample surface in Fig. 14. It can be noticed that the SAW velocity is reduced to 95% at a distance of roughly 10 µm from the surface. For both samples, the velocity recovers to almost 100% with distance from the surface. The recovery occurs at lower depths, roughly 80 µm, on the fine grain ground surface, as compared to about 140 µm on the surface ground with courser grain size. The results proove that by using a finer grain size a lower depth of damage can be achieved.

7. CONCLUSIONS

In this study, three different nondestructive techniques were employed to investigate the surface integrity of machined high-performance ceramics. Ground surfaces of Al_2O_3, SiSiC and SSiC were characterized with Confocal Laser Scanning Microscopy (CSLM), Atomic Force Microscopy (AFM), and Scanning Acoustic Microscopy (SAM).

CSLM topography maps revealed the fracture surface of Al_2O_3 ceramics. Brittle fracture occurs during grinding along the secondary alumina grains. CSLM x-z images showed the presence of a second phase particles in the alumina matrix resulting from the ceramics processing.

AFM topography maps of ground SiSiC ceramics resolved both amorphous Si and SiC-phase. It has been shown that the microroughness is smoother on the SiC-phase than on the amorphous Si. It can be concluded, that brittle fracture is observed on amorphous Si and plastic deformation is the material removal process on SiC.

SAM was employed extensively to investigate the influence of grinding parameters and techniques, as well as tribological load, on the surface integrity. Number and size of crack-like indications were determined from SAM iamges on longitudinal-face cross-ground SSiC and SiSiC. The crack density tends to decrease with the increasing size of abrasive particles. The crack density decreases with increasing cutting speed, whereas the cutting speed has little influence on the crack length. A reduction of crack density was observed after tribomechanical load. Creep-feed grinding of SiSiC ceramics leads to less surface damage than alternating grinding when the machining parameters are set for equal material removal rate in both the techniques.

With the SAM, V(z) curves were taken on cross sections of Al_2O_3 samples ground with two different abrasive grain sizes. The depth of damage is about 140 μm with >126 μm grain size. Using a finer grain size of <25 μm reduces the depth of damage to about 80 μm.

ACKNOWLEDGMENT

This work was supported by grants from Deutsche Forschungsgemeinschaft, joint research plan "Fine Grinding of Ceramics for Tribological Applications". Machining tests were performed at the Institute for Production Engineering and Machine Tools (IFW), University of Hannover. We tank Ms. D. Schroth of Fa. Leica, Bensheim, Germany, for the CSLM measurements.

REFERENCES

[1] T.R. Corle, G.S. Kino (eds.): *"Confocal scanning optical microscopy and related imaging systems"* Academic Press, San Diego, CA (1996)

[2] R. Wiesendanger, R. (ed.): *"Scanning probe microscopy and spectroscopy : methods and applications"* Cambridge University Press, Cambridge, UK (1994)

[3] B. Bhushan (ed.): *"Handbook of Micro/Nanotribology"*, CRC Press, Boca Raton, FL (1996)

[4] G.A.D. Briggs (ed.): *"Acoustic Microscopy"*, Clarendon Press, Oxford, UK (1992)

[5] S. Hirsekorn, S. Pangraz, G. Weides and W. Arnold: "Measurement of Elastic Impedance with High Spatial Resolution Using Acoustic Microscopy", *Appl. Phys. Lett.* **67**, (1995), pp. 745-747 and *Appl. Phys. Lett* **69**, (1996) (Errata), p. 2138

[6] S. Johanmir (ed.): *"Machining of advanced materials : Proceedings of the International Conference on Machining of Advanced Materials"*, July 20-22, 1993, Gaithersburg, MD, U.S. Dept. of Commerce, National Institute of Standards and Technology ; Washington, D.C., (1993)

[7] B. Karpuschewski, Th. Friemuth: "Feinschleifen tribologisch beanspruchter Oberflächen" in A. de Paoli (ed.): *Werkstoffwoche 96* , DGM Informationsgesellschaft (Publisher), Frankfurt a.M., in german (1996), pp. 249-254

[8] T. Lierse, B. Karpuschewski: "Thermomechanik des Schleifens keramischer Werkstoffe und röntgenographische sowie tribologische Pruefung der erzeugten Funktionsflächen" in A. de Paoli (ed.): *Werkstoffwoche 96* , DGM (Publisher), Frankfurt a.M., in german (1996), pp. 261-266

[9] S. Hirsekorn, S. Pangraz: "Materials Characterization with the Acoustic Microscope", *Appl. Phys. Lett.* **64**, (1994), pp. 1632-1634

[10] W. Arnold, U. Netzelmann, S. Pangraz: "Characterization of Surfaces by Acoustic Imaging Techniques" in: *Metals Handbook 10th ed., Vol. 18: Friction, Lubrication and Wear Technology*, ASM International, Materials Park, OH, (1993), pp. 406-413

[11] T.J. Reinhart: *"Engineered Materials Handbook, Vol. 4: Ceramics and Glasses"*, ASM International, Materials Park, OH, (1987)

ACOUSTIC SPECTRO-MICROSCOPY FOR QUANTITATIVE NONDESTRUCTIVE EVALUATION ON CERAMIC COATED MATERIALS

Tatsuhiko AIZAWA
Department of Metallurgy, University of Tokyo
7-3-1 Hongo, Bunkyo-ku, Tokyo 113, Japan

ABSTRACT

Acoustic spectro-microscopy (ASM) for coated materials is theoretically based on the multi-layered elastic medium model, so that the measured acoustic structure can be understood from the theoretical point of view. Interfacial defects and inhomogeneities can be detected non-destructively by using ASM; since the difference in the interfacial bonding condition reflects on the power intensity distribution in the acoustic structure, the adequate scanning conditions of incident angle and frequency can be rationally selected for linear scanning with successful nondestructive inspection of artificial defects. Due to good correlation between the measured and the theoretically estimated acoustic structure for multi-layered coatings, elastic characterization can be made to estimate the elastic properties via the inverse analysis where the unknown mechanical properties are parametrically varied to minimize the absolute difference between the measured and the theoretically calculated surface wave dispersion curves. Some model coated materials are employed here to experimentally demonstrate the validity and effectiveness of our developing acoustic spectro microscopy as a new methodology for quantitative nondestructive evaluation on the coated materials.

INTRODUCTION

Quantitative nondestructive evaluation on the multi-layered coating materials requires for the diagnosis of fabricated multi-layered coating system with rigid interlayer bonding, the inspection of interfacial defects, and the elastic characterization of each coating material (1). Nano-indentation method, atom force microscopy (AFM) or other surface contact methods (2) can afford to measure the pointwise displacement-load relation and to detect the surface geometry; however, they are insensitive to interfacial defects and irregular interlayer bonding conditions, and they have no means to estimate the elastic properties for each coating material in the whole multi-layered material system. Nondestructive inspection of interlayer defects and inhomogeneities and nondestructive elastic characterization of coatings are indispensable to make integrity assessment of thin and thick film coatings in practice. Use of X-ray and laser optical microscope enables us to make residual stress analysis and to detect various defects in the inside and on the surface of coatings; however, they have no ways to estimate elastic properties of each coating layer and to trace the geometric boundary of an interlayer defect. The acoustic spectro-microscopy (ASM) (3,4,5,6,7) has been developed as one of the most promising methodologies to provide the reliable answers to the above demands from industries.

In the present approach, the acoustic structure obtained by the reflectivity measurement via ASM is uniquely determined to an objective multi-layered coating system; for the prescribed injection of ultrasonic pulse into a material, the measured acoustic structure is one-to-one corresponding to kinetic energy transfer in reflection and transmission into the objective material system including a coupling matter, the multi-layered coatings and the substrate. Hence, the multi-layered elasto-dynamic model must be a strong theoretical base to calculate the above kinetic energy transfer and to describe the measured acoustic structure by ASM. This co-operation between theory and experiment (8,9) helps us to make nondestructive diagnosis of the multi-coated devices and parts by using the acoustic structure intrinsic to each system, to make nondestructive inspection of interfacial defects and irregularities, and to make elastic characterization of each layer material. In this study, two different classes of material system are employed to demonstrate the accurate sensitivity of the measured acoustic structure to the selected coating system and to explain good correlation between the experimentally measured and the theoretically predicted acoustic structures. As a model coating system, a PET-film coated glass substrate with an artificial interlayer defect is used to show that accurate geometry of this artificial interlayer defect can be detected and visualized via two dimensional scanning. Finally, Au-coated glass system is used for elastic characterization by the acoustic spectro-microscopy.

ACOUSTIC SPECTRO-MICROSCOPY

In AFM (Atomic Force Microscopy), very small contact force in the order of μ N is detected as the response of nano-indentation to describe the local geometry and deforming shapes of materials. Without additional knowledge or theoretical model, the measured force-displacement relation cannot be interpreted and transformed into mechanical properties. Furthermore, since every mechanical parameter and property is defined in elasto-statics, their spatial distribution and disturbance can never be described by those approaches. Our developing acoustic spectro-microscopy (ASM) is theoretically based on the elasto-dynamics, so that the acoustic structure can be one-to-one corresponding to the multi-layered elastic system. Hence, noticeable difference between the measured acoustic structure by ASM and the theoretically predicted acoustic response for multi-layered elastic system should be a proof of irregularities, inhomogeneities and defects located on the surface or in the interface between two layers. When these defects are absent in materials, the elastic properties and the geometric dimension of coatings can be estimated from direct comparison of experimental and theoretical data.

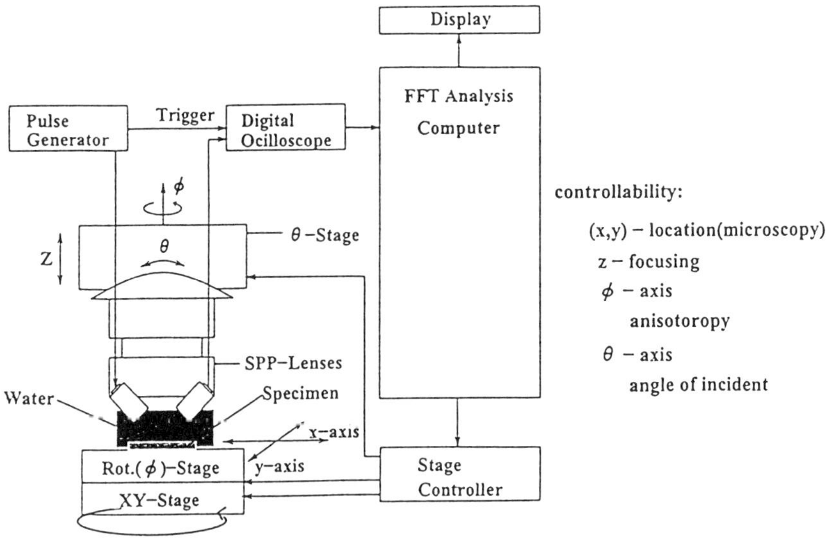

Fig. 1 Our developed acoustic spectro-microscopy working in the standard mode.

Figure 1 shows the standard type of our developing acoustic spectro-microscopy. Different from the conventional ultrasonic microscopy, the spherical-planar-pair lens (SPP lens) is used and controlled to move in five directions (in the parallel plane (x-y plane) to the surface of a sample, in the perpendicular direction (z-axis) to the surface, in the circumferential direction (θ - axis) and in the rotational direction (ϕ -axis)). The focus of this lens system is fixed at the specified point on the surface of a sample within the spatial mismatch less than 100 nm by using x-, y-, and z-axis driving units. The incident angle can be directly controlled by using the θ - axis driving unit; three different lens systems enable us to make measurement for $10° < \theta < 50°$. ϕ -axis control is necessary to make evaluation on the anisotropic properties of materials.

To be noted, since a ultrasonic pulse is injected into a sample through a coupler material (usually, water), data acquisition tools including the Fast-Fourier-Transform are available to make spectroscopic analysis of the reflected signals. From directly received signal, both the phase shift and the power amplitude distributions are obtained at every point of measurement in the function of the incident angle and the frequency. When the incident angle is fixed to be constant during measurement, the linear scanning of SPP lens system provides us the variation of these phase shift and intensity distributions along the specified direction. Any three axes among five can be selected and used simultaneously for reflectivity measurement; e.g. if x, y and θ axes are selected, the elastic characterization of layered materials can be made for a local spot specified in the x-y plane. In addition, two dimensional mapping of the measured intensity onto the x-y plane enables us to make visualization for computer tomography.

Through the data acquisition by varying the incident angle, the critical angle, where the surface wave is activated, can be detected as a significant amount of phase shift and a noticeable intensity reduction. From this measured critical angle (θ_c), the surface velocity (Vsaw) can be determined by

$$Vsaw = Vw / \sin (\theta_c) \tag{1}$$

where Vw is the acoustic velocity of a coupling material (usually, water). Since Vw is functional in temperature, the difference in temperature of the coupler is controlled to be less than 0.1 K during measurement. This Vsaw is a constant value indifferent to the ultrasonic frequency for the homogeneous material with the completely smooth surface. As have been reported in Refs. (8, 9), the multi-layered material system has a surface wave velocity dispersion, which is intrinsic to each system. The measured dispersion curve often becomes a basic information for elastic characterization and defect detection in the quantitative nondestructive evaluation of the structured materials. Since each layered coating system has its unique surface wave velocity dispersion, the measured Vsaw can be used to determine the elastic properties, the mass density and the coating thickness by the inverse analysis. First of all, various coating systems are employed to see that each system should have its unique acoustic structure and these acoustic structures can be successfully predicted by the multi-layered model (10).

ACOUSTIC STRUCTURE OF MULTI-LAYERED MATERIALS

In the ceramic coating, multi-layered system is often formed on the substrate or the matrix materials. Each layer might have its intrinsic function in the materials design; without any information of mechanical properties, it is difficult or nearly impossible to evaluate the validity and usefulness of these multi-layered coatings. As stated before, the elasto-dynamics provides us the multi-layered elastic model which is just corresponding to the measured acoustic structure by ASM. One-layer coating system is employed here to experimentally demonstrate the sensitivity of the measured acoustic structure to the layered elastic system. First, the sensitivity analysis was made by the elasto-dynamic model for 1] Au-coated glass substrate and 2] TiN-coated steel substrate. Table 1 lists the computational conditions for these two cases.

Table 1 Computational conditions to be used for theoretical model calculations.

	Young's Modulus (MPa)	Poisson's Ratio	Mass Density (kg/m3)	Longitudinal Velocity (m/s)	Shear Velocity (m/s)
Au	78. 5	0. 42	19300	3222	1196. 7
Glass	67. 5	0. 28	2760	5591. 5	3050. 8
TiN	411	0. 26	5300	9740. 8	5547. 8
Steel	210	0. 3	7800	6020. 2	3217. 9

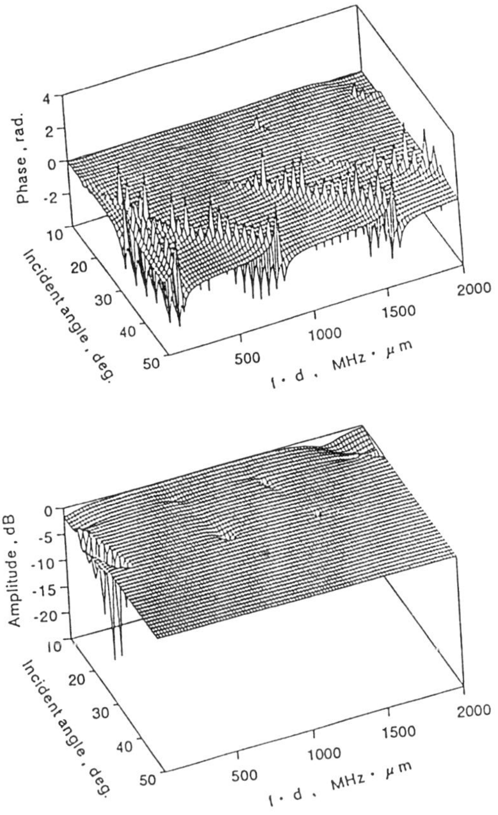

Fig. 2 Dependence of the reflectivity on the frequency for Au-coated glass substrate.

Figure 2 depicts the calculated phase shift and amplitude distributions in the function of the incident angle and f • d for Au-coated glass system. Here, f denotes for the frequency and d the thickness of coating. Since several regions have a large phase shift nearly from - π to π in Fig. 2, various surface waves are activated in this measurement. The acoustic structure analysis based on the elasto-dynamics told us that those activated surface waves can be identified in order from the lowest-frequency to the higher frequency as the Rayleigh wave, Sezawa waves and other higher surface waves. As explained before, these surface wave velocities can be determined from the measured dispersion curves in Fig. 2 in terms of Eq. (1). Here, the critical angle (θ_c) is functional in the frequency (f). Figure 3 shows the obtained surface wave velocity dispersion for Au-coated glass substrate. The pseudo Sezawa wave, whose kinetic energy is easy to leak into the substrate, was also depicted in this figure; in fact, abrupt reduction of power intensity in Fig. 2 corresponds to activation of this pseudo Sezawa wave.

This acoustic structure is qualitatively characterized by the following three features: 1] The Rayleigh wave velocity decreases monotonically with frequency down to the sound velocity of water, and 2] The first-order wave (Sezawa wave) velocity, also decreasing monotonically with f, has a cut-off point of f • d, in the lower range of which Sezawa wave turns to be a pseudo wave, and 3] Monotonically decreasing Vsaw dispersion curves are seen for the higher-order waves. These features are intrinsic to the layered system where the acoustic velocity of coating material is lower than that of substrate. In the following, PZT sol-gel coated MgO substrate system is employed to demonstrate that the above common features can be also measured by ASM.

Figure 4 depicts the cross-sectional view of sol-gel PZT coated MgO with the thickness of about 7 μ m. Since this coating is very dense, the relative density is assumed to be 100 % in the following calculation. Comparison between theoretically predicted and the experimentally measured power amplitude distributions is shown in Fig. 5. This good correlation must be a base to make integrity proof of this-type coatings and to perform further quantitative nondestructive evaluation on the coated system. Figure 6 depicts the comparison of both the calculated and measured surface wave velocity dispersion curves. As discussed later, optimization of elastic properties via the inverse analysis to minimize the absolute difference between two dispersion curves enables us to have accurate estimates of the elastic stiffness of a coating material.

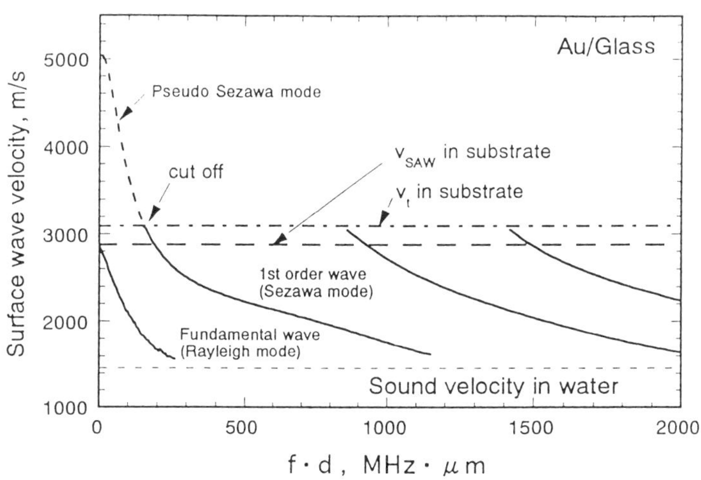

Fig. 3 Calculated surface wave velocity dispersion for Au-coated glass substrate.

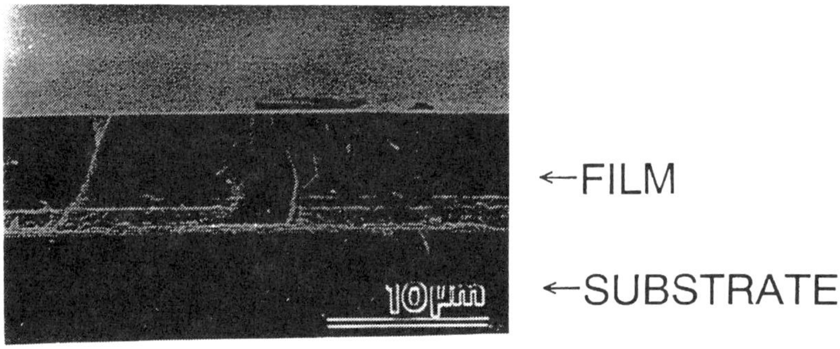

Fig. 4 Cross-sectional view of sol-gel PZT coated MgO.

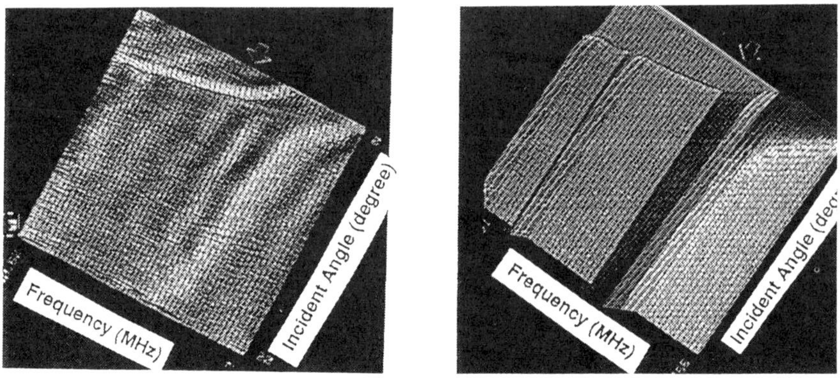

Fig. 5 Comparison between the theoretically predicted and the experimentally measured reflectivity distributions in the function of the incident angle and the frequency: (a) Theoretically predicted reflectivity, and (b) Measured reflectivity by ASM for Au-coated glass substrate.

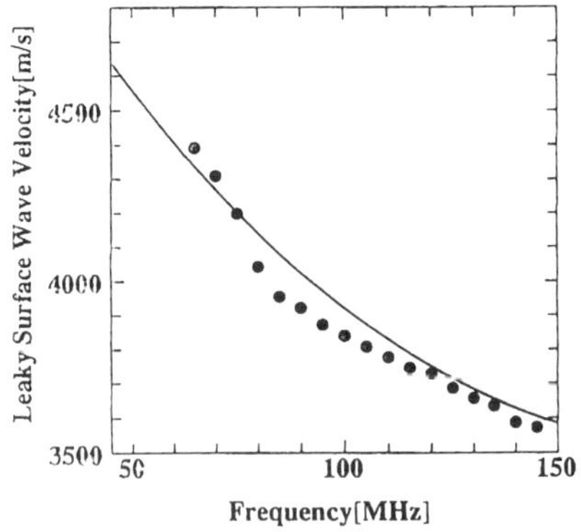

Fig. 6 Comparison between the theoretically calculated Vsaw dispersion and the measured Vsaw dispersion.

When the acoustic velocity of coating material becomes higher than that of substrate, the acoustic structure is expected to change itself drastically from Figures 2 and 3. In order to investigate this change in the acoustic structure, TiN coated steel substrate system was employed in what follows. Figure 7 depicts both the phase shift and the power amplitude distributions in the function of the incident angle and frequency. Although the Rayleigh surface wave is activated especially in the lower frequency, no higher-order surface waves were activated in the whole range. To be interested in the phase shift distribution, there seems to be a cut-off point where the amount of phase shift diminishes abruptly: for $f \cdot d > 500$ MHz $\cdot\ \mu$ m, no phase shift was observed.

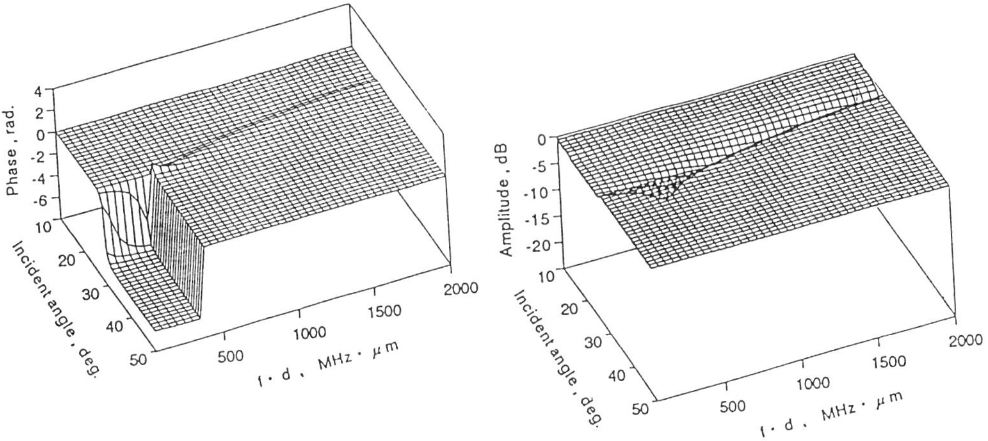

Fig. 7 Dependence of the reflectivity on the frequency for TiN coated steel substrate.

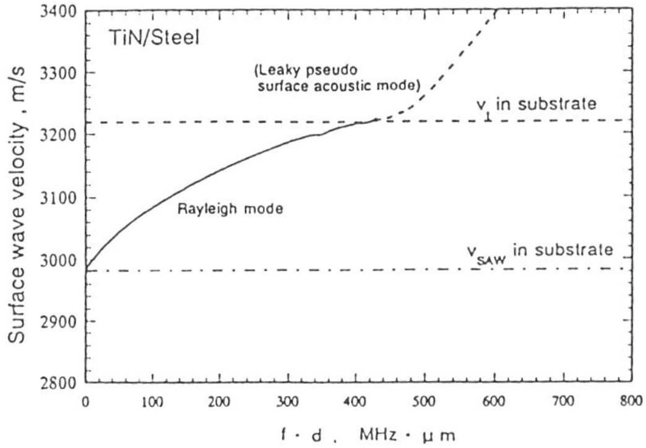

Fig. 8 Calculated surface wave velocity dispersion for TiN coated steel substrate.

TiN coated SKH steel substrate was used to experimentally demonstrate the above acoustic response predicted by the one-layer elastic model and to investigate the effect of TiN coating thickness (t) on the measured acoustic structure. Figure 9 depicts the measured phase shift distributions by ASM for two samples with different TiN coating thickness. No cut-off angles were seen for thin coating, but in the acoustic structure for thick coatings, distinct cut-off point appears. This cut-off phenomenon is so sensitive to the change of Young's modulus and coating thickness that the cut-off frequency is also available to simply estimate the elastic properties and to evaluate the dimensional accuracy of coating thickness.

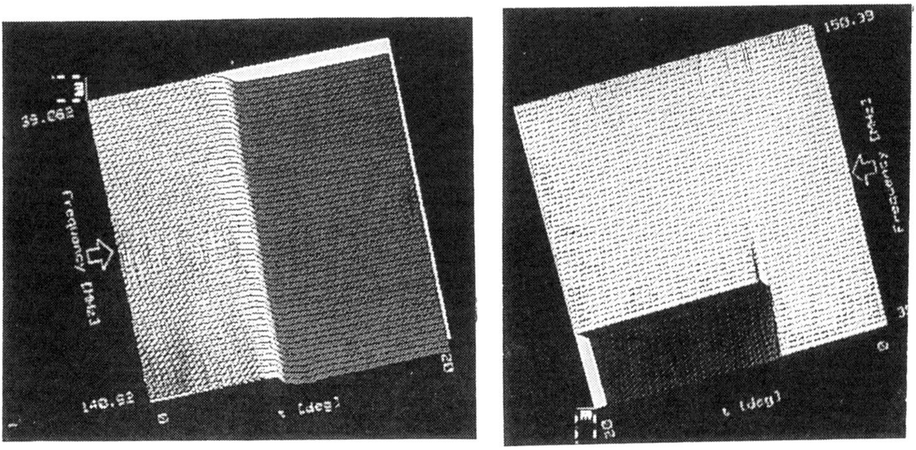

Fig. 9 measured phase shift distributions by ASM in the function of the incident angle and the frequency: (a) t = 2. 3 μ m, and (b) t = 6. 2 μ m.

NONDESTRUCTIVE INSPECTION OF INTERLAYER DEFECTS

The multi-layered coating system sometimes has interfacial defects and inhomogeneities originating from failure in coating procedure. Hence, nondestructive inspection is indispensable to improve the reliability of coated materials. Acoustic spectro-microscopy is also powerful to make nondestructive diagnosis of interfacial conditions between coatings or between coating and substrate. First of all, the elasto-dynamic theory for multi-layered elastic medium is used again to create three models with different interfacial conditions. TiN coated steel substrate system is taken for theoretical illustration of the sensitivity of acoustic structure to difference in the interfacial conditions. Figure 10 depicts the mechanical conditions of three models and the calculated phase shift distribution to each model. The acoustic structure to be measured by ASM is thought to have sufficient sensitivity to the change of interfacial conditions due to defects. In the following, let us consider the polymer-coated glass system for a model material to investigate whether ASM can detect the artificial defects between polymer and glass. The specimen is a PET (Polyethylene terephthalate) - coated glass: the PET film with the thickness of 12 μ m was glued on the glass substrate by using the adhesive agent. Non-adhesive area was made as an artificial defect as shown in Fig. 11. At the presence of the adhesive layer, the objective system might become a doubled-layer system: (PET-surface layer) - (adhesive layer) - (glass substrate). Measurement of surface roughness and micro-indentation testing revealed that 1] the maximum deflection of PET film was only about 0. 745 μ m, 2] no distinct differences were seen in the surface roughness both for adhesive and non-adhesive area, and 3] Deflection of PET film in micro indentation was comparable to surface roughness. Hence, the thickness of adhesive layer was found to be negligible small;

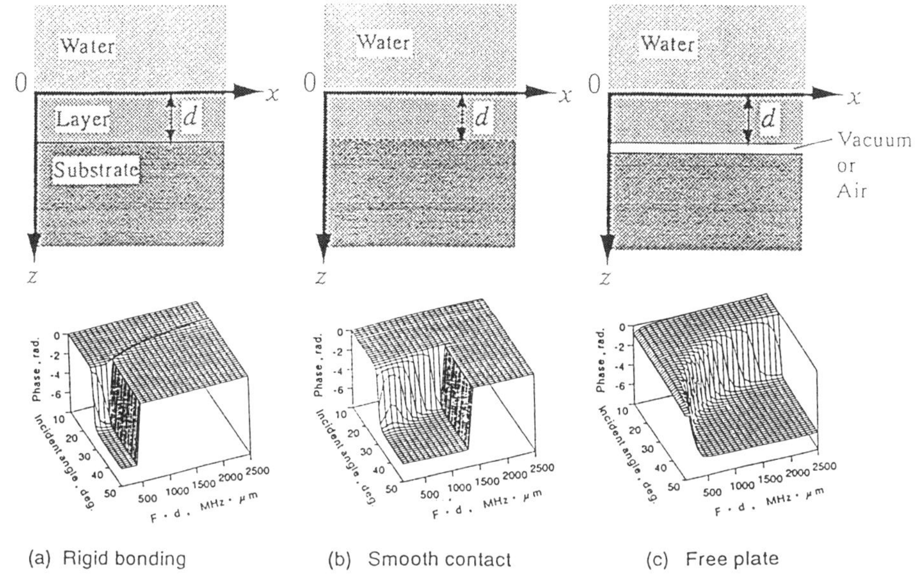

Fig. 10 Three interfacial models with different bonding conditions between coating and substrate and the calculated phase shift distribution in the function of the incident angle and the frequency for each interfacial model.

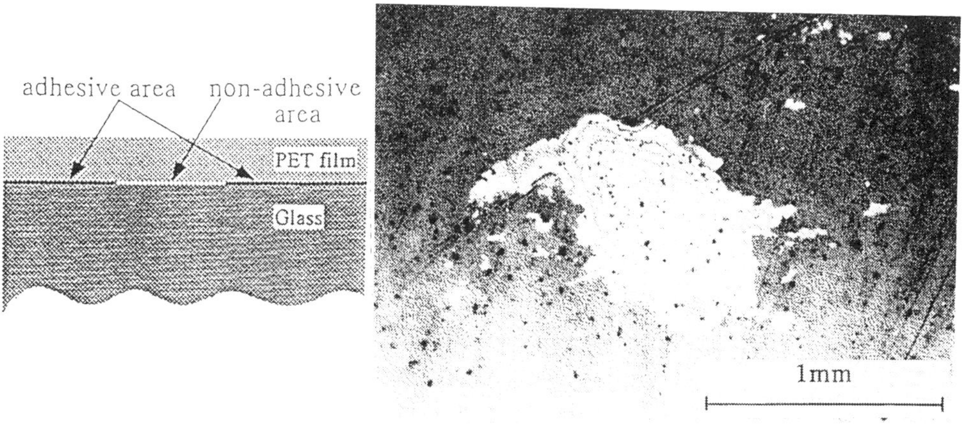

Fig. 11 Test specimen of PET-coated glass matrix with an artificial defect of non-adhesive area between PET coating and glass substrate.

single-layer model can be also available for theoretical understanding of the acoustic structure to be measured by ASM. The preliminary analysis told us that 1] Interfacial contact state was thought to satisfy the rigid bonding condition in the adhesive area, while the contact state in the non-adhesive region corresponded to the free-plate state rather than the smooth contact state due to the presence of an air in the interface, and 3] Phase shift distribution was too sensitive to the change of mechanical properties and inadequate for nondestructive inspection of defects. Then, the difference between adhesive and non-adhesive regions, or, the existence of postulated interfacial defects appears in the power amplitude with different dependency both on the incident angle and the frequency. Figure 12 depicts the measured acoustic structure for the whole sampling area including an artificial defect. The optimum sensing condition {Incident angle, Frequency} lies in the area with steep gradient of intensity distribution in this map: e.g. at the location A indicated in Fig. 12, a steep gradient was observed in the positive intensity distribution where the reflectivity from the non-adhesive region is expected to be higher in general. While, the reflectivity might be relatively weak at the location D in Fig. 12. From this understanding, sensing conditions can be directly determined as follows: [A] $\theta = 24.5^\circ$ and f = 100 MHz, [B] $\theta = 24.5^\circ$ and f = 35 MHz, [C] $\theta = 24.5^\circ$ and f = 125 MHz, and [D] $\theta = 34.5^\circ$ and f = 120 MHz. In actual nondestructive inspection, 2 mm x 2 mm region was specified on the specimen for two dimensional scanning with the sampling distance of 50 μ m. Through this scanning, the artificial defects are to be visualized as the difference of intensity just like the C-mode scanning. Figure 13 show two maps by using A and D sensing conditions. Comparing these maps to the optical image of a defect in Fig. 10, non-adhesive regions can be successfully described by the positive picture in case of A-scanning and by the negative picture in case of B-scanning. This fairly good agreement implies that 1) C-mode scanning by ASM can visualize the geometry of interfacial defects with high spatial resolution, 2) Detectability can be controlled by sensing conditions in the acoustic structure, and 3) the present nondestructive inspection method can be also utilized for the multi-layered coating system.

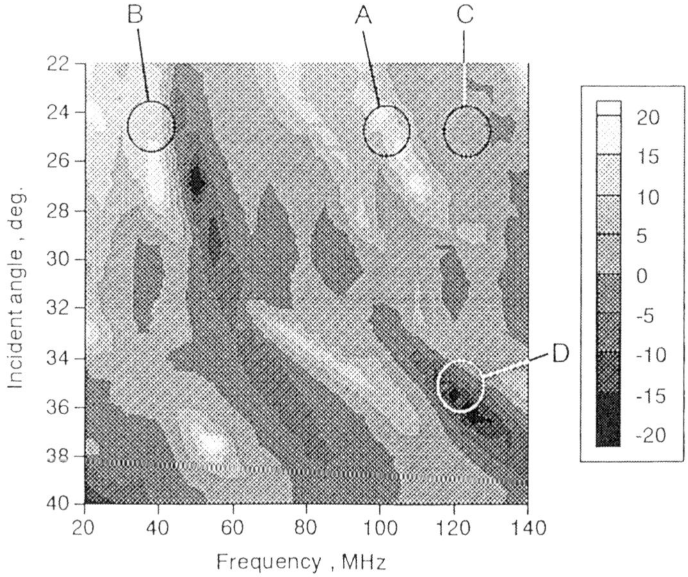

Fig. 12 Acoustic structure obtained by the reflectivity measurement for the whole regions including both the rigid-bonded interface and the non-adhesive area.

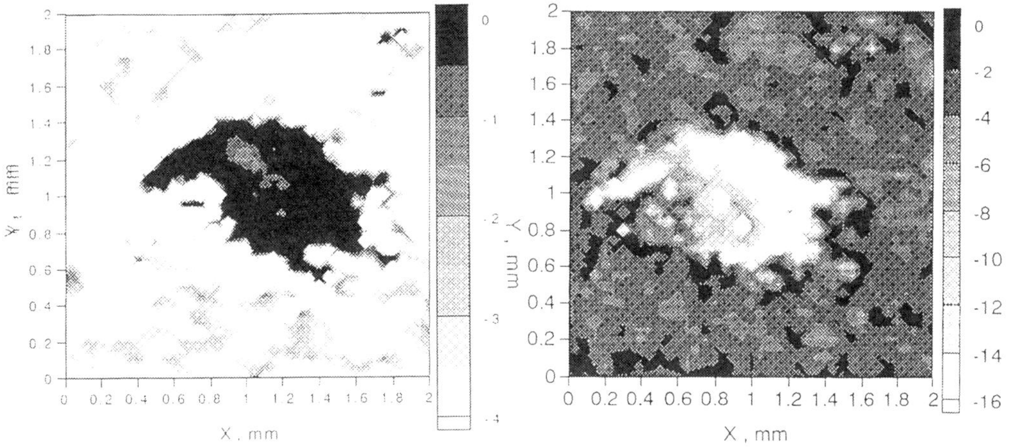

Fig. 13 Two dimensional scanning of amplitude with both the incident angle (θ) and the frequency (f) fixed at the specified values: A) θ = 24. 5 and f = 100 MHz, and D) θ = 34.5 and f = 120 MHz.

ELASTIC CHARACTRIZATION OF CERAMIC COATINGS

Correlation between the measured surface wave velocity dispersion and the calculated Vsaw model must be a base to estimate the elastic properties through the inverse analysis where the elastic properties are parametrically varied and optimized in order that the theoretical Vsaw dispersion curve should be in agreement with the measured one. Au-coated glass substrate system are taken for a example to validate the present approach to nondestructive elastic characterization of ceramic coatings. DC sputtering with 420 V and 1.0 A was used for Au coating onto a glass substrate at the presence of Ar with the pressure of 3 x 10-3 Torr as a carrier gas. Theoretical surface wave dispersion curve calculated by the optimized elastic properties is shown with the measured data: the estimated values are [E=78.5 GPa, ν =0.42].

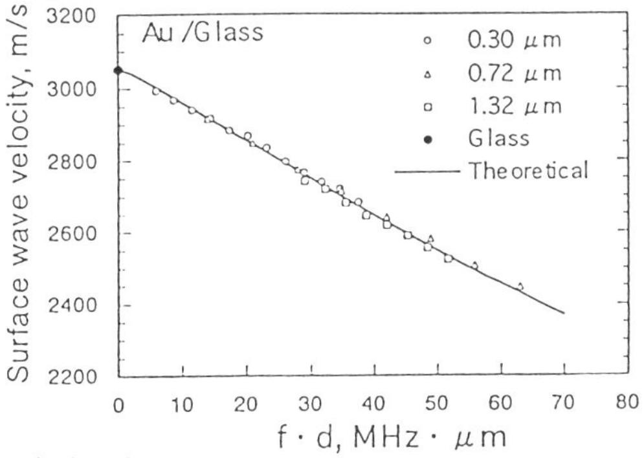

Fig. 14 Elastic characterization of Au-coated glass substrate system.

CONCLUSION

Acoustic spectro-microscopy is useful not only to make nondestructive inspection of interfacial defects but also to make elastic characterization of coating materials. Since acoustic structure obtained by the reflectivity measurement is just corresponding to the multi-layered elastic medium, various techniques can be available to quantitative nondestructive evaluation (QNDE) on the coated materials. Active use of the surface wave velocity dispersion curves enables us to make robust inverse analysis for quantitative determination of elastic properties, density and thickness of coatings; at present, three to four layered coating can be dealt with for elastic characterization. In the present paper, only the Rayleigh wave velocity dispersion was used for QNDE of coated materials; as discussed in Refs. (11), higher-order surface waves and pseudo Sezawa wave are useful not only for inspection of interlayer inhomogeneities but also for anisotropic elastic characterization. Further advancement will be reported in near future.

ACKNOWLEDGMENTS

Author would like to express his gratitude for discussion and advice to Dr. Y. Tsukahara (Toppan Print, Co. Ltd.) and Dr. I. Ihara (Nagaoka Institute of Science and Technology).

REFERENCES

1) A. Briggs: Acoustic Microscopy. (1992) Clarendon Press.
2) A.S. Edelstein and R.C. Cammarata (Eds.): Nanomaterials - Synthesis, Properties and Applications. (1996) Institute of Physics Publishing.
3) T. Aizawa, et al.: Nondestructive evaluation of PVD/CVD TiN coated WC/Co by acoustic spectro-microscopy. Acoustics-Optics and Acoustic Microscopy. ASME AMD-140 (1992) 55-70.
4) T. Aizawa, et al.: Nondestructive quality assurance of PVD/CVD ceramic coated WC/Co ceramics. ASME MD-44 (1993) 111-132.
5) I. Ihara, T. Aizawa and J. Kihara: Determination of ultrasonic attenuation in surface layer of materials by ultrasonic reflectivity measurement. Electronics Letters. 29-22 (1993) 1999-2003.
6) T. Aizawa and J. Kihara: Nondestructive evaluation of Co-agglomeration and WC growth in WC/Co superhard alloys. Nondestructive Characterization of Materials. VI (1994) 575-583.
7) T. Aizawa, et al.: Acoustic diagnosis for nondestructive evaluation of ceramic coatings on the steel substrate. ASME. PVP-302 (1995) 257-270.
8) I. Ihara: Surface wave spectroscopy by reflectivity measurement for quantitative nondestructive material evaluation. PhD Thesis of University of Tokyo (1996).
9) T. Aizawa and J. Kihara: Mechanical characterization of PVD/CVD TiN coated WC/Co by acoustic spectro microscopy. J. Faculty of Engineering, University of Tokyo. XLII (1993) 155-168.
10) L.M. Brekhovskikh: Waves in layered media. (1980) Academic Press.
11) Y. Xu, T. Aizawa and J. Kihara: Structure of surface wave on coarse-grained nickel observed by acoustic spectro-microscopy. Mater. Trans. JIM. 37-11 (1996) 1690-1698.

SURFACE AND SUBSURFACE FLAW DETECTION IN SILICON NITRIDE USING ACOUSTIC MICROSCOPY (AM)

Pramod Khandelwal
Allison Engine Company
P.O.Box 420, W-5
Indianapolis, IN 46202-0420

ABSTRACT

Silicon nitride is a leading candidate material system for potential use in automotive, industrial, and military applications for its unique high temperature material properties. Today's silicon nitride generally fails from surface defects caused by various processing steps. Reference standards with etched patterns in single crystal sapphire were fabricated. Simulated surface holes from 5 to 100 micron diameter size were excimer lasered in modulus of rupture (MOR) bars, button-head tensile specimens, and circular disks of PY6 hot isostatic press (HIP) processed silicon nitride. Subsurface holes were also produced in MOR bars and a billet to ascertain the detection sensitivity for volume defects. Acoustic microscopy resolved 70 micron apart line-pairs on the sapphire surface. In silicon nitride, 10 to 25 micron surface and 50 micron subsurface holes were detected up to a depth of 14 mm. Subsurface resolution was 220 to 315 micron in sapphire standards.

INTRODUCTION

Structural silicon nitride test specimens and components usually fail from both surface and volume defects. The flaws are introduced during the various stages of the fabrication process such as molding, drying, debinder removal (if any), sintering, and machining. The most common flaws present in today's high strength silicon nitride materials are iron (Fe) based inclusions, naturally occurring small voids, cracks, and machining induced surface damage. The first two types of flaws are found both on the surface and within the volume of the material, whereas the machining induced defects (e.g., cracks and grinding damage) occur only on the surface. The inclusions result from Fe impurities in the starting powder or from abrasion of the molding machinery. Nondestructive detection of these defects can screen out defective parts and substantially reduce variability, offering the potential for increased operating stresses. Additionally, the a priori knowledge of the defect size, distribution, and characteristics could potentially provide important information about the flaw population that affects the probability of detection (POD), intrinsic material strength, slow crack growth, and large scatter in time-to-failure during stress rupture testing and component evaluation.

In this paper, we will describe results of an acoustic microscopy study in which three different types of specimens with varying surface finishes and flaw sizes were used to ascertain flaw detection and resolution capability.

SPECIMENS

Three types of reference specimens, namely MOR bars, button-head tensile specimens, and biaxial disks, were designed and fabricated to optimize acoustic microscopy technique parameters. Calibration standards of single crystal C-axis sapphire were also fabricated for ultrasonic imaging to determine the surface and subsurface resolution of the acoustic microscopy technique.

Five MOR bars with 5 micron to 100 micron diameter and varying aspect ratio surface holes were fabricated. The holes were drilled by an excimer laser at Resonatics, Inc., using programmed x-y location of each hole. The laser parameters were developed for each size and depth of hole to create a quality straight hole. Optical micrography revealed chatter and damage on the surface around the hole. The surface finish of the MOR bars was between 0.4 to 0.5 microns (16 to 20 rms). In addition, the bottom of the holes was tapered, which decreased their effective depth and volume. Similar holes of 10 to 100 micron diameter and varying depth were induced in the gage area of the uniaxial button-head specimens. One specimen had 50 holes each of 25, 50, and 100 micron diameter and depth with an aspect ratio of 1. Another specimen, Figure 1, had 19 holes each of 10, 25, and 50 micron diameter with varying aspect ratios to ascertain the effect of the depth on flaw detection. Optical micrography, Figure 2, disclosed that the holes were irregular, chattered, damaged around the holes, and did not have flat bottoms in the button-head specimens. The surface finish of the tensile specimens was between 0.23 to 0.45 microns (9 to 18 rms). Biaxial circular disks were machined to the required surface finish using 400 grit wheel for multiaxial testing by Battelle. The surface finish was between 0.25 to 0.375 microns (10 to 15 rms). Subsequently, 50 laser-drilled surface holes of 25 to 100 micron diameter and depth (aspect ratio = 1) were induced to simulate a standard for disk.

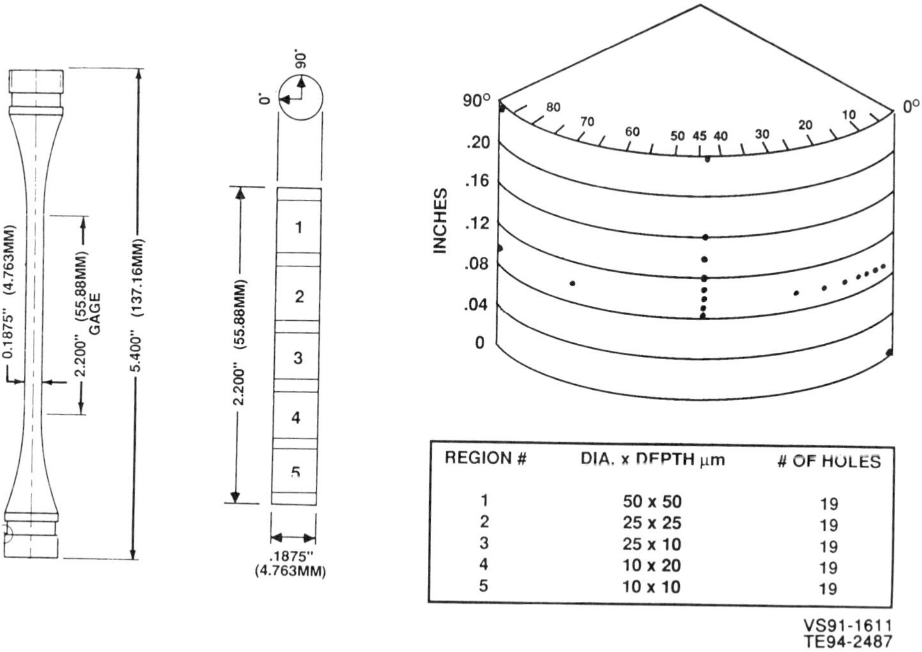

REGION #	DIA. x DEPTH μm	# OF HOLES
1	50 x 50	19
2	25 x 25	19
3	25 x 10	19
4	10 x 20	19
5	10 x 10	19

Figure 1. Layout of 10- to 50-micron surface drilled holes in a button-head specimen.

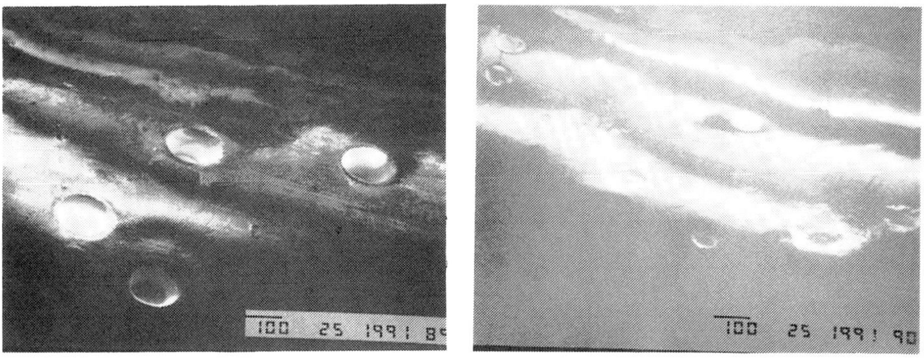

Figure 2. Optical photograph of laser drilled surface holes showing damage around the holes.

Lithographic methods have been used for creating surface and subsurface void arrays in many high temperature ceramic materials such as glasses, structural ceramics, and ceramic-metal interfaces. The combination of photolithographic methods with ion beam etching and hot pressing provides the methodology to fabricate reference standards with internal voids. These methods are described in detail elsewhere (Refs. 1, 2). Single crystal C-axis (perpendicular to flat faces) sapphire was used to make the resolution standards since its ultrasonic velocity is similar to that of silicon nitride. Single crystal wafers were acquired from Meller Optics. The specimens were prepared at the University of California, Berkeley. Two 2-in. x 2-in. square single crystal C-axis sapphire wafers (2-mm and 4-mm thick) were sectioned to produce 25.4-mm by 25.4-mm (±1.5 mm) pieces for lithographic processing, ion beam etching, and bonding. Two specimens with subsurface voids at 2 mm and 4 mm below the entry surface were produced. One 4-mm thick piece was used to produce a surface standard. Using a mask containing a 1951 Air Force Resolution Pattern, the desired defect pattern for nondestructive evaluation (NDE) studies was transferred to four 25.4-mm by 25.4-mm wafers of the 2-mm and 4-mm thickness for the standards (Figure 3). A profilometer was used to determine the etched feature depth, to assure a pattern depth within the 0.2- to 0.9-μm range as planned.

Fabrication of reference standards with isolated internal or volume defects has historically been a major problem in structural ceramic materials. Internal defects in Type-B MOR bars and rectangular 15-mm thick billets were simulated with laser-drilled and pilot holes. The pilot holes were induced by Bullen Ultrasonics, Eaton, Ohio, and laser-drilled holes by an excimer laser at Resonatics, Inc, respectively. The MOR specimens with laser-drilled holes of 10- to 100-micron diameters and 250-micron to 2-mm depth were fabricated with an aspect ratio ranging from 1 to 290 to assess the subsurface detection capability. Reference specimens with pilot holes and laser-drilled holes in flat rectangular billets were fabricated ranging from 0.5 to 3 mm. Excimer laser-drilled holes, each with a 1-mm depth and 50-, 100-, or 250-μm diameter, were drilled into the pilot holes.

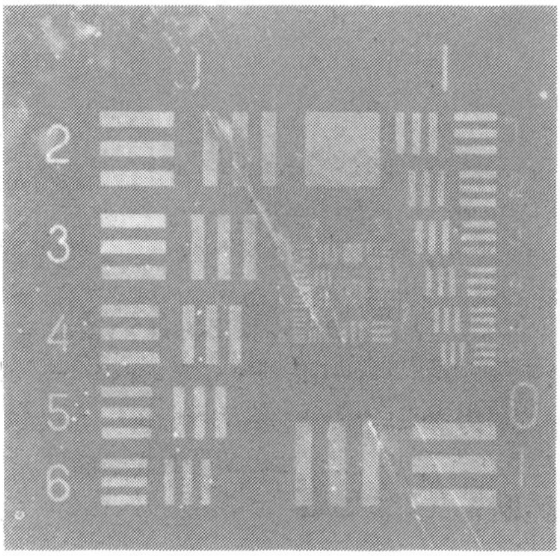

USAF TARGET 1951

NUMBER OF LINES PER MILLIMETER IN USAF
RESOLVING POER TEST TARGET 1951

ELMT NO.	GROUP NUMBER 0	1	2	3	4	5	6
1	1.00	2.00	4.00	8.0	32.0	32.0	64.0
2	1.12	2.24	4.49	8.98	17.95	36.0	71.8
3	1.26	2.52	5.04	10.1	20.16	40.3	80.6
4	1.41	2.83	5.66	11.3	22.62	45.3	90.5
5	1.59	3.17	6.35	12.7	25.39	50.8	102.
6	1.78	3.56	7.13	14.3	28.51	57.0	114.

VS93-3952 TE94-468

Figure 3. Sapphire reference standard for ultrasonic image calibration.

EQUIPMENT

Failure of high strength ceramic material normally is initiated from naturally occurring and machining induced surface defects. Secondary failures occur from internal defects such as inclusions. One of the most sensitive NDE techniques to characterize such surfaces is acoustic microscopy. The fracture controlling flaw size in today's silicon nitride is very small (1 to 50 μm). A 100 MHz computer-controlled acoustic microscopy (AM) system was developed and implemented at Allison to characterize PY-6 ceramic test specimens prior to testing, as shown in Figure 4. This system utilizes Panametrics 100 MHz pulser/receiver, single shot 100 MHz peak detector with zero-span auxiliary equipment, which increases the dynamic range of the peak detector output, 50 MHz transducers, and a 486/33 MHz Dell computer control data acquisition and imaging software. The system utilizes a PixelWorks imaging board (1280 x 768), Panametrics Multiscan version 2.1 software for data acquisition, image processing and display on Mitsuishi 15-in. Diamond monitor Model HL6605 (1280 x 1024). The system is equipped with a 232 megabyte hard drive, a 44 megabyte Bernoulli removal cartridge drive system, and a Fujitsu 128 megabyte magneto-optical disk drive with removable disks to store large amounts of C-scan imaging data. The image storage capability of the system enables precision scanning of specimens to detect minute defects, as necessary. The immersion tank, 16.5 x 15.5 x 11.5 in., was fabricated from anodized aluminum. The scanner has x, y, z, and _ axis of scanning. All the axes are programmable and computer controlled. The scanning speed of the system can be varied up to 6 in./sec with a resolution of at least 5 mm in x, y, and z directions, respectively. Specimens can be inspected at a rotary speed of 360 deg/sec with 0.005 deg resolution. C-scan setup and image data can be acquired and stored on the disk in real time. It can be presented on the screen or hard copy generated using a Seiko Model 5504 color printer. A 50 MHz 0.2-in. (5.08 mm) focal length surface wave transducer (F# = 0.8) was predominantly used for the inspection of specimen surfaces. Two other 50 MHz transducers with F# 2 and F# 6 were used in this study to assess surface and subsurface flaw detectability. A 25 MHz transducer was used for deep subsurface defects.

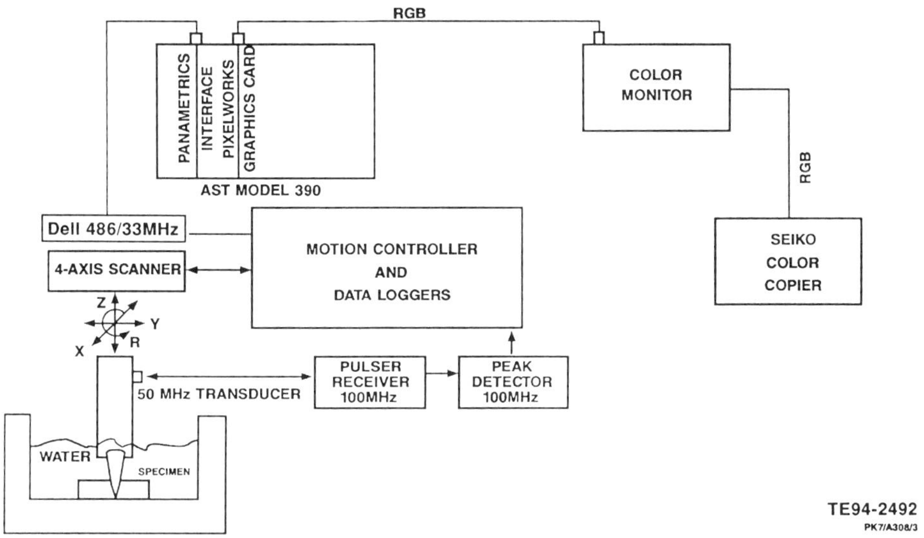

Figure 4. Schematic of a 100 MHz acoustic microscopy system.

RESULTS AND DISCUSSION

Surface Flaw Detection

Acoustic microscopy (AM) examination of the MOR bar, button-head tensile specimen, and biaxial disk reference standards with surface holes was conducted. MOR specimens with surface holes were examined using a 50 MHz transducer with a focal length of 5.08 mm and crystal diameter of 6.35 mm (F# 0.8). This transducer was predominantly used for surface characterization. The specimens were scanned parallel to their length and machining direction. The transducer was incremented in the width direction. Acoustic microscopy detected 25 x 25 μm size holes with clarity (Figure 5). Smaller 10 micron holes could not be discerned from the background microstructure signal of the longitudinally machined surface with a surface finish of about 0.5 microns.

The button-head tensile specimens with 10- to 100-micron surface holes were also evaluated by acoustic microscopy using the F# 0.8 transducer. The specimens were radially scanned by rotating the specimen and keeping the transducer stationary. Subsequently, the transducer was incremented to a new axial location and the specimen was rotated and inspected. Acoustic microscopy detected all the 50 holes in a tensile reference standard each of 25-, 50-, and 100-μm diameter in size. Another tensile specimen standard, Figure 1, with 10- to 50-micron size hole specimens was also characterized. All the 10-micron diameter x 20-micron deep holes were detected (Figure 6). Again, surface machining damage inhibited detection of shallower 10-micron deep defects.

Acoustic microscopy detected 25-μm diameter holes in the biaxial disk specimen. The results for all three types of specimens are summarized in Table I. Microfocus X-ray detected 50 to 100 micron surface flaws whereas acoustic microscopy detected 10 micron surface voids. Acoustic microscopy detected 25 x 25 micron surface holes in the biaxial circular disk using the setup for MOR bars. Comparative analysis of the detectability of surface holes in three different surface finish specimens by acoustic microscopy was significantly affected by the surface finish (Table I) as expected. The surface finish of the tensile and biaxial specimens was better than MOR bar specimens. Ten micron diameter holes were detected in tensile specimens compared with 25 micron holes in MOR bars.

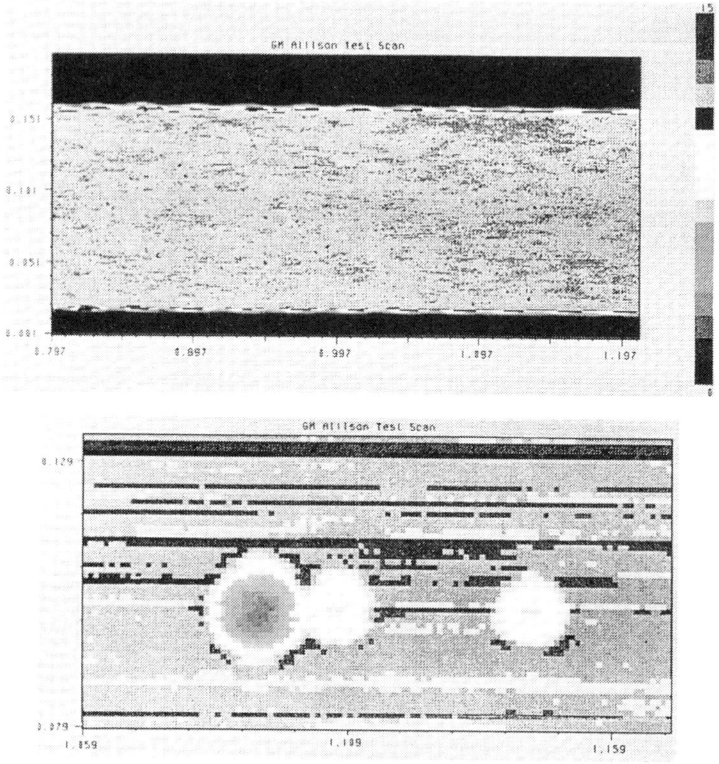

TE94--2494

Figure 5. Detection of 25 x 25 micron surface holes in MOR bars.

AM Characterization of Sapphire Resolution Standards

Sapphire resolution standards (25.4 mm x 25.4 mm) were examined by acoustic microscopy to determine the spatial resolution capability of separating surface and subsurface defects. Three reference specimens were evaluated: (i) surface-etched pattern on a 4-mm thick wafer; (ii) 2-mm/2-mm thick specimens with pattern at the interface; and (iii) 4-mm/2-mm specimens with pattern on the 4-mm wafer. Optical micrography measured the depth of the surface-etched pattern on the 4-mm wafer to be between 0.2 μm and 0.9 μm. The specimen was evaluated using the 50.0 MHz, F# 0.8 transducer at a water path of 5.04 mm with an increment of 12.5 micron. The image data were collected by focusing the beam at the water-sapphire interface and gating the reflected signal. The -6 dB focal spot diameter was approximately 30 micron. Magnification of the image on screen revealed that Group 3 Element 6 line pairs 70 μm apart were resolved, which is consistent with the literature (Ref. 3). Gilmore et. al (Ref. 3) has reported that line-pair resolution can be significantly improved by using a Weiner filter. The shallow depth of the pattern made it difficult to resolve smaller line-pairs.

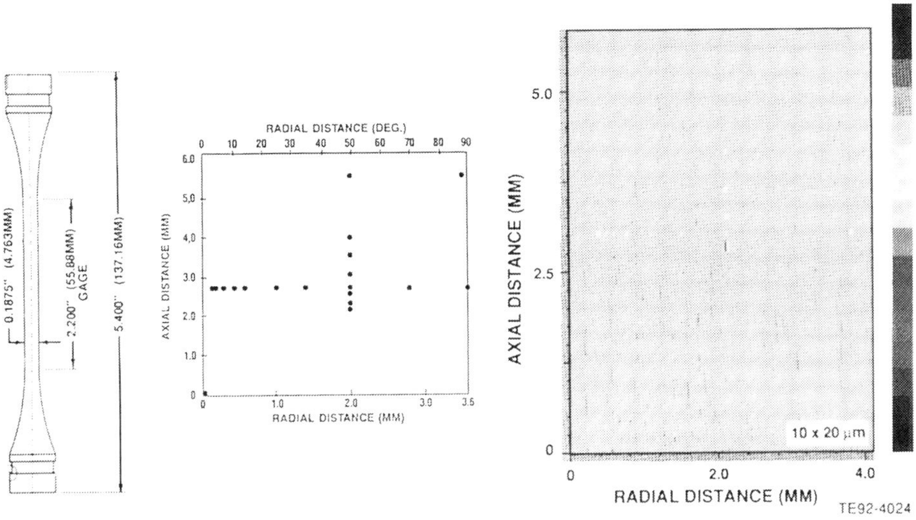

Figure 6. Detection of 10-micron diameter surface holes in the gage area of the tensile specimen.

Table I. Detecting of surface holes by acoustic microscopy.*

Specimen type	Nominal hole size** Diameter—μm	Depth—μm	Flaws present	Flaws detected
MOR bar	100	100	100	100
Surface finish	50	50	50	50
Avg: 0.45 μm	25	25	50	49
Tensile	100	100	50	50
Surface finish	50	50	50	50
Avg: 0.32 μm	25	25	50	50
	25	10	19	19
	10	20	19	18
	10	10	19	9
Biaxial disk	25	25	49	49
Surface finish				
Avg: 0.3 μm				

*Transducer: 50 MHz, F# 0.8

**Actual flaw shape/size were uncontrolled.

The second sapphire standard (2 mm/2 mm) with the pattern 2 mm below the surface was evaluated using a 50 MHz transducer with a 12.7-mm focal length and crystal diameter of 6.35 mm (F# 2.0). The transducer was focused at the bonded interface of the two wafers where the etched pattern was located. The scanning increment was 12.5 micron at a water path of 1 mm, which was experimentally optimized for best results. The line pair in Group 2 Element 2 with 223 micron separation were resolved (Figure 7). This is consistent with -6 dB focal spot diameter of

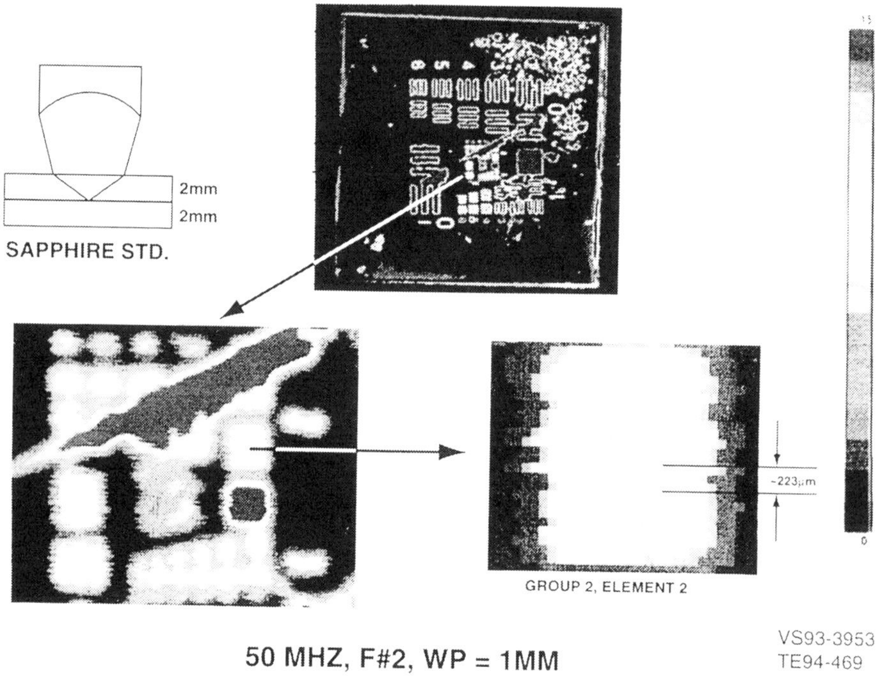

Figure 7. Acoustic microscopy detected 223-micron apart line-pair 2 mm below the surface in a sapphire standard.

around 120 micron. Many other voids were also detected because of somewhat poor bonding of the sapphire wafer in the unetched area. In addition, the twin developed during fabrication also was detected. The third sapphire target (4 mm x 2 mm) was interrogated by entering the acoustic beam from the 4 mm wafer side. Here, a 50 MHz nominally 3.81 cm focal length transducer was focused 4 mm below the surface at the two wafer bonded interface. The scanning increment was 25 micron and the water path was 12.7 mm. The line-pair in Group 1 Element 5 were resolved at a distance of 315 microns.

Subsurface Flaw Detection

Acoustic microscopy detected 25-micron diameter and deep subsurface holes readily in MOR bars using a 50 MHz transducer with a focal length (F) of 12.7 mm and a diameter (d) of 6.35 mm (F# 2.0) at a water path of 12.7 mm (Figure 8). Smaller holes of 10 micron nominal diameter could not be discerned from the background microstructure. Again, the holes were imperfectly drilled as described previously. The rectangular billet with deeper subsurface holes was examined using a 25 MHz nominal frequency 3 in. in focal length transducer with a water path of about 2.75 mm. The 50-mm diameter holes were readily detected at a subsurface depth of 14 mm (0.55 in.) (Figure 9).

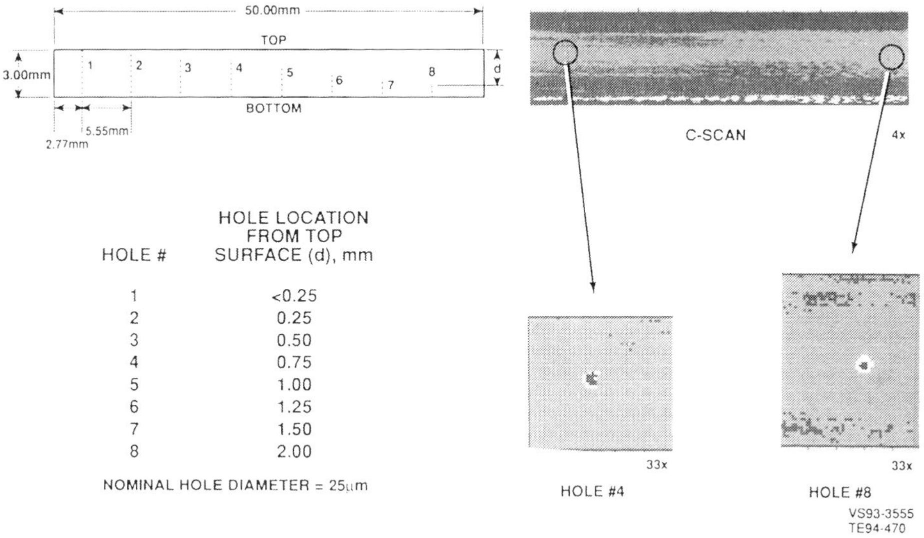

Figure 8. Acoustic microscopy detected 25-micron internal holes 2 mm below the surface.

CONCLUSIONS

Reference standards with surface holes in various silicon nitride specimens were fabricated. Acoustic microscopy detected 25-micron diameter surface holes in both the MOR bars and biaxial disks. Ten (10) micron surface holes were detected in tensile specimens that had a better surface finish. Detection of defects below 25 micron depends largely on the surface finish. The surface image resolution was 70 micron whereas the subsurface image resolution varied between 220 and 315 micron. Subsurface 50-micron holes were detected to a depth of 14 mm.

ACKNOWLEDGMENTS

This research effort was sponsored by the U.S. Department of Energy, Assistant Secretary for Energy Efficiency and Renewable Energy, Office of Transportation Technologies, as part of the Ceramic Technology Project of the Propulsion System Materials Program, under contract DE-AC05840R21400 with Martin Marietta Energy Systems, Inc. The author would like to thank Mr. M. McCormick for conducting the experiments. Thanks are also due to Dr. A. Glaeser, University of California, Berkeley, for fabricating the sapphire standards.

REFERENCES

[1]J. Rodel and A. M. Glaeser, "Production of Controlled Morphology Pore Arrays: Implications and Opportunities," *J. Am. Cer. Soc.*, Vol. 70, No. 8, pp C172-C176, 1987.

[2]J. Rodel and A. M. Glaeser, "Photolithography: A New Tool for Ceramic Science," *Materials Research Symposium Proceedings*, Vol. 155, pp. 293-306.

[3]R. G. Gilmore, A. M. Glaeser, and J. C. Wade, "Calibrating Ultrasonic Images for the NDE of Structural Materials," ASME Paper 93-GT- 300.

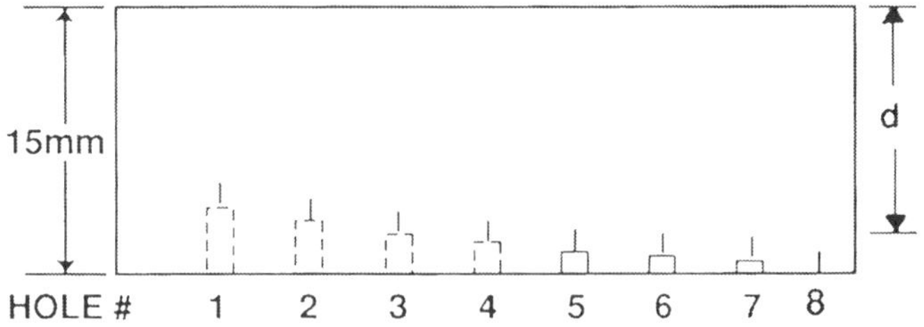

HOLE #	HOLE DEPTH, mm COUNTERBORE	LASER	DISTANCE FROM SURFACE (d)
1	3.00 mm	1.00	11.00 mm
2	2.75 mm	1.00	11.25 mm
3	2.50 mm	1.00	11.50 mm
4	2.00 mm	1.00	12.00 mm
5	1.50 mm	1.00	12.50 mm
6	1.00 mm	1.00	13.00 mm
7	0.50 mm	1.00	13.50 mm
8	0.00 mm	1.00	14.00 mm

NOMINAL COUNTERBORE DIAMETER = .500 mm
NOMINAL LASER HOLE DIAMETER = 0.050 mm

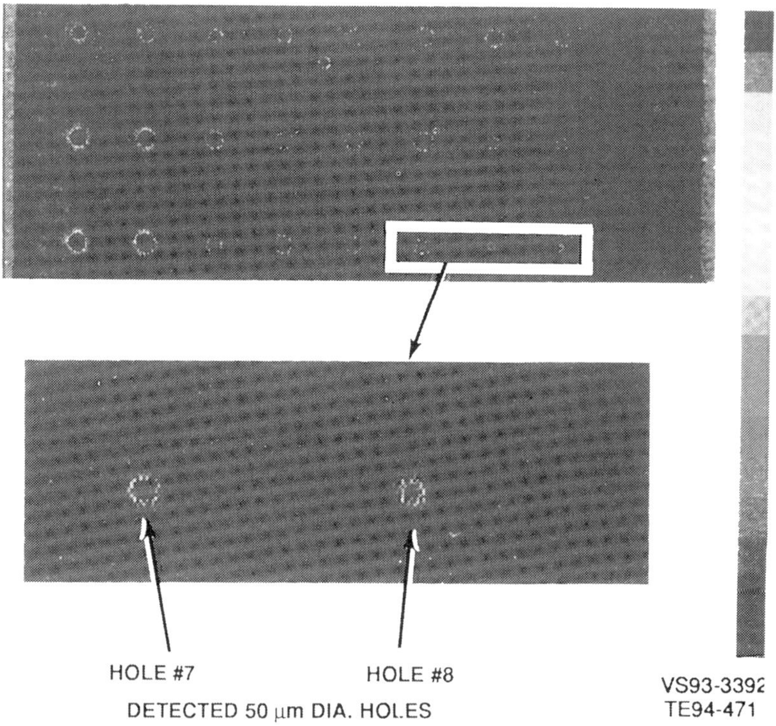

Figure 9. Acoustic microscopy detected 50-micron subsurface holes 14 mm below the surface.

GRAIN BOUNDARY CHARACTERIZATION AND EVALUATION BY ACOUSTIC SPECTRO-MICROSCOPY

Y. Xu and T. Aizawa
Department of Materials Science, Faculty of Engineering,
University of Tokyo,
7-3-1 Hongo, Bunkyo-ku, Tokyo 113, JAPAN

ABSTRACT

Acoustic spectro-microscopy was applied to nondestructive evaluation of the straight grain boundary between two nickel crystals(grains). Although the intrinsic spatial resolution to ASM is 50 to 100μm, the acoustic interaction across the grain boundary can be described by accurate linear sensing. Linear sensing with different directions and incident angles with spectroscopic analysis provides us a new method to make grain boundary characterization.

INTRODUCTION

Grain boundary characterization has been mainly performed by various high-resolution microscopes because microscopic evaluation of lattice structure change across the grain boundary and precise analysis of constitutive and dopant elements on/along the grain boundary is very important to discuss the role of the grain boundary on the functional and structural response of materials.[1] However, there is still a large gap between the above microscope observation/analysis and the functional/mechanical properties macroscopically measured. In other words, a mesoscopic method is awaited to make quantitative evaluation of the grain boundary with sufficient resolution and to provide the homogenized response of materials for a macroscopic understanding of their mechanical and functional behavior. The ultrasonic method must be one of the candidate methodologies for mesoscopic grain boundary characterization.

Since the first invention of scanning acoustic microscope by Lemons and Quate in 1973,[2] various studies have been reported for materials evaluation in theory and experiment.[3-6] In recent years, research activities have been shifted from nondestructive imaging of defects and inhomogeneities in materials to quantitative nondestructive elastic characterization both in isotropic and anisotropic materials.[7-11] Authors[12, 13] have succeeded in theoretical and experimental description of Rayleigh surface wave(RSW) and pseudo Rayleigh type surface wave(pseudo-RSW) structure for various nickel crystals with different crystallographic orientations together with simultaneous determination of elastic stiffness tensor and crystallographic orientation of a crystal(grain) embedded in polycrystal by using the acoustic spectro-microscopy(ASM).

In the present paper, a potential possibility of ASM as a tool to make grain boundary characterization is experimentally demonstrated with consideration of an intrinsic spatial resolution to ASM. Special notes are stressed on the acoustic interaction and its change experienced in the linear sensing across the grain boundary and the characteristic vibration of the reflected wave power at the vicinity of the grain boundary. These characteristics are discussed as a promising approach to quantitative grain boundary evaluation.

ACOUSTIC SPECTRO-MICROSCOPY (ASM)

Our developed acoustic spectro-microscopy is briefly introduced with a special equipment for sensing the grain boundary. An ultrasonic pulse injected from a planar lens reflects from a sample in the coupler and is received by a spherical lens.[12] The reflected signal is data-processed into the power and the phase spectra in the function of the incident angle(θ) and the frequency(f). To be noted, these spectra data can be obtained at any specified point on the surface of sample material and in any incident directional angle by moving the above spherical-planar-pair(SPP) lenses through the driving stage units.

This system has six stages for spatial control: X and Y stages for location of a focusing area in measurement, a Z stage for focusing depth-wise to a specimen, a θ stage for varying the incident angle, a φ stage for rotating the specimen, and a χ stage for transportation of a specimen from the acoustic spectro-microscope to the optical microscope and vice versa.

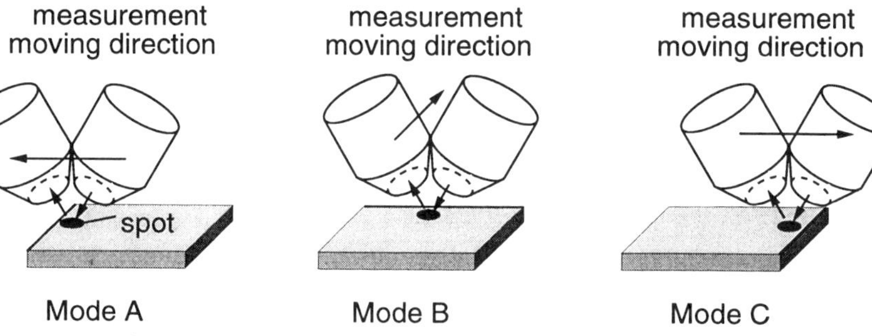

Fig. 1 Three edge-scanning modes for determination of spatial resolution on the material surface

Both the RSW and the pseudo-RSW velocities at the specified point and the measurement direction on the sample can be directly determined from the measured critical incident angles where the surface waves are activated. The spatial resolution by ASM has close relationship with the transducer spot size in the function of the frequency. Along the depth direction from the material surface, the resolution can be estimated to be one-wavelength of the surface wave for an isotropic material. This is because most of kinetic energy of RSW is imposed in this surface layer. On the other hand, the spatial resolution on the material surface must be experimentally determined by the edge-scanning method. Figure 1 depicts three edge-scanning modes with the different measurement moving directions. To be noted, the incident angle must be shifted from the Rayleigh critical angle. Figure 2 shows a typical normalized power of the reflected wave in the function of the frequency and the measurement position. The spatial resolution $\triangle x$ can be determined as the distance where the normalized power decreases from 90% down to 10%; *e.g.* $\triangle x \sim 100\mu m$ at f=40MHz in Fig. 3. The obtained spatial resolution by three different edge-scanning modes are illustrated in Fig. 4 as a function of frequency. Since the eye-spot is shaped nearly circular for the SPP lens, the spot diameter is considered to be about 50~100μm for the frequency range from 40 to 140MHz.

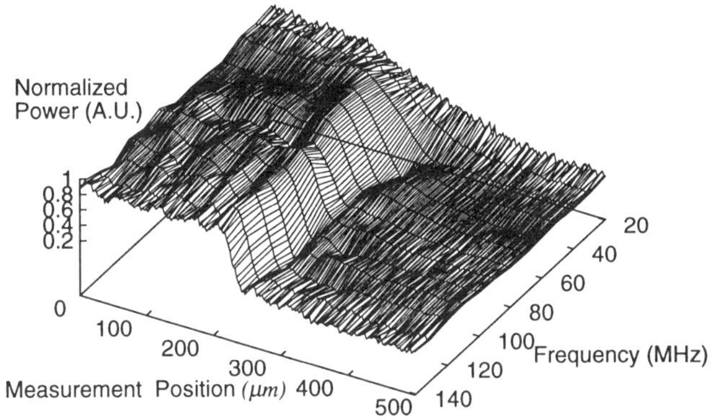

Fig. 2 Normalized reflected wave power in the function of the measurement position and the frequency in case of the mode C.

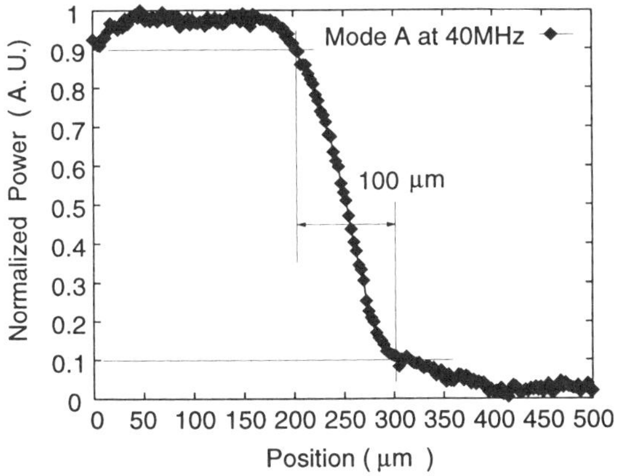

Fig. 3 Typical change of the normalized power with the position.

EXPERIMENTAL RESULTS AND DISCUSSION

In this experiment, the incident angle from the planar lens is incrementally varied by $\triangle\theta$=0.1°, and the temperature change $\triangle T$ of coupler (water) is controlled to be

$|\Delta T| \leq 0.1K$. The measurement error in this condition is about $10m/sec$.

Neighboring coarse grains (call it bi-crystal system) in the pure nickel polycrystal were employed as a model specimen to investigate the effect of grain boundary on the surface wave structure. The grain size was controlled to be larger than 1mm, and the target grain boundary was selected to be a right plane to the surface of the specimen. After careful polishing, the crystallographic orientation of the selected bi-crystal system was determined by the Laue back reflection method. Table I listed the model bi-crystal systems to be used in the following.

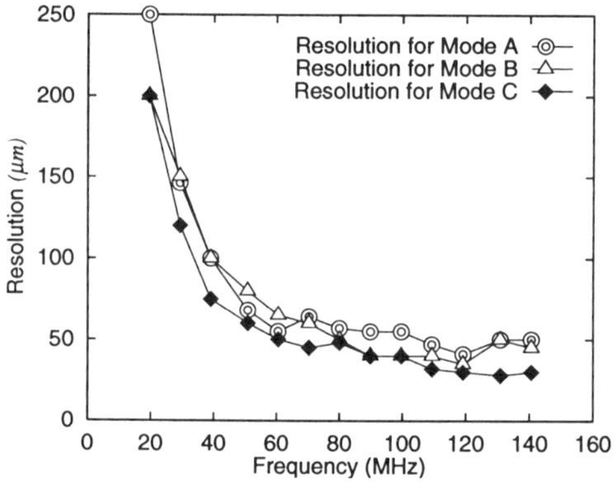

Fig. 4 Frequency dependence of resolution for three modes

Table I. Crystallographic orientation of the selected bi-crystal system measured by the Laue back reflection method: α, β and γ are the angle of normal direction of grain surface with the base crystal axes [100], [010] and [001], respectively

Number of Specimen	Number of Grain	Crystallographic Orientation		
		α	β	γ
D1	D1-left	(29.0°)	66.0°	74.7°
	D1-right	66.3°	55.3°	(44.2°)
D2	D2-left	85.9°	(30.5°)	59.8°
	D2-right	55.0°	(38.0°)	77.0°

Table II. Point symbol corresponding to relative position in the experiments (unit: μm)

a	-570	**b**	-430	**c**	-285	**d**	-140	**e**	-70	**f**	-35	**g**	-17.5	**h**	-8	**i**	0
q	570	**p**	430	**o**	285	**n**	140	**m**	70	**l**	35	**k**	17.5	**j**	8		

Figure 5 shows the schematic diagram of spatial scanning direction across the grain boundary. 17 points(marked by letters **a** to **q**) were selected for measurement between the two centers of each coarse grain in the perpendicular direction of the grain boundary, as listed in Table II. At each point, three directions φ =90°, -45°

and +45° to the grain boundary direction were selected to change the scanning angle.

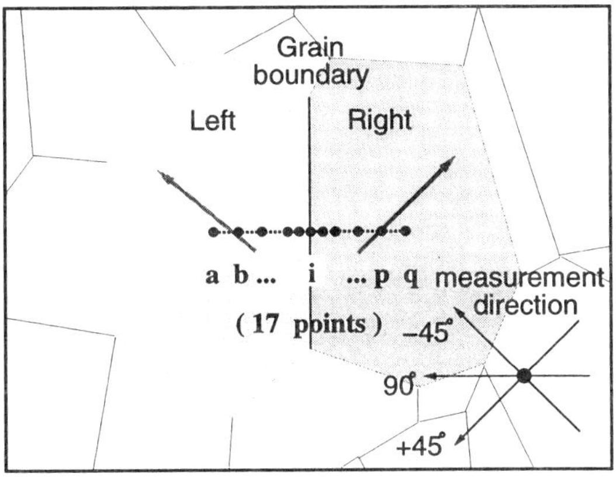

Fig. 5 Schematic diagram of spatial scanning direction across the grain boundary in a selected bi-crystal system.

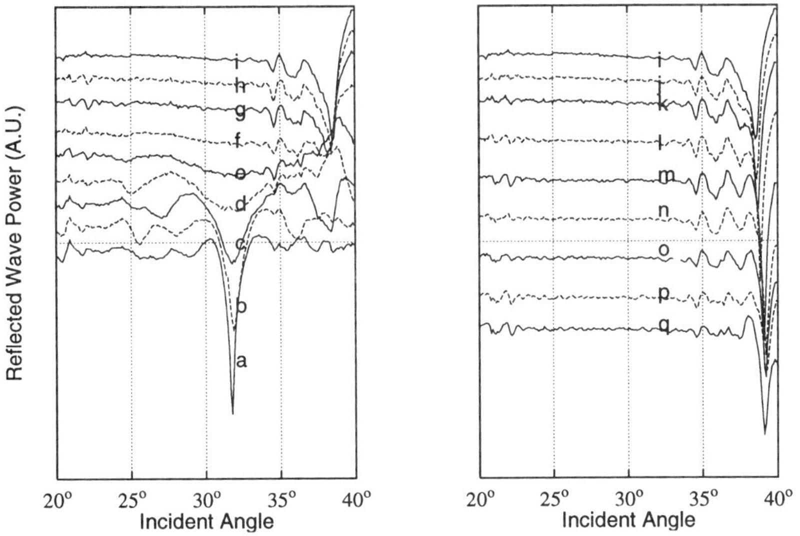

Fig. 6 Variation of reflected wave power distribution in the incident angle with the linear scanning from **a** to **q** along +45° direction for D2 bi-crystal system.

Figure 6 shows the reflected wave power changing with the incident angle in the +45° direction by f=40MHz for different measurement positions in the bi-crystal system D2. At the point **a**, since the eye-spot of SPP lens lies in the inside of the left-hand side grain of D2(D2-left), the measured power dip is just corresponding to the Rayleigh critical angle. Coming from **a** to **d**, this power dip becomes shallow, and no dips were seen at the point **e**. Just on the grain boundary, at the point **i**, the measured power dip was located at the angle smaller than the Rayleigh critical angle measured at points from **o** to **q** for the right-hand side grain of D2(D2-right). To be interested, although the eye-spot at 40MHz is still in the inside of D2-left, no reflected wave dip was observed at **e**.

The above phenomenon might be because the reflected surface wave coming back from the grain boundary superposes on its original activated surface wave. This attenuation behavior is intrinsic to the prescribed bi-crystal system; in other words, dependent on the grain boundary characteristics, this attenuation behavior has a drastic change. In fact, in case of the linear scanning along 90° direction for the D1 bi-crystal system, no attenuation of the reflected signals were observed for every received power at any point, as shown in Fig. 7. This reveals that the surface wave might better transmit through the grain boundary rather than reflect from it. Instead of the change of the surface wave attenuation, a gradual change of the dip angle, or, a gradual decrease of the surface wave velocity was observed from **a** to **q**. As shown in Fig. 8, the surface wave velocity continuously changes from D1-left to D1-right across the grain boundary. Roughly speaking, as suggested by Ref. 12, the measured surface wave velocity change by the linear scanning has no frequency dependence, but the surface wave velocity change in the D1-left here is enhanced with increasing frequency. This might be a typical grain boundary effect.

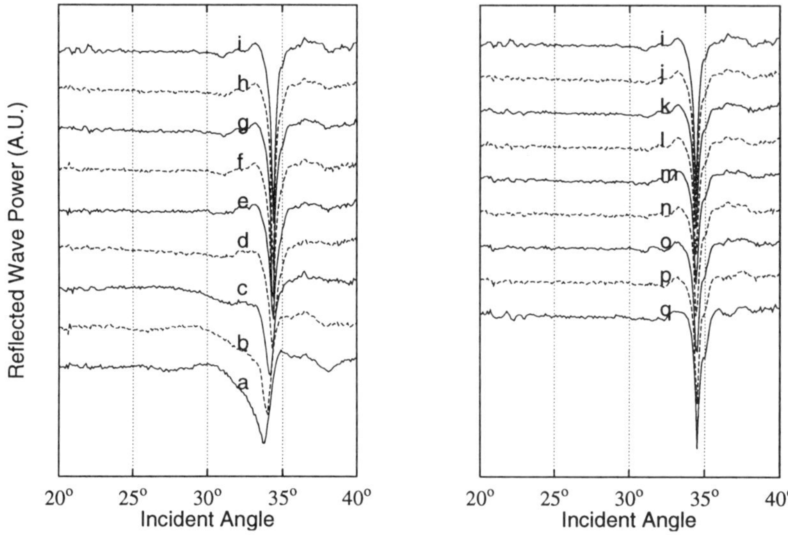

Fig. 7 Variation of reflected wave power distribution in the incident angle with the linear scanning from **a** to **q** along 90° direction for D1 bi-crystal system.

Theoretically, the surface wave velocity converges to the true Rayleigh surface wave velocity of D1-left crystal $v_r = 2710m/s$ approaching from the point **i** on the grain boundary to the point **a** in the D1-left; while it goes to the true Rayleigh surface wave velocity $v_r = 2617m/s$ approaching from the point **i** to **q** in the

D1-right. Hence, if the grain boundary had nothing to do with the surface wave propagation across the grain boundary, the above measured frequency dependence should disappear. When the scanning direction was reversed by 180°, or, D1-right to D1-left, this grain boundary effect diminished in Fig. 9.

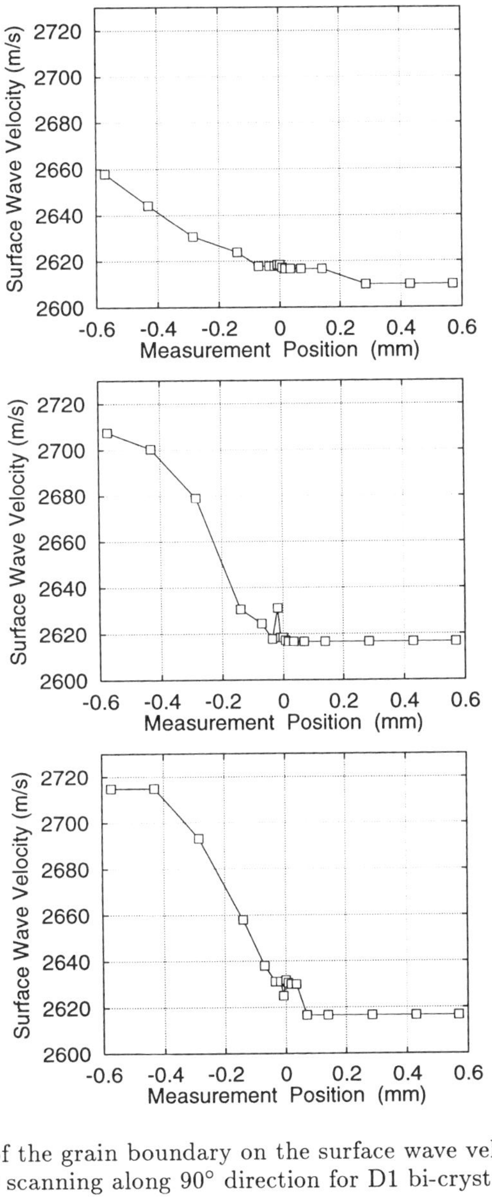

(a) f=40MHz

(b) f=90MHz

(c) f=130MHz

Fig. 8 Effect of the grain boundary on the surface wave velocity change in the linear scanning along 90° direction for D1 bi-crystal system.

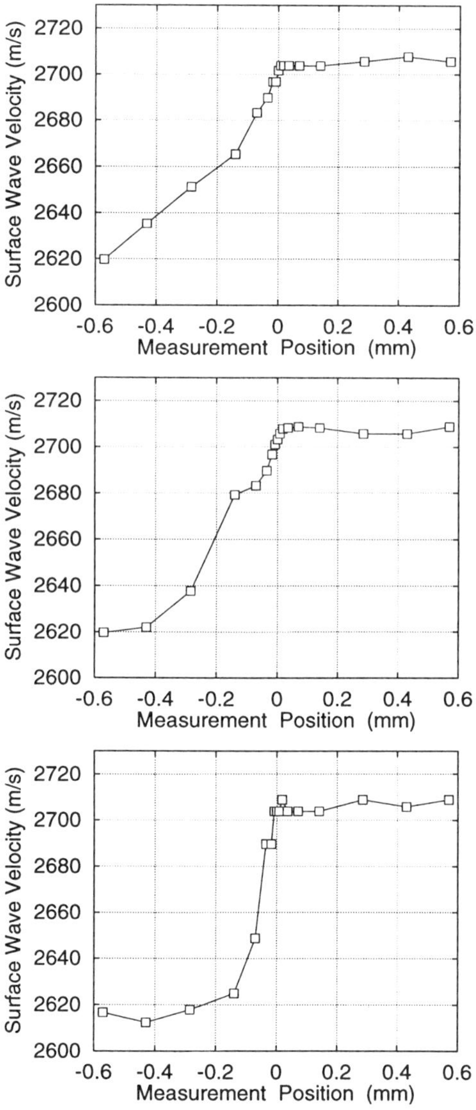

(a) f=40MHz

(b) f=90MHz

(c) f=130MHz

Fig. 9 Surface wave velocity change in the linear scanning along -90° direction for D1 (in this case D1-right denotes for the left-hand side grain, and D1-left the right-hand side one) bi-crystal system.

In measurement of the surface wave velocity change by linear scanning along -45° direction for D2 bi-crystal system, the other surface wave structure appeared as shown in Fig. 10. To be noticed, two intrinsic surface waves to both grains with different crystallographic orientation were simultaneously activated and received by ASM. This may be caused by the "Schoch displacement" phenomenon existing in Rayleigh critical angle, in which angle the Rayleigh type surface wave is excited on the liquid-solid interface.[14–20]

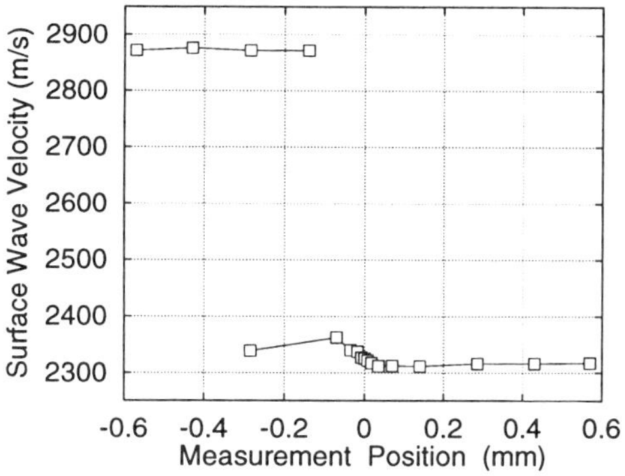

(a) f=40MHz

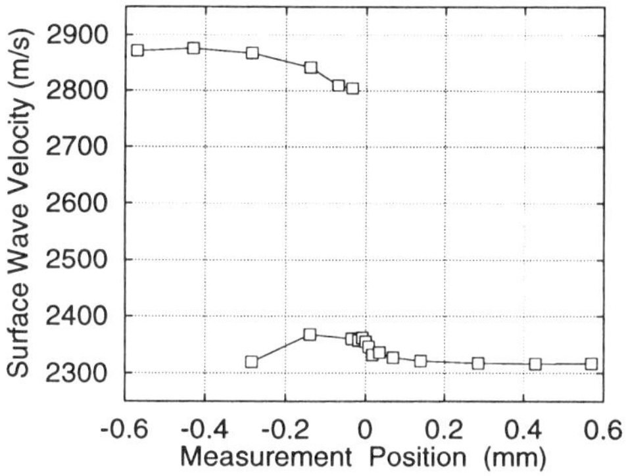

(b) f=90MHz

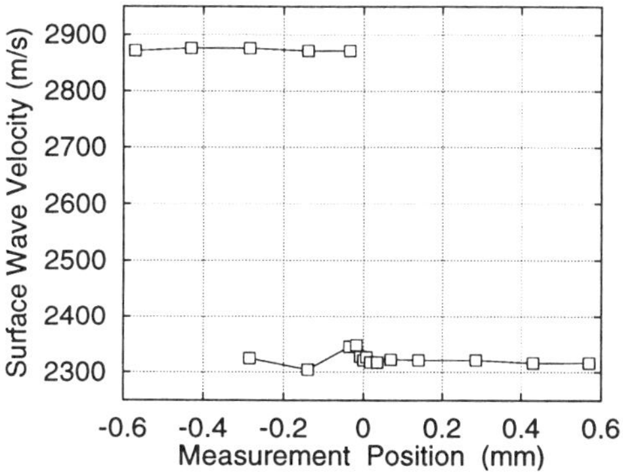

(c) f=140MHz

Fig. 10 Schoch displacement effect on the surface wave velocity change observed in the linear scanning along -45° direction for D2 bi-crystal system.

Owing to this "Schoch displacement" effect, the measured v_r in the right-hand side grain is constant irrespective of the frequency in Figs. 8 to 10: the left-hand side grain in the bi-crystal system has nothing to do with the right-hand side grain in the acoustic structure.

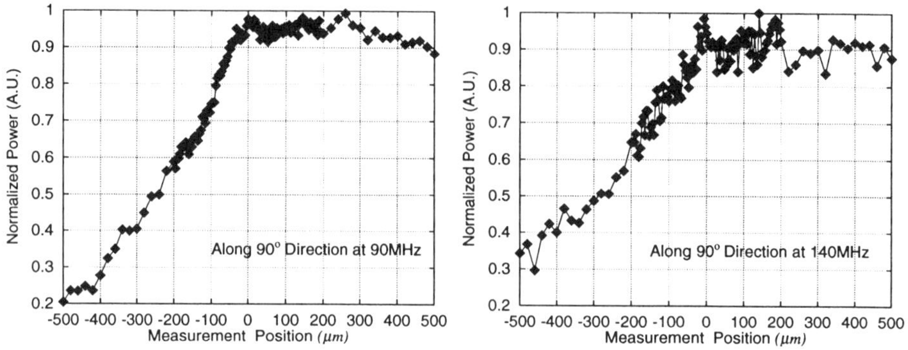

(a) (b)

Fig. 11 Normalized power change in the linear scanning with $\theta = 33.2°$ along 90° direction for D1 bi-crystal system: (a) f=90MHz and (b) f=140MHz.

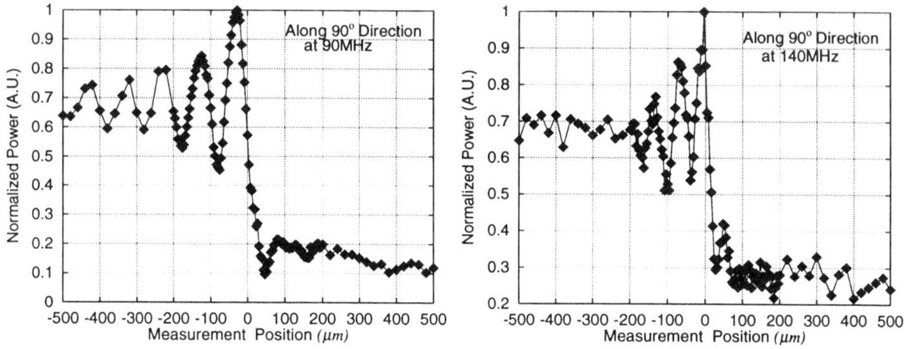

(a) (b)

Fig. 12 Normalized power change in the linear scanning with $\theta = 31.0°$ along -45° direction for D2 bi-crystal system: (a) f=90MHz and (b) f=140MHz.

In order to make close sensitivity analysis of the grain boundary to the surface wave structure, the incident angle θ is fixed by the Rayleigh critical angle to activate the surface wave for either bi-crystal system, and the grain boundary is controlled to be located in the parallel to the Y-axis in the moving stage to make linear scanning measurement only by using the X-stage. Figure 11 shows the normalized power change when $\theta = 33.2°$(Rayleigh critical angle for D1-left) in the linear scanning along the 90° direction for D1 bi-crystal system. No distinct differences were seen by varying the frequency. The kink where the linearly increasing reflected wave power becomes constant is just corresponding to the grain boundary. Then, when detecting this kink point, the location of the grain boundary can be determined with the spatial error less than $10\mu m$. Figure 12 shows the normalized power change when $\theta = 31.0°$ (Rayleigh critical angle for D2-right) in the linear scanning along -45° direction for D2 bi-crystal system. Approaching the grain boundary, the normalized power becomes vibrational and its amplitude enlarges itself. This might be considered an acoustic resonance activated by the incident wave. To be noted here, the point where the steepest gradient is observed in the vibration is closely corresponding to the location of grain boundary. Furthermore, when increasing the frequency, the spatial pitch of vibration narrows and the enhancement of vibration is only activated at the vicinity of the grain boundary. Owing to the above high sensitivity of the acoustic structure to the grain boundary, acoustic spectro microscopy must be available as a new tool for nondestructive grain boundary characterization.

CONCLUSIONS

Acoustic spectro-microscopy has significant utility not only for determining the velocities of RSW and pseudo-RSW, but it also can be used for linear sensing analysis to evaluate the effect of grain boundary on the propagation of surface wave, and to situate the position of the grain boundary. This technology can be used on non-metal materials such as ceramics. The more circumstantial research on the effects of grain boundary and grain size on the propagation of surface wave will be reported elsewhere.

REFERENCES

[1] Z. Horita, D.J. Smith, M. Furukawa, M. Nemoto, R.Z. Valiev and T.G.Langdon, "An Investigation of grain boundaries in submicrometer-grained Al-Mg Solid Solution Alloys Using High-Resolution Electron Microscopy" *J. Mater. Res.*, **11** 1880-1890 (1996).

[2] R.A. Lemons and C.F. Quate, "A Scanning Acoustic Microscope," Proc. IEEE Ultrasonic Symposium, 18-21 (1973).

[3] M.G. Somekh, G.A.D. Briggs and C. Ilett, "The Effect of Elastic Anisotropy on Contrast in the Scanning Acoustic Microscope," *Philosophical Magazine A*, **49** [2] 179-204 (1984).

[4] R.D. Weglein, "Acoustic Micro-Metrology," *IEEE Trans. Sonics Ultrason.*, **SU-32** 225-234 (1985).

[5] A. Briggs, "Acoustic Microscopy," Clarendon Press·Oxford, New York, 1992.

[6] J. Kushibiki and N. Chubachi, "Material Characterization by Line-Focus-Beam Acoustic Microscope," *IEEE Trans. Sonics Ultrason.*, **SU-32** 189-212 (1985).

[7] K.K. Liang, G.S. Kino and B.T. Khuri-Yakub, "Material Characterization by the Inversion of V(z)" *IEEE Trans. Sonics Ultrason.*, **SU-32** 213-224 (1985).

[8] Y. Tsukahara, N. Nakaso and K. Ohira, "Angular Spectral Approach to Reflection of Focused Beams with Oblique Incidence in Spherical-Planar-Pair Lenses," *IEEE. Trans. Ultrason. Ferroelec. Freq. Contr.*, **UFFC-38** 468-480 (1991).

[9] I. Ihara, H. Koguchi, T. Aizawa and J. Kihara, "Determination of Ultrasonic Attenuation in Surface Layer of Materials by Ultrasonic Reflectivity Measurement," *Electron. Lett.*, **29** 1999-2000 (1993).

[10] N. Nakaso, K. Ohira, M. Yanaka and Y. Tsukahara, "Measurement of Acoustic Reflection Coefficients by an Ultrasonic Microspectrometer," *IEEE. Trans. Ultrason. Ferroelec. Freq. Contr.*, **UFFC-41** 494-502 (1994).

[11] I. Ihara, H. Koguchi, T. Aizawa and J. Kihara, "Estimation of Mean Acoustic Reflection Coefficient of Polycrystalline Aggregate," *Appl. Phys. Lett.*, **66** [8] 956-958 (1995).

[12] Y. Xu, T. Aizawa and J. Kihara, "Structure of Surface Wave Structure on Coarse-Grained Nickel Observed by Acoustic Spectro-Microscopy," *Materials Transactions, JIM*, **37** 1690-1698 (1996).

[13] Y. Xu, T. Aizawa and J. Kihara, "Simultaneous Determination of Elastic Constants and Crystallographic Orientation in Coarse-Grained Nickel by Acoustic Spectro-Microscopy," *Materials Transactions, JIM*, to be published.

[14] A. Schoch, "Seitliche Versetzung Eines Total Reflektierten Strahls Bei Ultraschallwellen," *Acustica,* **2** 18-19 (1952).

[15] O.I. Diachok and W.G. Mayer, "Velocity Measurements of Lateral Beam Displacement Upon Reflection," *IEEE Trans.*, **SU-16** 219-221 (1969).

[16] H.L. Bertoni and T. Tamir, "Unified Theory of Rayleigh-Angle Phenomena for Acoustic Beams at Liquid-Solid Interfaces," *Appl. Phys.*, **2** 157-172 (1973).

[17] W.G. Neubauer, "Ultrasonic reflection of a Bounded Beam at Rayleigh and Critical Angles for a Plane Liquid-Solid Interface," *J. Appl..Phys.*, **44** [1] 48-55 (1973).

[18] W.G. Neubauer and L.R. Dragonette, "Measurement of Rayleigh Phase Velocity and Estimates of Shear Speed by Schlieren Visualization," *J. Appl. Phys.*, **45** [2] 618-622 (1974).

[19] T.J. Plona, M. Behravesh and W.G. Mayer, "Ultrasonic Bounded Beam Reflection Effects at a Liquid-Anisotropic-Solid Interface," *J. Acoust. Soc. Am.*, **56** [6] 1773-1775 (1974).

[20] M.A. Breazeale, L. Adler and G.W. Scott, "Interaction of Ultrasonic Waves Incident at the Rayleigh Angle onto a Liquid-Solid Interface," *J. Appl. Phys.*, **48** [2] 530-537 (1977).

PROCESS CONTROL OF CVD FIBER COATINGS

J.G. Jones*, P.D. Jero, and D.J. Kent, Wright Laboratory Materials Directorate, Wright Patterson AFB, OH 45433
*TMC, Inc.

This work seeks to develop an MOCVD system using *in situ* sensors and automated process control to deposit controlled, reproducible oxide fiber coatings (e.g. $LaAl_{11}O_{18}$). A CVD system capable of continuous fiber coating has been assembled in which all of the system parameters are logged by and controllable by computer. The system employs liquid precursor delivery for precise precursor stoichiometry control and inert gas seals for near atmospheric operation. *In situ* measurements are accomplished via a mass spectrometer, laser micrometer, thermocouples, and a Raman spectrometer which are integrated into the CVD system. A fuzzy logic controller has been developed to control the O_2 flow rate based on the mass spectrometer data. The CVD system, process control system, *in situ* sensors, and fuzzy logic controller are described in detail. An initial example of process control is reviewed.

I. INTRODUCTION

Most ceramic matrix composites (CMC's) require a fiber interface coating in order to exhibit the graceful failure or pseudo-ductility that make them potentially useful high temperature structural materials.[1,2] Although interface coatings are widely recognized as a key to desirable properties in (most) CMC's, coating technology has received relatively little attention. As a result, coating technology has lagged composite development, hindering successful application of CMC's. Consequently, in today's search for new oxidation resistant interfaces, the inability to deposit coatings of controlled chemistry, thickness, and morphology is a significant problem.[3]

Chemical vapor deposition (CVD) is one of the processes used to deposit interface coatings on fibers for CMC's. CVD is a well known process used commercially to deposit coatings and bulk materials for a variety of industrial applications. A voluminous CVD literature exists [see, for instance, Ref 4]. Control of CVD processes can be very complex, and significant efforts in modeling, *in situ* sensing, and intelligent process control of various CVD processes has been undertaken [see, for instance, Refs. 5 & 6] Various sensors, models, and control algorithms have been used, some of which may be relevant to fiber coatings.

The application of CVD to fiber coatings for CMC's presents unique challenges (i.e. small, nonplanar, moving substrate; complex coating chemistries). This, combined with the fledgling size of the CMC industry, has resulted in CVD fiber coating technology remaining a mixture of science and art. This work seeks to advance the state of the art in CVD fiber coating such that coating properties (e.g. chemistry, crystallinity, thickness, topography) can be measured in real-time during the coating deposition. The objectives of this (continuing) work are to 1) improve the CVD fiber coating process control/repeatability, thereby improving the uniformity of the coatings, 2) implement *in situ* sensors so as to measure coating or gas phase properties in real-time, and 3) implement closed-loop process control based on the sensor data to produce coatings with the desired chemistry, crystallinity, thickness, and topography. A secondary goal, at this point, is to demonstrate and evaluate the coatings in a CMC.

The composite system chosen for study was a lanthanum hexaluminate ($LaAl_{11}O_{18}$) coating on a sapphire monofilament (in an alumina matrix, eventually). Lanthanum hexaluminate is one member of the β-alumina/magnetoplumbite family of compounds that has been identified as potentially valuable as interface coatings in CMC's.[7-9] A complex (multi-cation) oxide coating was deliberately chosen in order to test the system, sensor, and control algorithm capabilities. The use of a sapphire monofilament represents a simplification from the tow fibers ultimately desired, but it simplifies coating evaluation, obviates (temporarily) issues associated with coating thickness uniformity throughout a tow, and allows for coating heat treatments at temperatures that would degrade polycrystalline oxide fibers. These are, admittedly, significant simplifications.

II. CVD SYSTEM

A hot wall CVD system has been implemented which is capable of coating fiber in a continuous manner, Figure 1. Use of inert gas seals obviates the need for a sealed system with large spool boxes and their requisite complications. The CVD system operates at near atmospheric pressure and uses liquid precursor delivery for precise control of coating stoichiometry. Separate in-line furnaces (not shown in Fig. 1) are used for desizing the fiber prior to deposition and for heat treatment of the coating after deposition. A supply spool and motorized take-up spool (also not shown in Fig. 1) complete the system.

The hot walled reactor consists of a 2.54 cm diameter quartz tube ~30 cm long. The reactor furnace is capable of operation to 1200°C, although depositions are typically accomplished in the 500-700°C range. At the front of the reactor, a 4-way stainless steel cross connects the reactor to the vaporizer, inert gas seal, and the process gas inlet. The cross, seal, and process gases are all heated to inhibit condensation of the vaporized precursor. At the exhaust, a 5-way heated cross is used to connect the reactor to a second inert gas seal, the mass spectrometer capillary, reactor exhaust tubing, and to an access port for *in situ* thermocouples. The 1 m long capillary tube which connects the reactor exhaust gas to the mass spectrometer is also heated. A single zone furnace capable of operation to 1200°C is in-line ahead of the reactor to desize and clean the fiber. A three zone furnace capable of operation to 1500°C is in-line after the reactor. It is used for heat treatment of the fiber coating.

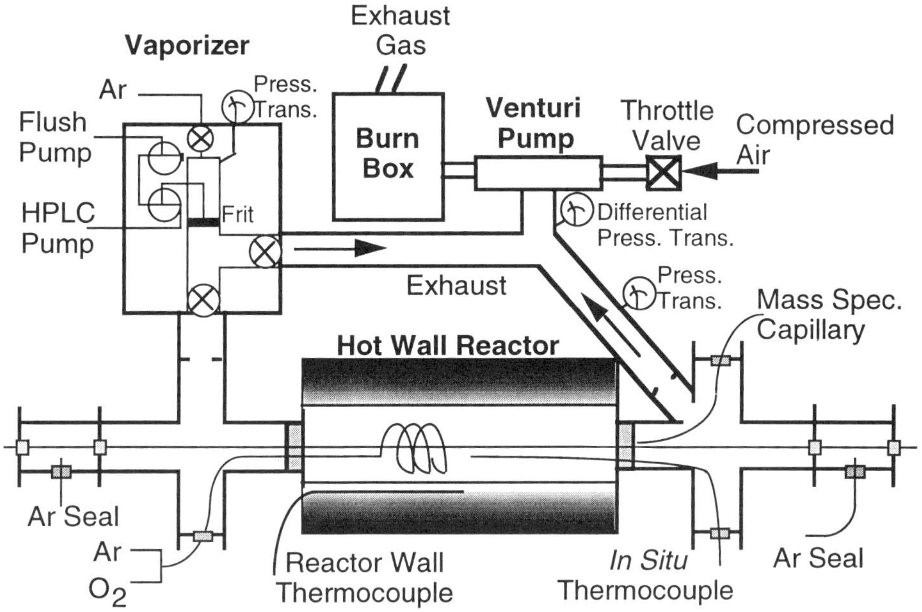

Figure 1. Hot Wall CVD Reactor System.

Gas flows are handled by seven mass flow controllers including five for Ar and two for O_2. Of the Ar mass flow controllers, one maintains a constant inert environment for the precursor HPLC pump, the second controls the carrier gas flow through the vaporizer, and the third controls the Ar process gas flow (used to dilute the O_2 process gas). The remaining two Ar controllers feed the inert gas seals at the system inlet and outlet. The size of the seal orifices and the Ar flow are adjusted such that the seals are slightly pressurized. Ar from the seals is forced both into the reactor and out of the system, ensuring that the ambient atmosphere is excluded from the system. The O_2 mass flow controllers regulate the flow of O_2 to the in-line desize furnace, and the O_2 process gas entering the reactor.

The reactor pressure is maintained slightly (~1 torr, typically) below atmospheric pressure through the use of a venturi pump system. Pressure control is obtained by regulating the amount of compressed air allowed into the pump, which in turn controls the amount of effluent which is pulled through the pump. A differential pressure transducer with a 5 volt range for a 10 torr span is used to precisely measure the difference between atmospheric pressure and the reactor exhaust pressure. An MKS 250B pressure controller regulates the flow of compressed air through the venturi via a throttle valve. Regulation of the exhaust pressure has been demonstrated to within 0.1 torr of the set point.

The effluent flows through a burn box which is heated to >500°C before exiting into the fume hood. This ensures that any residual hydrocarbons are fully oxidized. The effluent is diluted with compressed air in the venturi pump, so there is always excess oxygen available if it is needed.

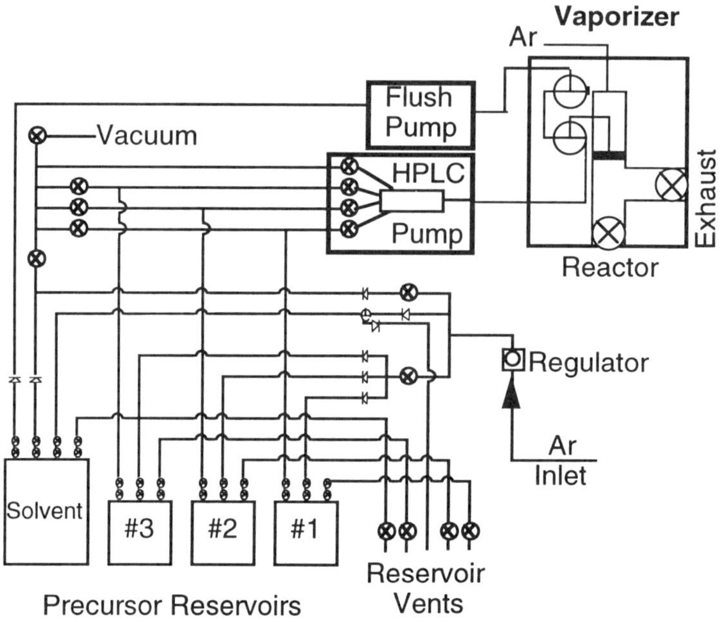

Figure 2. CVD Liquid Delivery System.

The liquid delivery system (LDS) which was implemented, Figure 2, consists of the Sparta 450 LDS components from Advanced Technology Materials Inc. which include an HPLC pump with 4 computer controlled inlet valves (for mixing precursors), a 4 zone, resistively heated vaporization chamber (vaporizer), and a flush pump. A precursor delivery manifold was designed and built which employs 3 precursor reservoirs and a solvent reservoir and allows for four different modes of operation: 1) pumping precursor(s) or solvent through the HPLC pump, 2) pushing solvent through any line to clean it out, 3) pushing argon through any line to remove the solvent, and 4) pulling a vacuum on any line before introducing a new precursor. The computer controlled inlet valves on the HPLC pump allow for mixing the precursors with an accuracy of 0.1%. The manifold is constructed entirely of stainless steel tubing and is sealed, so even air-sensitive precursor can remain in the lines for extended periods of time. The flush pump is used to clean precursor out of the vaporizer and the short capillary tube which feeds precursor to the vaporizer, at the end of each run.

The vaporizer components include a gate valve for carrier gas, two liquid valves, a stainless steel frit, two gate valves for the exit path of the precursor vapor, and a pressure transducer to verify operation. The liquid valves are used to select between the HPLC pump and the flush pump. Using the pressure transducer above the frit and the reactor pressure transducer (which is below the frit), the pressure drop across the frit can be monitored and used to determine the efficiency of the frit. For this work, the vaporizer was heated to 275°C.

The lanthanum hexaluminate precursors being employed are lanthanum methoxyethoxide (La(mee)3) and aluminum *sec*-butoxide (Al(O-s-Bu)3) in excess methoxyethanol (mee). The Al(O-s-Bu)3 is viscous (like honey) as received and is diluted with mee. This results in significant heat generation, presumably due to an exchange reaction which forms Al(mee)3 and liberates *sec*-butanol. This has not yet been verified. The total (La + Al) solution concentration is adjusted to be ~0.25 M, as recommended by the LDS manufacturer. Precursor mixtures with La:Al ratios of 1:4 and 1:18 were made. The HPLC inlet valves can be controlled so as to allow vaporization of mixtures with compositions anywhere between these end points. This approach is more precise than using separate Al and La precursors because of the small amount of La relative to Al present in the desired coating (1 La to 11 Al). The precursors are water sensitive, so they are prepared in a dry box under a nitrogen atmosphere.

III. PROCESS CONTROL SYSTEM

The CVD system was automated using LabVIEW® 4.0 from National Instruments, a Power Macintosh 8100/100 computer system, an N400 expansion chassis from Second Wave Inc., two acquisition boards from National Instruments, and two four port serial interface cards from Applied Engineering. Building on these fundamental components, the entire CVD system has been automated and can be operated entirely from the computer platform. The automated components include the seven mass flow controllers and corresponding gate valves, the venturi pressure system, 13 PID temperature controllers, 4 thermocouples, 3 pressure transducers, all LDS system functions, and the fiber take-up spool velocity. *In situ* sensors were also integrated into the computer platform including a mass spectrometer, laser micrometer, and thermocouples inside the reactor.

The seven mass flow controllers and venturi pressure system are controlled using two MKS-232 interfaces which support four analog channels each. These two interfaces each provide the ability to set and measure the analog channels with 12 bit resolution using a serial port. Hence the mass flow rate for each channel can be set and the actual flow rate of each channel can be monitored.

The 10 reactor system temperature controllers are controlled and monitored using an RS-485 2 wire bus. These controllers are connected in series using only 2 wires and additionally, only one subroutine needed to be written to access the controllers as the only difference between them is their address number, zero through nine. Each controller can be tuned for operation at a typical setpoint either manually or by using the computer software to initiate the built-in tuning algorithm. Three additional PID controllers are used for the three-zone heat treatment furnace and can be accessed in a similar fashion. They are accessible using an RS-485 4 wire bus.

Fiber velocity control is accomplished using a stepper motor and worm gear combination in conjunction with a custom designed stepper motor control circuit which includes an RS-232 interface. The fiber velocity can be specified in either the forward or reverse directions from 0.01 to 30 cm/min.

The flush pump controller requires 4 to 20 mA for zero to 100% flow rate. The flush pump is controlled by the computer using a control voltage from an acquisition card and a custom made voltage to current converter circuit. Using this design, the flush pump will automatically shut off when the computer is turned off.

The HPLC pump is a serial device and was automated by simply recoding the PC source code (supplied by ATMI) into LabVIEW. The software provides for complete functionality including flow rate commands, mixing ratio control, pressurization limits, and feedback on pump status. The pump status indicates how long the pump has been in operation since the last command and the current precursor line pressure. Software was also written to allow for semi-automatic purging of gas from the precursor inlet lines.

The 25 pneumatic valves were automated using relays configured in four banks of 8. Each relay was wired in a Boolean OR configuration with an existing panel switch and thus can be activated either remotely or manually. In addition, the state of 10 of the valves, 5 in the vaporizer and 5 in the precursor delivery system, can be monitored using digital inputs configured in two banks of 8. This feedback was instrumental in developing the software and ensuring that the correct valves really were being selected.

IV. *IN SITU* SENSORS

In situ measurements of the gas stream are made using a mass spectrometer at the exhaust of the reactor and thermocouples inside the reactor tube. *In situ* measurements of the fiber are accomplished using a Raman spectrometer and a laser micrometer. Using the information recorded from the gas stream and fiber provides for the opportunity of process discovery. By discovering discrepancies between what is expected based on the theoretical models and what actually occurs, improvements in the fidelity of the models used for CVD can be made. Efforts to date have relied primarily on data from the mass spectrometer and thermocouples.

The laser scanner employed is a Model 200 dual axis scanner and Model 2021 SPC processor from LaserMike Inc. The scanner is in-line, at the reactor exit. An averaged diameter measurement can be collected approximately each second. Due to the diameter variations of the monofilament fiber, the value of the coating thickness data collected to date is extremely limited. However, if a second laser scanner were installed in front of the reactor, the outputs could be correlated to yield a reliable coating thickness measurement. The Raman spectrometer is an EIC VIS500 Echelle Spectrograph, from EIC Laboratories, which is coupled by fiber optics to a remote microprobe. The microprobe resides in a stainless steel box which is in-line and can be placed at the reactor exit or after the heat treatment furnace. The work on the Raman spectroscopy of the coating is still in its infancy and will not be described further in this paper.

The exhaust gas from the CVD reactor is analyzed using a Balzers QMS 420 mass spectrometer. It utilizes a non-discriminating gas inlet system which includes a 1m resistively heated 0.150 mm ID capillary tube which connects the reactor exhaust to a flange on the mass spectrometer vacuum column. This configuration results in a two-stage reduction of the pressure from atmospheric to 10^{-4} torr with the partial pressure composition of the gas mixture remaining unchanged. The actual analysis is performed at high vacuum , typically 10^{-6} torr, using a quadrupole mass filter. Spectrum analyses can be recorded at intervals as small as ~5 seconds.

Three type K thermocouples are located at selected locations within the reactor tube. Another type K thermocouple resides outside the tube and is used for control.

V. FUZZY LOGIC *IN SITU* CONTROL

Data from a number of runs showed that the temperature in the reactor as well as the effluent gas composition varied substantially over time due to the ebb and flow of exothermic reactions (combustion) occurring in the reactor [the reactor atmosphere is essentially a mixture of alcohol vapors and O_2 in excess Ar]. This despite the fact that the reactor wall temperature (the process variable) remained essentially at the setpoint. In an attempt to stabilize the process [and as an initial attempt at process control,] a scheme was developed in which the O_2 input flow rate was linked to the relative amounts of O_2 and another species in the effluent gas stream, as measured by the mass spectrometer. It should be noted that even if this approach successfully stabilizes the system, optimization of coating properties may dictate a different oxygen level. Nonetheless, the approach serves as a useful control demonstration tool. A controller, Figure 3, which was written in LabVIEW® and based on fuzzy logic[10], was developed which adjusts the O_2 flow rate to cause the intensity of the (effluent) O_2 peak, 32 AMU, from the mass spectrometer to track (equal) that of water, 18 AMU. The controller error is the difference in the intensities of O_2 and H_2O. The fuzzy controller uses two inputs, error and error rate, to generate a change in output flow rate of O_2 command. Based on the fuzzy variables shown in Table I, triangular membership functions were used for both the fuzzification of the error and error rate, Figure 4. Defuzzification was accomplished through the use of a 3 by 5 fuzzy associative memory (FAM) rule base defining the output change of O_2 in terms of five different fuzzy variables, as indicated in Table II.[11]

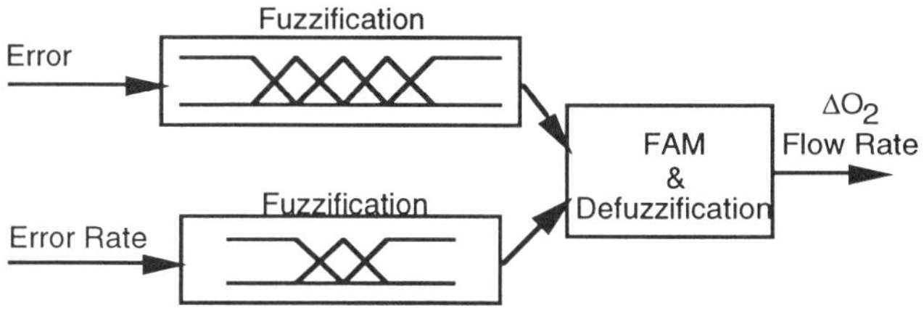

Figure 3. Fuzzy Controller Block Diagram.

Table I. Fuzzy Variables For Fuzzy Associative Memory.

Error		Error Rate		Change In O_2 Flow Rate	
PL	Positive Large	P	Positive	PF	Positive Full
PS	Positive Small	Z	Zero	PH	Positive Half
Z	Zero	N	Negative	Z	Zero
NS	Negative Small			NH	Negative Half
NL	Negative Large			NF	Negative Full

The degree of membership of the input fuzzy sets which correspond to each FAM rule are combined in an AND fashion to produce the degree of clipping of each specified output fuzzy set. In fuzzy logic an AND corresponds to taking the minimum value of the degree of membership of each corresponding fuzzy membership function. The output membership functions were chosen to be triangular, Figure 5.

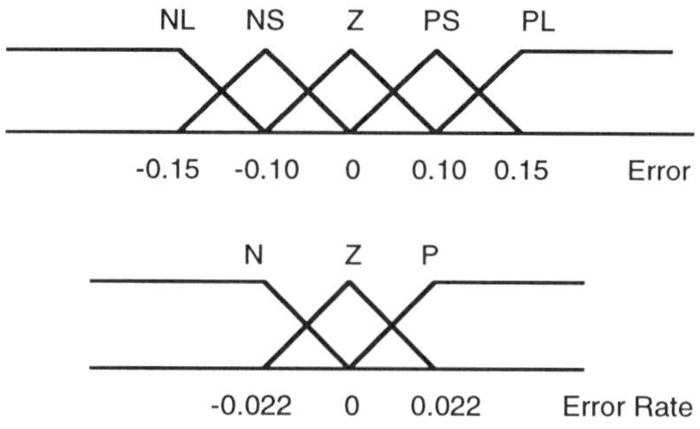

Figure 4. Fuzzy Input Membership Functions.

Table II. Fuzzy Associative Memory Bank.

Error Rate	P	PF	PH	NH	NH	NF
	Z	PF	PH	Z	NH	NF
	N	PF	PH	PH	NH	NF
		NL	NS	Z	PS	PL
		Error				

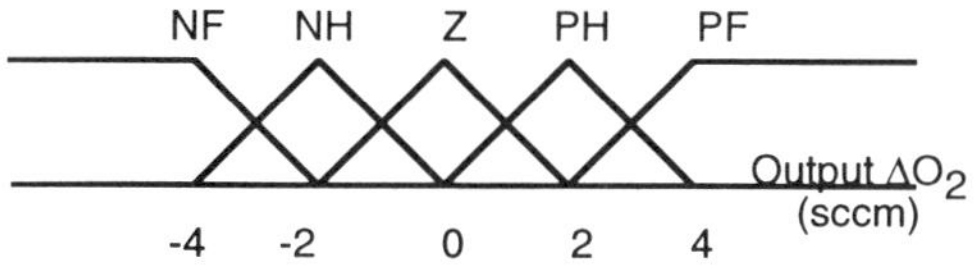

Figure 5. Fuzzy Output Membership

The correlation-minimum inference method with centroid defuzzification was used to generate the output command, Figure 6.[11] Using this method the resulting output value is the center of gravity of the clipped output fuzzy sets. Letting X_i and Y_i correspond to the degree of membership of the FAM input fuzzy sets, the antecedents, and Z_i correspond to the output fuzzy sets, the consequents, the output is calculated as:

$$O = \frac{\sum_{i=1}^{NRules} Z_i \min(X_i, Y_i)}{\sum_{i=1}^{NRules} \min(X_i, Y_i)} \qquad (1)$$

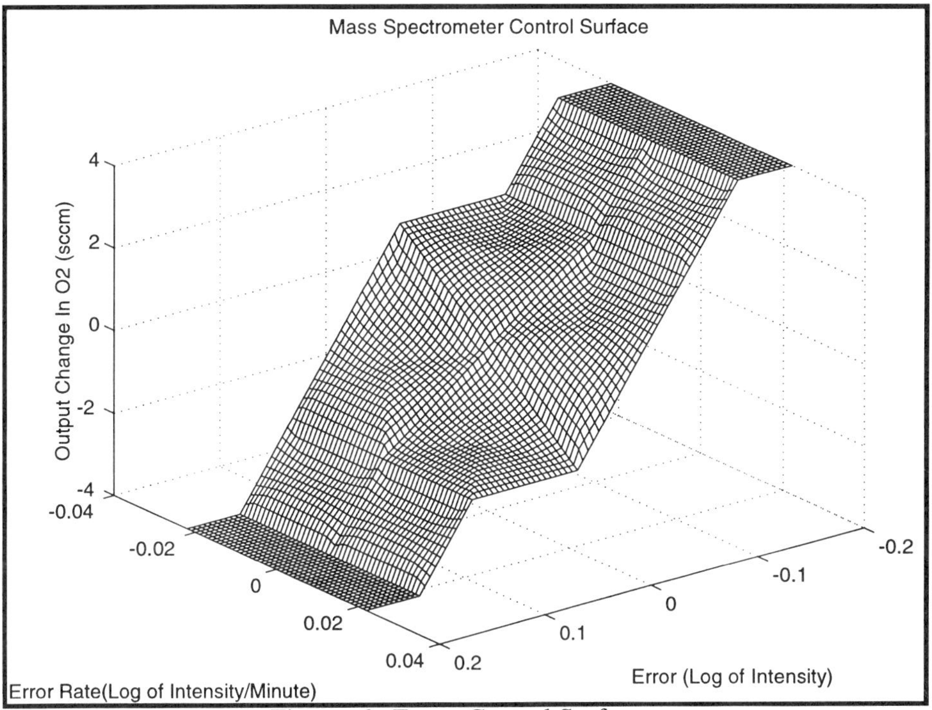

Figure 6. Fuzzy Control Surface.

VI. PROCESS CONTROL EXAMPLE

The fuzzy controller was exercised to cause the O_2 peak intensity from the mass spectrometer to track the H_2O peak intensity, by automatically adjusting the O_2 flow into the reactor. In this particular operational regime (e.g. reactor temperature, precursor & gas flow settings, etc.) the system is unstable in the "uncontrolled" state; the reactor temperature and effluent gas composition fluctuate significantly. Figures 7 and 8 show the mass spectra and temperature data, respectively, collected over the course of a deposition using the fuzzy control algorithm. The control algorithm successfully correlated the O_2 and H_2O peaks, as desired, however the system remained unstable. While this particular correlation (O_2 to H_2O) proved useless for stabilizing the operation of the system, subsequent work has demonstrated stabilization of the system using the same fuzzy control algorithm with different correlation parameters.[12] Patent issues prevent further discussion at this time.

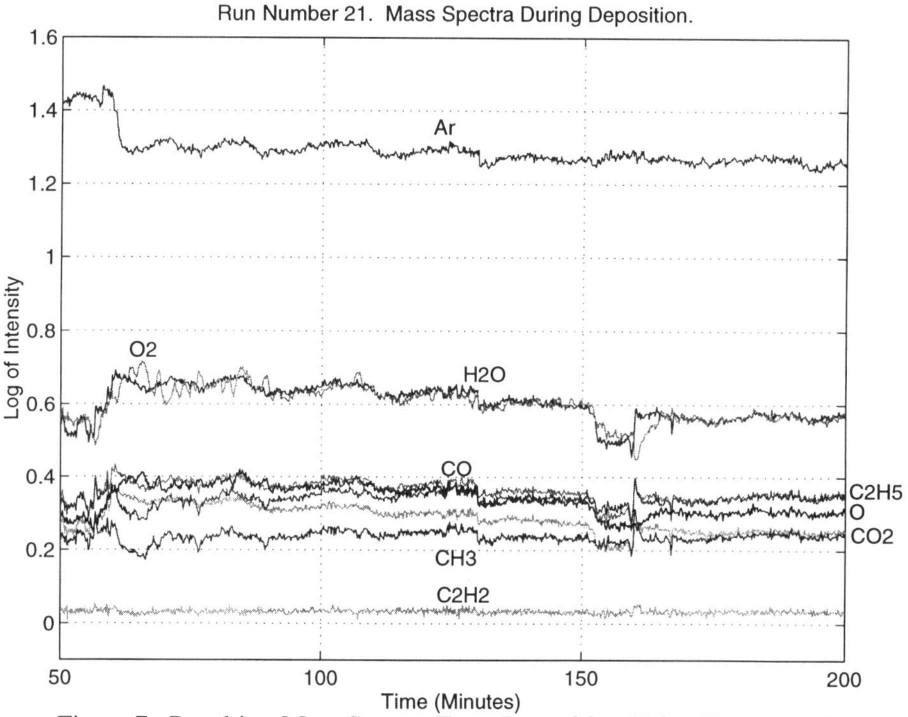

Figure 7. Resulting Mass Spectra From Deposition Using Fuzzy Logic.

There are two other points to be made with regard to this example. First, the furnace temperature measured (and controlled) just outside the reactor wall was maintained at the 500°C setpoint to within 5°C while the actual reactor temperature fluctuated in a temperature range significantly higher. Second, while the intensities of the dominant effluent species drift somewhat over the ~150 minutes of deposition, they are not obviously indicative of the significant reactor temperature [hence deposition rate, morphology, and composition] variations which are occurring. Without using an *in situ* thermocouple for monitoring, the reactor temperature variability would go virtually unnoticed.

VII. SUMMARY

An MOCVD system capable of continuous fiber coating has been assembled in which all of the system parameters are logged by and controlled by computer. The system is designed for coating of continuous fibers for CMC's. The system employs liquid precursor delivery for precise precursor stoichiometry control, inert gas seals for near atmospheric operation, and a mass spectrometer, laser micrometer, thermocouples, and a Raman spectrometer for *in situ* measurements. Process control experiments employing a fuzzy logic algorithm have been initiated.

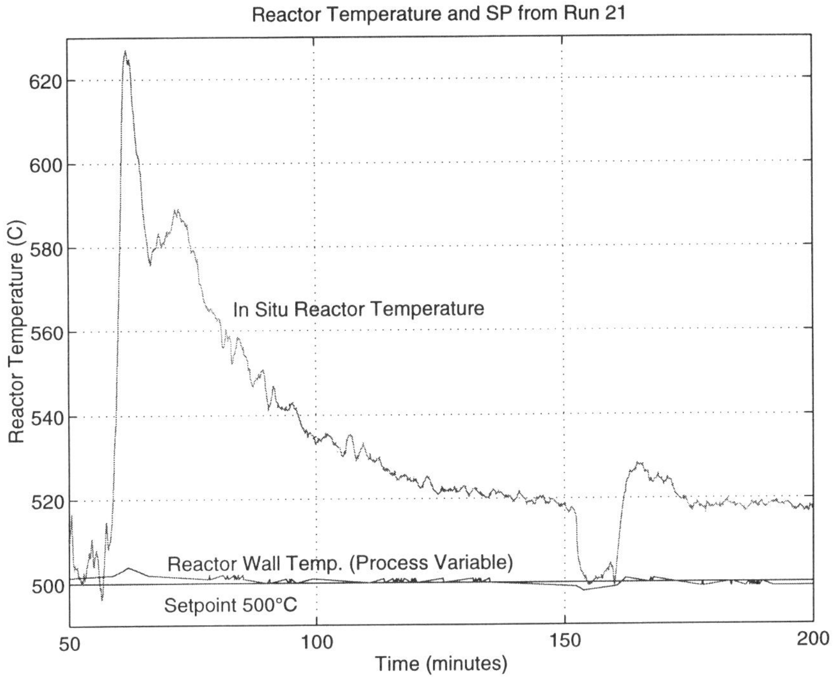

Figure 8. Reactor Temperature Resulting From Deposition With Fuzzy Controller.

VIII. REFERENCES

1. D.B. Marshall and A.G. Evans, "Failure Mechanisms in Ceramic Fiber/Ceramic Matrix Composites," J. Am. Cer. Soc., **68**, [5], 225-231 (1985).
2. R.J. Kerans, R.S. Hay, N.J. Pagano, and T.A. Parthasarathy, "The Role of the Fiber-Matrix Interface in Ceramic Composites," Bull. A. Cer. Soc., **68**, [2], 429-442 (1989).
3. R. J. Kerans, private communication.
4. Proceedings of the 7th International Conference on Chemical Vapor Deposition, ed. T.O. Sedgewick and H. Lydtin, The Electrochemical Society, Princeton, NJ (1979).
5. T.M.Besmann, R.H.Lowden, D.P. Stinton, J. C. McLaughlin, B. W. Sheldon, T. L. Starr, and A. W. Smith, "Processing Science for Chemical Vapor Deposition," Report WL-TR-94-4044, Wright Laboratory, Air Force Materiel Command, WPAFB, OH (1994).
6. J.R. Strife and R. J. Kee, "Manufacturing Science of Silicon Nitride Chemical Vapor Deposition," Report WL-TR-96-4091, Wright Laboratory, Air Force Materiel Command, WPAFB, OH (1996).
7. P.E.D. Morgan and D.B. Marshall, "Functional Interfaces for Oxide/Oxide

Composites," Mat. Sci. Eng., **A162**, 15-25 (1993).

8. M.K. Cinibulk, "Magnetoplumbite Compounds as a Fiber Coating in Oxide/Oxide Composites," Cer. Eng. Sci. Proc., **15**, [5], 721-730 (1994).
9. S. Sambasivan, J.A. Moris, and W.T. Petusky, "Rb β-Alumina as an Interface Coating in Oxide CMC's," Cer. Eng. Sci. Proc., **17**, [4], 250-257 (1996).
10. G.J. Klir, B. Yuan, Fuzzy Sets And Fuzzy Logic Theory and Applications, Prentice-Hall, NJ, 1995.
11. B. Kosko, "FUZZY ASSOCIATIVE MEMORIES"; pp. 314-322 in Neural Networks And Fuzzy Systems A Dynamical Systems Approach To Machine Intelligence, Prentice-Hall, NJ, 1992.
12. J.G. Jones, "Intelligent Process Control Of Fiber Chemical Vapor Deposition," PhD dissertation, University of Cincinnati, 1997.

PIEZOSPECTROSCOPIC CHARACTERIZATION OF THERMAL BARRIER COATINGS

Michael J. Lance[a,b], J. Allen Haynes[a,c], W. Roger Cannon[b], Mattison K. Ferber[a]

[a] Oak Ridge National Laboratory, P. O. Box 2008, Oak Ridge, TN 37831

[b] Department of Ceramics, Rutgers University, Piscataway, NJ 08855

[c] Department of Materials and Mechanical Engineering,
University of Alabama at Birmingham, Birmingham, AL 35294

ABSTRACT

A technique to measure stress non-destructively with a high spatial resolution using incoherent light scattering and luminescence is presented. The technique employs a Raman microprobe to measure small peak shifts in vibrational and electronic spectra which correlate to applied or residual stresses in the ceramic. The predicted frequency shift for a non-degenerate peak of polycrystalline alumina is estimated from single crystal phonon deformation potentials (PDPs) and secular equations by averaging over all crystal orientations. The predicted shift compares well to the experimentally measured shift. Thermal barrier coatings are analyzed to demonstrate the usefulness of the technique for solving unique mechanics problems in ceramics.

INTRODUCTION

The ability to measure stress with a high spatial resolution can be a valuable tool in research on advanced ceramics. The mechanical response of ceramic materials are significantly affected by local stresses around and within individual grains, whiskers and fibers. Characterization of these local stresses is not possible with techniques currently available. Neutron diffraction, x-ray methods and strain-gages can measure stresses only with macroscopic dimensions.

Raman spectroscopy has been successfully used to measure stress in ceramic materials such as diamond,[1] silicon nitride,[2] zirconia[3,4] and alumina.[5] The technique relies on the stress-induced peak shift of vibrational energies owing to the anharmonicity of atomic bonding. The Raman effect is initiated by irradiating a sample with a monochromatic light source (usually a laser). The laser can be focused using a microscope to a spot size of 1 μm which allows for the measurement of local stresses with a high spatial resolution. Raman microscopy has also been used to detect phase changes in zirconia.[4,6]

A corollary technique to piezo-Raman is piezo-luminescence. Here, the characteristic luminescence peaks of a dopant ion within a ceramic crystal are measured. Strain distorts the unit cell and the positions of the electron orbitals relative to the dopant ion which causes the luminescence peaks to shift with stress. As with piezo-Raman, the spatial resolution is very high (1 μm). This technique has been successful for measuring stress in alumina where the ruby luminescence peaks (R-lines) are used.[7,8] Piezo-spectroscopy has been used to solve a number of unique mechanics problems: stresses around indents,[8] crack-tip stress fields,[9] oxide scales[10] and fiber composites.[11]

Recently, thermal barrier coatings (TBCs) have been applied to air-cooled superalloy components in gas turbine engines in order to decrease their operating temperatures, thus improving engine durability and efficiency. A typical plasma-sprayed TBC system consists of a porous, thermally-insulating ceramic top coating (usually zirconia) applied over an oxidation-resistant metallic bond coating (usually an MCrAlY).[12] During engine operation a thin oxide scale (predominately Al_2O_3) forms along the rough metal/ceramic interface due to the reaction of oxygen with Al in the bond coat. The dense, continuous alumina scale provides a diffusion barrier which protects the underlying metal from rapid oxidation attack. However, during thermal cycling, stresses are induced in the YSZ due to (1) interfacial scale growth, (2) the rough bond coat interface, and (3) metal/ceramic coefficient of thermal expansion (CTE) mismatches. These stresses initiate localized fracture in the YSZ top coating near the rough bond coat interface. Eventual coalescence of these near-interface cracks leads to failure of the TBC by spallation of the top coating. The specific effects of interfacial scale growth on the stress state of the TBC, as well as the general evolution of YSZ stress within the near-interface region, are poorly understood. Piezo-spectroscopy has been used to analyze the development of stress in the oxide scale for TBCs formed by physical vapor deposition (PVD).[13] In this paper, piezo-spectroscopy is used to measure stress and phase content within the zirconia thermal barrier and the interfacial oxide scale for TBCs formed by plasma spraying.

THEORY

All crystals have unique normal vibrations that are dependent upon the masses of the constituent atoms in the lattice and the strength of the bonding between the atoms. The number of normal vibrations and their degeneracy is limited by the symmetry of the crystal. Furthermore, only normal crystal vibrations that result in a change in the polarizability of the unit cell will be Raman active. For example, group theory predicts 27 normal vibrations for sapphire but only 12 of these are Raman active (five doubly degenerate and two non-degenerate peaks). For a harmonic (Hookean) bond, the force constant does not vary with strain. In reality, all bonds are anharmonic and the force constant will vary with strain. Since the frequency at which a bond vibrates is dependent upon its force constant, Raman peaks will shift with stress.

The frequency shift can be determined by solving the secular equation:

$$|\Delta K - \lambda I| = 0 \tag{1}$$

where I is the unit matrix, ΔK is the change in the spring constant with strain and λ is:

$$\lambda = \omega_f^2 - \omega_i^2 \approx 2\omega_i \Delta\omega \tag{2}$$

with $\Delta\omega$ as the stress-induced peak shift. For a non-degenerate Raman peak of sapphire, the crystal symmetry limits the eigenvalue (λ) to:

$$\lambda = K_{11}(\varepsilon_{11} + \varepsilon_{22}) + K_{33}\varepsilon_{33} \tag{3}$$

where K_{11} and K_{33} are material constants called the phonon deformation potentials (PDPs) and ε_{ii} are strains in the crystal frame of reference. By substituting stresses for strains using the compliance matrix of sapphire and integrating over all crystal directions, the effect of a generalized macro-stress on the peak positions of polycrystalline alumina is given as:

$$\lambda_{poly} = C * \sigma'_h \tag{4}$$

with the constant C equal to:

$$C = 2K_{11}(S_{11} + S_{12} + S_{13}) + K_{33}(2S_{13} + S_{33}) \tag{5}$$

Therefore, the peak position is affected only by the mean hydrostatic stress (σ'_h) and not the shear stresses. Also, the stress tensor can not be resolved since the trace of the stress tensor is collected as one term. Substituting known values of the compliance matrix and the PDPs for the 417 cm^{-1} peak of alumina predicts the constant in Eq (5) to be 1.12 cm^{-1}/GPa for equi-biaxial stress. This value compares well with our experimental result of 1.2 cm^{-1}/GPa. A similar result is obtained for Raman peaks of any poly-crystal with only the constant in Eq. (5) changing.

The theory for piezo-luminescence has already been derived[14] and the result is similar to piezo-Raman. Both techniques measure the mean hydrostatic stress in polycrystalline ceramics. Luminescence is common only in a few ceramics whereas Raman peaks can be obtained from most ceramics. However, piezo-luminescence from Cr^{3+} in alumina is very sensitive to stress and the peaks are much more intense than the Raman technique.

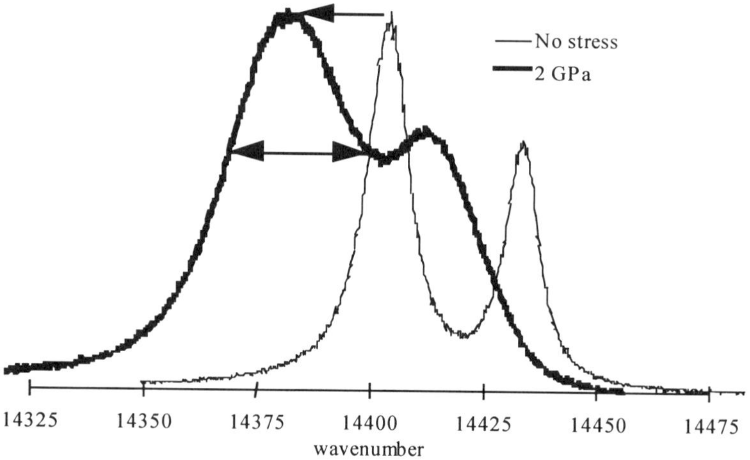

Fig. 1 The peak shift of R-lines of alumina under 2 GPa. The broadening represents a stress gradient across the oxide scale.

EXPERIMENTAL PROCEDURE

The spectra were obtained using an ISA U1000 Raman microprobe with an attached CCD (charged coupled device) detector. The CCD has a position sensitive chip that can resolve spectral features without rotating the gratings in the monochromator. The 514.532 nm line of a 5-watt argon ion laser was used to excite the spectra. The microprobe incorporates an optical microscope with a TV camera and screen which allow the user to visually probe on the surface. The spatial resolution was approximately 1 micron. A neon lamp was used to simultaneously calibrate all peak positions of the spectra and a curvefit program on the software package LabCalc was used to measure peak positions, widths and heights.

The coatings consisted of an air plasma sprayed (APS) 8 w/o Y_2O_3 partially stabilized ZrO_2 thermal barrier deposited on a vacuum plasma sprayed (VPS) Ni-22Cr-10Al-1Y bond coat. Deposition conditions have been reported elsewhere.[12] The substrate was Rene N5, a single crystal Ni-based superalloy. One specimen was exposed to 270 one hour cycles at 1150°C in order to induce damage within the YSZ top coating. Under these test conditions, the average lifetime of this TBC system was 360 cycles.[12]

CHARACTERIZATION OF STRESS WITHIN THERMAL BARRIER COATINGS

The stress within the aluminum oxide interfacial scale was determined by measuring the shift and broadening of the two ruby fluorescence peaks of alumina. Fig. 1 shows the peak shift of the luminescence lines under 2 GPa of compressive stress. In addition to shifting position, the peaks also broaden which indicates a stress gradient across the scale.

Piezo-Raman was used to measure stress in the zirconia thermal barrier. Fig. 2 shows the stress shift of a tetragonal zirconia peak under 2 GPa of compressive stress. Comparing to Fig. 1, one can see that the peak shift is much less for Raman peaks given the same stress. A neon lamp was used as an absolute standard to calibrate all peak positions.

Fig. 3 shows a typical cross-section of a plasma sprayed TBC after thermal cycling. Plasma sprayed TBCs require a rough bond coating surface for mechanical adhesion of the APS YSZ top coating. This often results in a torturous metal/ceramic interface with large semi-spherical asperities. The interfacial Al_2O_3 scale is visible in Fig. 3 as the dark-contrast layer between the lamellar APS ceramic top coating and the dense VPS NiCrAlY bond coat. After 270 cycles, cracking and layering of the interfacial scale was observed in localized areas (most often on convex surfaces). In adjacent regions the Al_2O_3 scale often remained intact with no visible cracking or decohesion. The thicker regions of scale on the tops of the semi-spherical asperities in Fig. 3 consist of heavily cracked Al_2O_3.

The amount of cracking in the porous YSZ top coating varied from location to location along the interface. A stress-induced crack in the YSZ can be observed in Fig. 3. This prominent crack lies near the plane of the thicker scales on the tips of the bond coat asperities, and runs parallel to the bond coat surface. Some regions of the interface were more heavily cracked, while other regions displayed no visible damage in the plane of polish.

Since the CTE of the metallic bond coat is larger than that of the oxide scale, on planar surfaces oxide scales are typically subject to in-plane compressive stresses after cooling to room temperature. However, on convex surfaces of relatively small radius (such as the asperities in Fig. 3), significant radial tensile

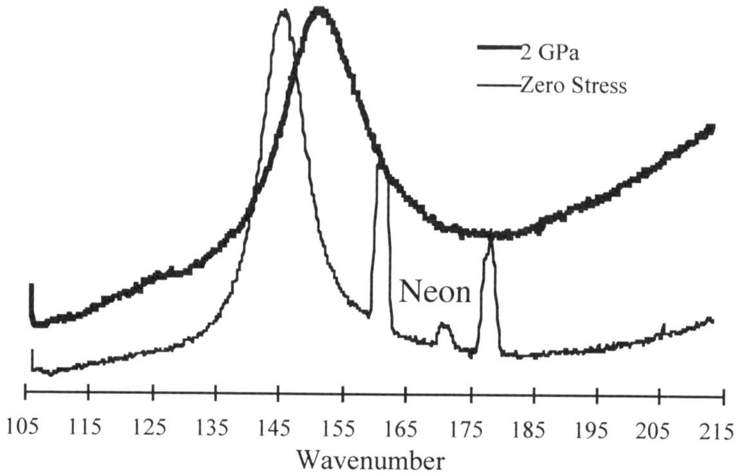

Fig. 2 Stress-induced peak shift for tetragonal zirconia under 2 GPa of stress. Neon lines are used to calibrate every scan.

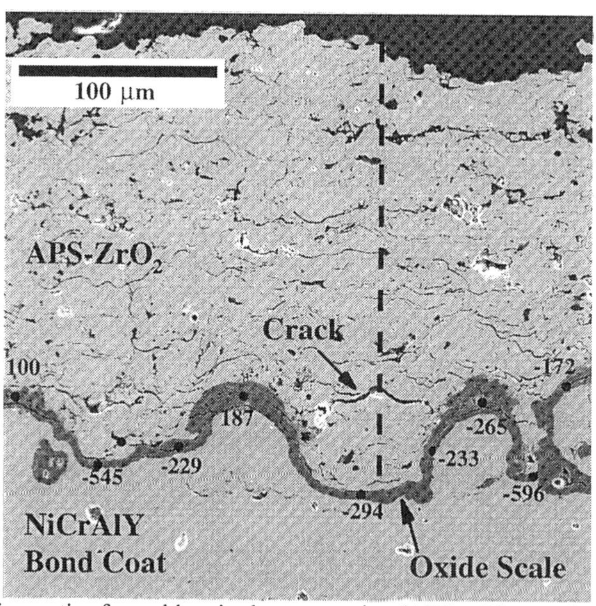

Fig. 3 Thermal barrier coating formed by air-plasma spraying (APS). The coating consists of a NiCrAlY bond coat and a stabilized zirconia thermal barrier. A dark oxide scale can be seen between the zirconia and the bond coat which forms at high temperatures. This sample had undergone 270 cycles in air at 1150°C. Stresses in the scale are labeled.

stresses can be induced in the thin Al_2O_3 film. These radial tensile stresses can induce scale fracture or delamination, with cracks typically propagating parallel to the metal surface. Cracking or delamination typically relieves much of the stress within the scale. Stress measurements of the Al_2O_3 made with piezo-luminescence at specific locations along the interfacial oxide scale are labeled on Fig. 3. It must be emphasized that the measured values are hydrostatic stresses and thus are not representative of in-plane stresses on the rough surfaces. The magnitude of hydrostatic compressive stress in the Al_2O_3 was greatest in the flat or concave regions in which no scale cracking was observed. As expected, on the tips of the large asperities the magnitude of the compressive stresses was reduced, and in some of these convex areas tensile stresses were measured (relative to the Al_2O_3 standard). In all of the regions which displayed reduced compressive stresses (or tension) scale cracking and/or delamination was observed (with most cracks propagating parallel to the metal surface), indicating the capacity of this technique to measure stress relief in Al_2O_3 scales due to cracking.

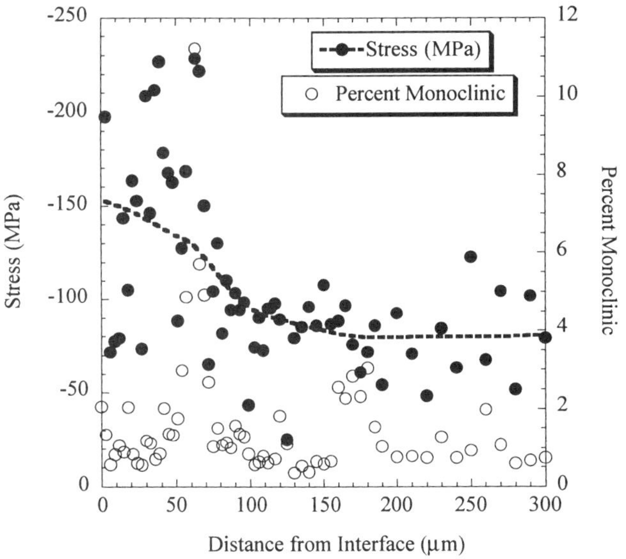

Fig. 4 Stress and phase distribution in the zirconia thermal barrier pictured in Fig. 3. Near the interface the stress in the zirconia is more compressive.

Fig. 4 shows the stress distribution and phase content measured in the zirconia along the dashed line in Fig. 3. The YSZ stresses are compressive due to the CTE mismatch of the zirconia and bond coat. The magnitude of the compressive stress in the APS YSZ is much greater near the bond coat interface.

Above the crack the stress drops significantly and remains constant at about 100 MPa. The ceramic that is adjacent to the rough metal surface maintains a greater residual stress state since residual stresses due to CTE mismatch have not been relieved by localized cracking/delamination. Also, an increase of monoclinic zirconia is observed around the crack, as shown in Fig. 4. This suggests a transformation occurs either during fracture or once the material in the plane of the crack is no longer constrained.

These coatings eventually fail by YSZ fracture along the plane of the crack in Fig. 3, leaving a thin layer of YSZ covering the bond coat surface.[12] These results suggest that during thermal cycling localized stresses induced by bond coat roughness and out-of-plane volume expansion due to interfacial scale growth contribute to development of tensile stresses in the YSZ normal to the surface. These tensile stresses cause the ceramic top coating to fracture, and spallation eventually occurs due to progressive crack growth and link-up. Piezo-spectroscopy shows that the observed cracking causes stress to be relieved in both the zirconia and the alumina scale.

CONCLUSION

The theory of piezo-spectroscopy has been developed. The mean hydrostatic stress can be measured within polycrystalline ceramics with the technique. The technique has successfully been used to characterize the failure mechanisms of plasma sprayed thermal barrier coatings.

ACKNOWLEDGMENTS

This research is sponsored by the Assistant Secretary for Energy Efficiency and Renewable Energy, Office of Transportation Technologies, as part of the High Temperature Materials Laboratory Fellowship Program, Oak Ridge National Laboratory, managed by Lockheed Martin Energy Research Corp. for the U. S. Department of Energy under contract number DE-AC05-96OR22464.

REFERENCES

[1]J. A. Baglio, B. C. Farnsworth, S. Hankin, G. Hamill, and D. O'Neil, "Studies of Stress Related Issues in Microwave CVD Diamond on <100> Silicon Substrates," *Thin Solid Films*, **212**, 180-185 (1992).

[2]T. D. Fang, "Lattice Dynamics of Stress-Induced Shifts in the Raman Microprobe Spectra of Beta Silicon Nitride," M. S. Thesis in Electro-Optics, University of Dayton, December 1993.

[3]V. Sergo, D. R. Clarke, and W. Pompe, "Deformation Bands in Ceria-Stabilized Tetragonal Zirconia/Alumina: I, Measurement of Internal Stresses," *J. Am. Ceram. Soc.*, **78** [3] 633-40 (1995).

[4]J. F. Tsai, J. D. Belnap, and D. K. Shetty, "Crack Shielding in Ce-

TZP/Al2O3 Composites: Comparison of Fatigue and Sustained Load Crack Growth Specimens," *J. Am. Ceram. Soc.*, **77** [1] 105-17 (1994).

[5]S. H. Shin, F. H. Pollak, and P. M. Raccah, "Effects of Uniaxial Stress on the Raman Frequencies of Ti2O3 and Al2O3," *Proceedings of the Third International Conference on Light Scattering in Solids*, edited by M. Balkanski, R. C. C. Leite, and S. P. S. Porto, Campinas, Brazil, 1975.

[6]H. Toraya, M. Yoshimura, and S. Somiya, "Calibration Curve for Quantitative Analysis of the Monoclinic-Tetragonal ZrO2 System by X-ray Diffraction," *J. Am. Ceram. Soc.*, **67** [6] C119-C121 (1984).

[7]Q. Ma and D. R. Clarke, "Stress Measurement in Single-Crystal and Polycrystalline Ceramics Using Their Optical Fluorescence," *J. Am. Ceram. Soc.*, **76** [6] 1433-40 (1993).

[8]S. E. Molis and D. R. Clarke, "Measurement of Stresses Using Fluorescence in an Optical Microprobe: Stresses around Indentations in a Chromium-Doped Sapphire," *J. Am. Ceram. Soc.*, **73** [11] 3189-94 (1990).

[9]A. G. Haerle, W. R. Cannon, and M. Denda, "Direct Measurement of Crack-Tip Stresses," *J. Am. Ceram. Soc.*, **74** [11] 2897-901 (1991).

[10]J. Bernie, C. Craggs, D. J. Gardiner and P. R. Graves, "*Ex Situ* and *In Situ* Determination of Stress Distributions in Chromium Oxide Films By Raman Microscopy," *Corr. Sci.*, **33** [1] 1-12 (1992).

[11]Y. Huang and R. J. Young, "Analysis of the Fragmentation Test for Carbon-Fibre/Epoxy Model Composites by Means of Raman Spectroscopy," *Composites Sci. Tech.*, **52** 505-517 (1994).

[12]J. A. Haynes, E. D. Rigney, M. K. Ferber, and W. D. Porter, "Oxidation and degradation of a plasma-sprayed thermal barrier coating system", *Surface and Coatings Technology*, **86-87** 102-108 (1996).

[13]R. J. Christensen, D. M. Lipkin, D. R. Clarke and K. Murphy, "Nondestructive evaluation of the oxidation stresses through thermal barrier coatings using Cr^{3+} piezospectroscopy," *Appl. Phys. Lett.*, **69** [24] 3754-56 (1996).

[14]J. He and D. R. Clarke, "Determination of the Piezospectroscopic Coefficients for Chromium-Doped Sapphire," *J. Am. Ceram. Soc.*, **78** [5] 1347-53 (1995).

Advanced Characterization Techniques

ULTRASONIC VISCOSITY MEASUREMENT OF MOLTEN LIQUIDS

R. Daniel Costley, W.M. Ingham, Jason A. Simpson
Diagnostic Instrumentation and Analysis Laboratory
Mississippi State University
P.O. Drawer MM, Mississippi State, MS 39762-5932
Phone: (601)325-2105 FAX: (601)325-8465
costley@dial.msstate.edu

ABSTRACT

A torsional waveguide and its use as a sensor to measure viscosity of liquids is described. Measurements were made in the 7 to 500 poise range with two different calibration liquids. The viscosity of the liquids was varied by varying the temperature from 20°C to 60°C. The viscosity measurements of the two liquids correlated very well with each other. The measurements were not compensated for temperature effects (other than the change in viscosity). The results were very encouraging. Experiments were also performed to see if the sensor could be used to measure liquid level. A few potential problems were identified with this use of the sensor, but several solutions to these problems were suggested. A similar waveguide was used to make measurements in molten glass, at temperatures up to 1060°C. These results were also very encouraging. Modifications to the design of the sensor were proposed for future high temperature experiments.

INTRODUCTION

Viscosity of the molten glass is a key variable in determining the quality of the final glass product. At low viscosities the melt can be highly corrosive. At high viscosities the melter can become plugged. "Melt viscosity is the most important processing property; it controls processing rate, product homogeneity, and heat transfer within the molten glass."[1] Thus, the viscosity is an important parameter which can be used by the vitrification industry for the processing of waste material and by the glass industry for production of high quality glass products. The major problem in measuring the viscosity of the molten waste product is the extremely hot and corrosive environment.

Although the primary motivation for this work is the development of a high temperature viscometer for use in the vitrification of hazardous and radioactive wastes, it has application for monitoring other molten liquids as well. However, at this time there is no on-line method for measuring the viscosity of the molten liquids. The work described here addresses this shortcoming by developing a robust technique to measure viscosity of glass as it is being processed. The temperature of the molten material can be over 1500°C, depending on the type of glass. The goal of any viscometer would vary for any particular applications. For the vitrification of hazardous and radioactive wastes into waste glass it would be desirable to measure the molten glass in the range of 20 to 100 poise within +/- 10 poise. The temperatures will reach up to 1300°C. For commercial glass applications these criteria could be more demanding, i.e., higher temperatures, larger viscosity range, and higher resolution.

This paper describes a torsional waveguide and demonstrates its use to measure the viscosity of liquids, in the 1 to 500 poise range. Its use as a level meter is also discussed. It is also shown that a waveguide can be used to measure viscosity of liquids at very high temperatures. Necessary modifications and improvements are presented and discussed.

BACKGROUND

A versatile system has been developed in which either torsional or extensional waves can be excited and detected in thin rods, or wires, made of magnetostrictive materials.[2] Magnetostrictive materials deform under the influence of a magnetic field. The type of deformation depends on the polarization of the magnetic domains within the material and the alignment of the external magnetic field. When a current is passed through a magnetostrictive rod the domains are aligned circumferentially within the rod. A twist in the rod is produced if the rod is then subjected to an external, axial magnetic field. If the external field is transient, then the twist propagates as a torsional wave down the length of the rod. This effect is referred to as the Wiedemann effect.[3] The axial magnetic field is produced by passing current through a coil that is wrapped around the wire. The detection of torsional waves is accomplished via the inverse-Wiedemann effect, in which the torsional wave produces a current in the wire.

The magnetostrictive rod can be attached, by either solder or glue, to another rod made of a different material, such as stainless steel. It is this second material which usually comes into contact with the liquid. Waveguides of noncircular cross-section are used to determine liquid density.[2,4] In two phase fluids, such as particulate suspensions or liquid-vapor mixtures, they have also been used to measure equivalent density. Waveguides of circular cross-section are used to determine viscosity.[5] Both types of waveguides can be used to determine temperature and liquid level.

Since the work described here deals with measuring viscosity, the discussion will be confined to torsional waveguides of circular cross-sections. Of course, many different modes of torsional waves are possible, depending on the frequency content of the torsional wave and the diameter of the rod. The frequency content of the torsional wave pulses used in this work is under 100 kHz and the diameter of the waveguides is less than 2.4 mm (0.1 inch). Only the lowest mode of propagation of torsional waves is allowed under these conditions. Thus, the discussion that follows is restricted to the lowest torsional mode.

Mathematical solutions are available that describe torsional wave propagation under various conditions. These conditions range from the case in which the sides of the circular rod are stress free[6] to that in which the rod is immersed in a viscous liquid.[5] For the latter case, the numerical solutions show that both the phase velocity and attenuation of torsional waves depend on the ratio of the density to the viscosity. The motion of the surface of a rod due to a torsional wave is parallel to the surface and in a circumferential direction. When the surface is in contact with a liquid, the motion of the solid will induce motion within the liquid. If the liquid is viscous, shear stresses will be produced at the surface which oppose the motion in the solid rod. These stresses, or viscous drag, will cause the torsional wave to attenuate. The phase velocity of the wave will decrease as well.

Dimensional analysis has been used to obtain similar, albeit qualitative results. Although a quantitative relationship was not obtained as in the numerical solutions mentioned earlier, insights were gained and the effects of temperature were considered. It was shown that the decay in amplitude could be represented as a function of dimensionless numbers according to the following expression:

$$\frac{A_2}{A_1} = f\left(\frac{\lambda}{L}, \frac{a}{L}, \frac{c_r}{Lf}, \frac{\alpha T}{L}, \beta L, \frac{\eta}{L^2 f \rho}\right) \qquad (1)$$

where A_1 and A_2 are the amplitude of the torsional wave at two different points, a and L are the radius and the length of the rod, c_r is the speed of the waves in the rod and the liquid, T is the temperature, α is the coefficient of thermal expansion, β is the attenuation in nepers or dB per meter, λ is the wavelength, f is the frequency, and η and ρ are the viscosity and density of the liquid, respectively. The wave speed in the rod, c_r, refers to the lowest mode of the torsional wave, which is nondispersive.[6]

Of course, this set of dimensionless numbers is not unique. However, they do confirm the dependence of the amplitude decay on the ratio of η to ρ. They also serve to aid in the design of a device that can measure the viscosity, or more correctly, the ratio of viscosity to density. This is also referred to as the kinematic viscosity.

These numbers were derived using the Buckingham Pi Theorem which, as mentioned, does not determine the functional relationship between the parameters.[7] However, it is believed that some of these parameters will have only little influence on the output. For instance, the wavelength should not have much influence if it is long enough so that only the fundamental mode is excited. Once the sensor is designed, its radius and length will be fixed so that they will not affect the outcome of the experiment. However, it is important to consider all three of these parameters in designing a waveguide. The coefficient of thermal expansion will only play a role in experiments involving changes in temperature. The attenuation of torsional waves in the rod, β, should be negligible at room temperature. However, it is a function of temperature as well and will also have to be accounted for in any calibration at high temperature.

It has been shown that the speed of the torsional wave in the waveguide, in addition to the attenuation, depends on the viscosity of the adjacent liquid.[5] However, the temperature of the waveguide affects the wave speed as well, in several ways. First, the speed of torsional waves in the material is a function of temperature. Secondly, the rod will elongate as the temperature increases so that the torsional pulse will have a larger distance to travel. At this point in the development of the sensor it was decided that it would be easier to relate the viscosity of the surrounding liquid to the attenuation of the torsional wave rather than try to unravel the travel time information.

EXPERIMENTAL CONFIGURATION

Calibration Liquid Experiments

A waveguide was composed of three rods of different diameters silver soldered together. The top section was a remendur wire 910 mm (36 inches) long. The middle section was a stainless steel rod 2.4 mm (0.093 inches) in diameter and 258 mm (10 inches) in length. The bottom section was another stainless steel rod of 1.6 mm (0.062 inch) diameter and 150 mm (5.9 inches) in length. The coil which generated and detected the torsional waves was fitted over the top end of the remendur rod, as shown in Figure 1.

The viscous liquids used were N15000 and N4000 calibration liquids. The viscosities of the liquids are tabulated at several discrete temperatures: 20 °C, 25 °C, 40 °C, and 60 °C and range in viscosity from approximately 7 to 500 poise. An exponential curve was fit through these points to estimate the viscosity of the liquids at intermediate temperatures. These curves are shown in Figure 2. The

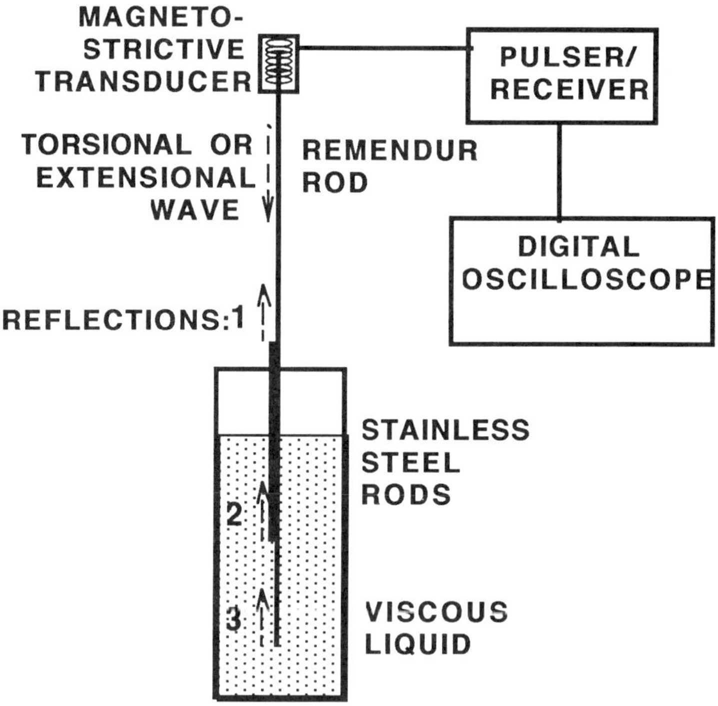

Figure 1. Experimental arrangement for measuring viscosity of liquids with torsional waveguide.

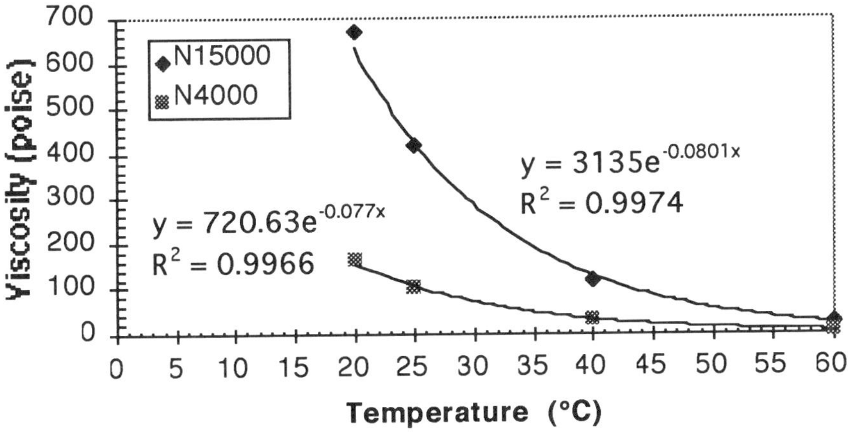

Figure 2. Viscosity as a function of temperature for the two calibration liquids used. The curves are exponential curve fits. The points represent the calibration data supplied with the liquids.

viscous liquid was poured into a standard 100 ml graduated cylinder and the cylinder was placed into a temperature controlled water bath. The bottom end of the waveguide (150 mm stainless steel section) was inserted entirely into the viscous liquid so that the level was somewhere along the middle section of the waveguide (258 mm stainless steel section).

A torsional wave was generated in the remendur by exciting the coil with a Pulser/Receiver unit. The torsional wave traveled down the waveguide and was partially reflected at each of the soldered joints. The remaining energy was totally reflected at the end of the waveguide. One such waveform is shown in Figure 3. The pulses labeled 1, 2, and 3 in the figure are the reflections from joint between the top and middle section, between the middle and bottom section, and from the end, respectively.

The ratio of the amplitudes of the third to the second pulse should depend only on the viscosity of the liquid, and not on the level of the liquid. This is because both the end of the waveguide and the 2nd soldered joint are submerged in the liquid. The ratio between the second to the first amplitude should depend on the level of the liquid as well as its viscosity. However, since the viscosity is determined by the ratio of the 3rd to the 2nd amplitudes the level can be determined as well.

Molten Glass Experiment

A preliminary experiment was performed in molten glass. This experiment was similar to the one described above with a couple important differences. The waveguide was configured differently. The remendur wire was attached to a 1.83 m (6 ft.) length of Inconel, 1.6 mm (0.062 inch) in diameter. A bead of silver solder was placed 610 mm (24 inches) and at 200 mm (8 inches) from the bottom.

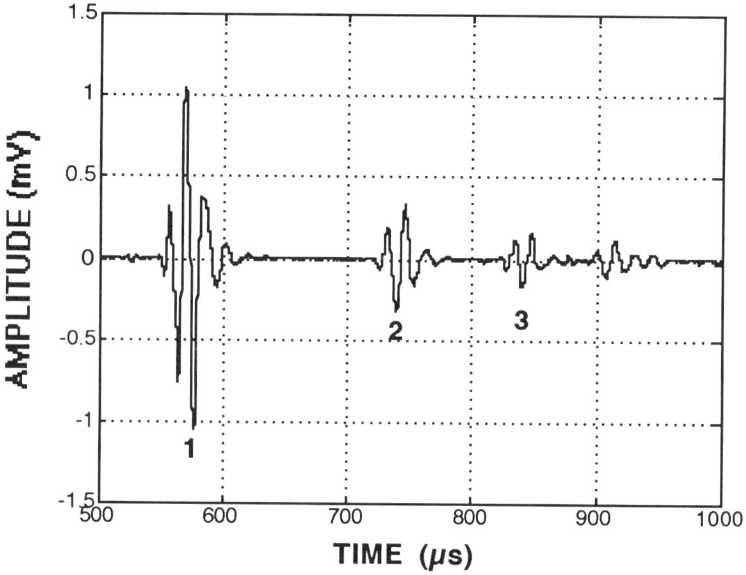

Figure 3. Signal from torsional waveguide inserted in N15000 calibration liquid. The temperature, viscosity and depth were 60°C, 27 poise and 20.5 cm.

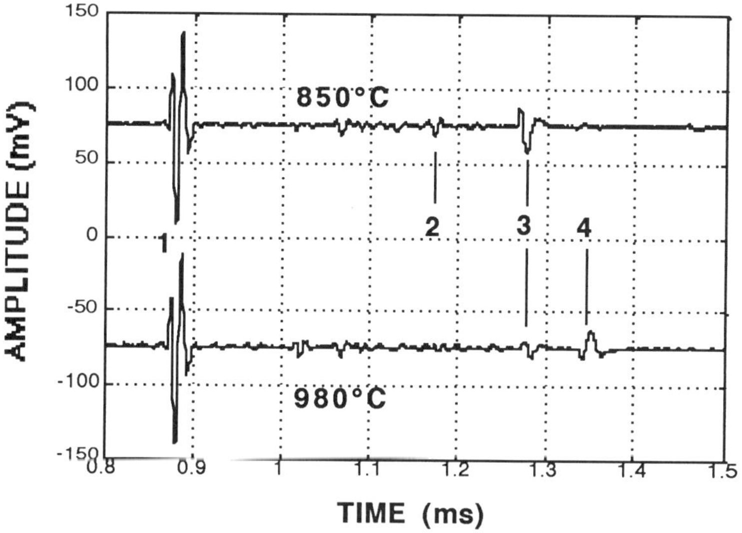

Figure 4. Signals from Inconel waveguide inserted in glass melt. Temperatures indicated are oven temperatures.

A lead based glass was melted in a top loading oven in a 10 cm (4 inch) high crucible. The softening point of the glass was 565°C. The depth of the glass within the crucible was approximately 75 mm (3 inches). The temperature was varied from 800°C to 1060°C. The bottom solder bead (at 200 mm) was within the oven, while the upper one was not. The rod remained cool to the touch within a few inches of the oven without any external cooling.

Waveforms recorded at two different temperatures are shown in Figure 4. The large pulse just before 0.9 ms in both the top and bottom plots, and labeled 1 in the figure, is the reflection of the torsional wave from the top solder bead. The amplitude and arrival time of the reflection remained essentially constant during the entire experiment. This is because this part of the rod was sufficiently far from the oven that its temperature did not change significantly. The reflection at approximately 1.15 ms in the top plot (labeled 2) was from the bottom solder bead. This reflection decreased in amplitude as the temperature increased. It can be seen that this reflection is no longer observable in the bottom plot. The reflection just before 1.3 ms (labeled 3) was from the air-glass interface, as seen in the top plot in the figure. This reflection also diminished in amplitude as the temperature increased. Below 920°C no reflection was observed from the bottom of the waveguide. However, at temperatures above 920°C this reflection was observed. This reflection, labeled 4 in the bottom plot of Figure 4, is approximately at 1.35 ms.

RESULTS

Calibration Liquid Experiments

As mentioned earlier, the various amplitude ratios were correlated to the viscosity of two different viscosity calibration liquids, N4000 and N15000. The ratio of the 3rd to the 2nd amplitude is plotted in Figure 5 against viscosity. Viscosity is used as the abscissa in this graph, instead of the ratio of viscosity to density, since the densities of the two liquids are so similar. The viscosity was varied by varying the temperature, in 5°C increments, from room temperature to 60°C. Several different experiments were performed to show that the results were repeatable. Also, the level of the liquid was varied, as noted in the legend, to show that this ratio was indeed independent of liquid level. The results from the two different liquids agree as well. There is some discrepancy between the two liquids at approximately 120 poise. It was determined that this is not due to the difference in density of the liquids. It could be due to the difference in temperature. The N4000 liquid is at room temperature at this viscosity, while the N15000 liquid is at 40°C. The difference could be due to the elongation or differences in attenuation in the stainless steel waveguide.

The ratio of the 2nd amplitude to the 1st amplitude is plotted against viscosity in Figure 6. Again, as noted in the legend, the experiments were performed in the 2 different liquids and at different liquid levels. The depth referred to in the legend is from the from the bottom of the waveguide to the liquid surface. As can be seen in the figure, the ratios calculated from data taken at similar depths roughly fall along the same curves. For a given viscosity, the ratio corresponding to a shallow liquid level is greater than the ratio from a deeper level. Thus, this could be a technique to determine liquid level in addition to viscosity. This is, of course, assuming that the viscosity is uniform along the entire length of the waveguide. It

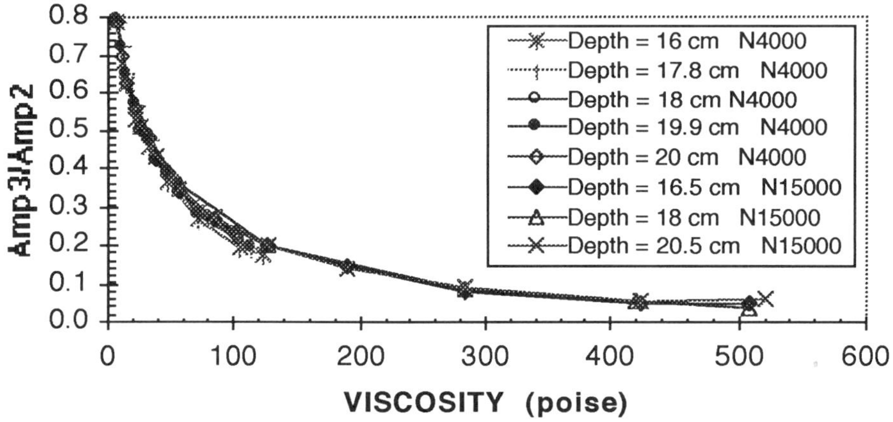

Figure 5. Ratio of the amplitudes of the 3rd to the 2nd pulses plotted against the viscosity of the calibration liquids.

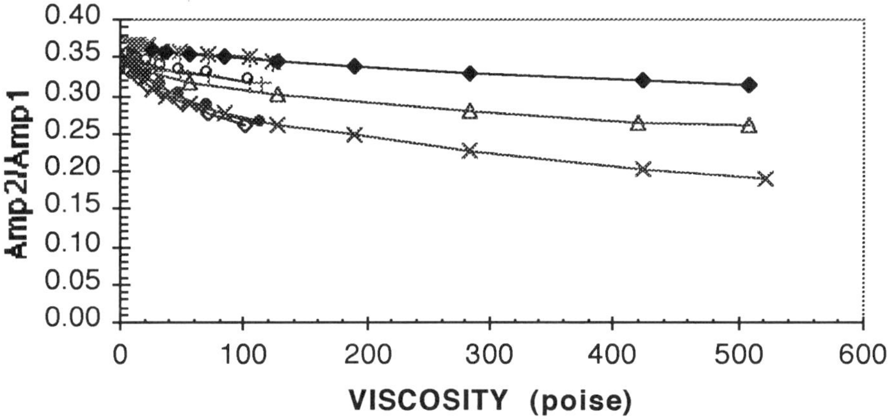

Figure 6. Ratio of the amplitudes of the 2nd to the 1st pulses plotted against the viscosity of the calibration liquids. The legend in Figure 5 applies to this figure.

is interesting to note that the curves converge as the viscosity decreases. This is to be expected because there is no attenuation of the torsional wave when the viscosity is zero, i.e., the amplitude of the reflection would be the same as if the rod were immersed in air.

There was concern that as the liquid level drops that the liquid would adhere to the sides of the waveguide and adversely affect the amplitude reading. This was investigated in two simple experiments with the two different liquids. The waveguide was inserted into the liquid at a given depth and a reading was taken. It was then inserted to a deeper depth (4 cm difference) and another reading was taken. The waveguide was then pulled out to the original depth and readings were taken periodically over the next hour or two. The ratio of the 2nd to the 1st amplitudes over this period are plotted in Figures 7 and 8. Unfortunately, the ratio never fully recovered to its original value. However, as expected, it did rebound faster for the less viscous liquid. Also, the dip within the first ten minutes after the waveguide was not expected. It could be because there is a smooth, gradual taper between the rod and the fluid during this time in contrast to the abrupt change in impedance when the surface is flat. This would be analogous to an acoustic horn or tapered buffer rod.

Molten Glass Experiment

For the high temperature measurements, the amplitude ratio of the reflections from the top solder bead and the bottom of the rod were plotted against temperature in Figure 9. We know the viscosity decreases as the temperature increases, however viscosity data is not available for this glass. As seen in Figure 9, the amplitude ratio increases as the temperature increases up to 1000°C. This is due to the decrease in the viscosity. At temperatures over 1000°C the amplitude ratio decreases. This is believed to be because the attenuation of the torsional wave in the Inconel rod dominates at these temperatures. The melting point of Inconel is approximately 1300°C so that this explanation is plausible.

CONCLUSION

The design of a device for measuring viscosity of highly viscous liquids was presented and discussed. As a viscometer this device appears to be very robust and simple to implement. The data from this device is processed in a straightforward and uncomplicated manner. The use of this device as a level detector was also discussed. However, it was shown that determining the level of the liquid after the level had dropped would be difficult due to the viscous liquid adhering to the sides of the rod. Coating the waveguide with a nonwetting material might improve its performance as a level detector. However, the sensor would still be able to accurately measure a rising level as it is now configured. It could be used to detect a dropping level, which would be a useful way to warn of leaks. The actual level possibly could be determined from the curves shown in Figures 7 and 8, once the viscosity had been determined from the other amplitude ratio. However, this would make the data processing a little more complicated.

Data was taken with a similar device in molten glass at high temperatures with encouraging results. However, several design changes will have to be implemented. In particular, another high temperature material needs to used instead of Inconel. One suggestion is molybdenum, which has a melting point of over 2600°C. The attenuation of torsional waves in this material below 1500°C is not expected to be as great as it is for Inconel. A different technique for sectioning the rod to produce reflections is needed as well. The softening of the silver solder at high temperatures affected the reflection from the solder bead within the oven. A ferrule made of molybdenum could be welded to the rod instead of using the solder bead. If the rod is made of molybdenum, it could be plated with nickel or platinum to reduce corrosion and/or oxidation of the rod.

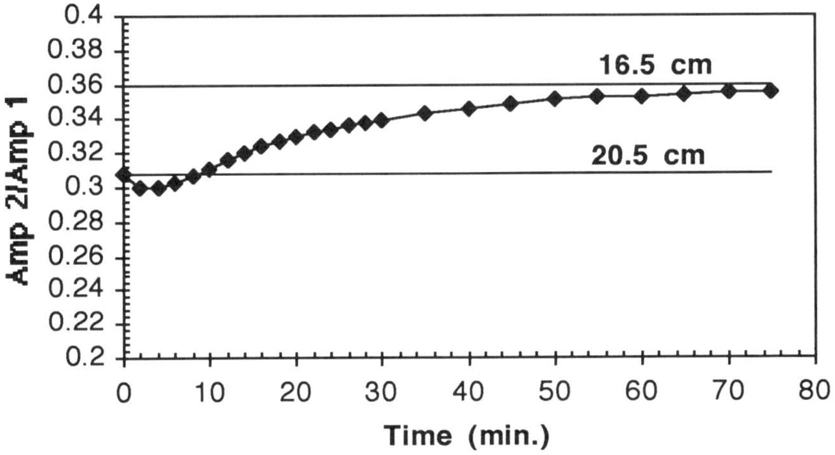

Figure 7. Amplitude ratio as a function of time after waveguide was lifted 4 cm out of N4000 liquid. The temperature of the liquid was 40°C and the viscosity was 27 poise. The horizontal lines are the initial readings taken at the depths indicated.

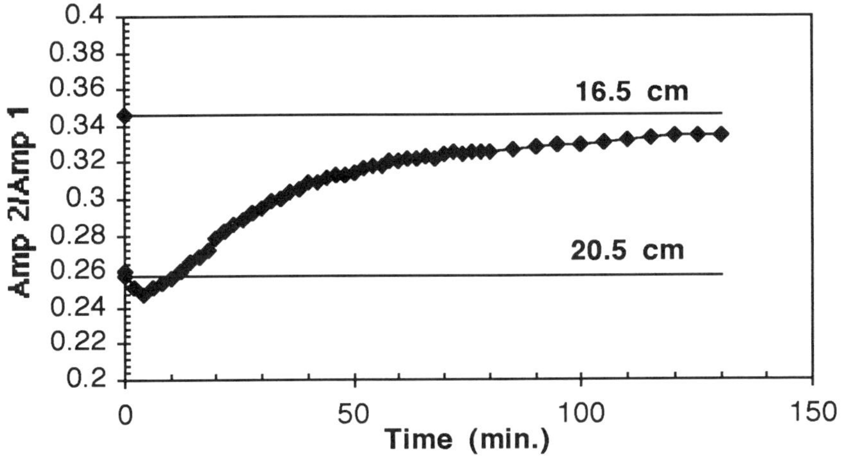

Figure 8. Amplitude ratio as a function of time after waveguide was lifted 4 cm out of N15000 liquid. The temperature of the liquid was 40°C and the viscosity was 115 poise. The horizontal lines are the initial readings taken at the depths indicated.

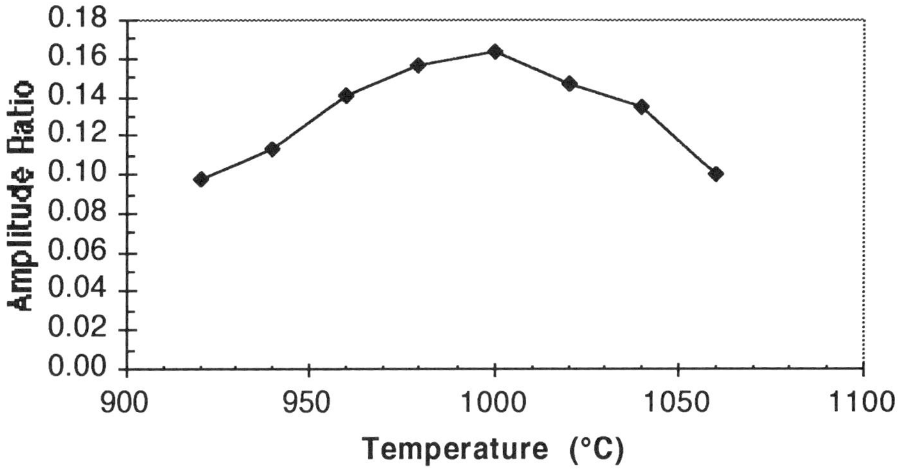

Figure 9. The ratio of the amplitudes of the reflections from the top solder bead and the end of the rod as a function of temperature for a lead based glass.

REFERENCES

[1]U.S. Environmental Protection Agency, *HANDBOOK: Vitrification Technologies for Treatment of Hazardous and Radioactive Waste*, Office of Research and Development, EPA/625/R-92/002, 1992.

[2]L.C. Lynnworth, *ULTRASONIC MEASUREMENTS FOR PROCESS CONTROL: Theory, Techniques, Applications*, pp. 216-221 and 432-445, Academic Press, Boston, 1989.

[3]N.S.Tzannes, "Joule and Wiedemann effects - The simultaneous generation of longitudinal and torsional stress pulses in magnetostrictive materials," *IEEE Transactions on Sonics and Ultrasonics*, **13**(2) 33-41, 1966.

[4]J.O. Kim, *et. al.*, "Torsional sensor applications in two-phase fluids," *IEEE Transactions on Ultrasonics, Ferroelectrics, and Frequency Control.* **40**(5) 563-576, 1993.

[5]J.O. Kim, W. Yuzhou and H.H. Bau, "The effect of an adjacent viscous fluid on the transmission of torsional stress waves in a submerged waveguide," *Journal of the Acoustical Society of America*, **89**(3) 1414-1422, 1991.

[6]K.F. Graff, *Wave Motion in Elastic Solids*. pp. 464-470, Dover, New York, 1975.

[7]R.W. Fox and A.T. McDonald, *Introduction to Fluid Mechanics*. 2nd ed. pp. 305-315, John Wiley & Sons, New York, 1978.

THICKNESS-INDEPENDENT ULTRASONIC IMAGING APPLIED TO CERAMIC MATERIALS

Don J. Roth

Dorothy V. Carney

NASA Lewis Research Center
21000 Brookpark Rd.
Cleveland, Ohio 44135
ph: 216-433-6017
fax: 216-433-8300
email: don.j.roth@lerc.nasa.gov

ABSTRACT

A single transducer ultrasonic imaging method that eliminates the effect of thickness variation in the image, and its application to plate and hollow tubular ceramic materials, are described. The method thus isolates ultrasonic variations due to material microstructure. Its use can result in significant cost savings, because the ultrasonic image, and consequently assessment of material and manufacturing quality, can be interpreted correctly without the need for precision thickness machining during nondestructive evaluation stages of material development. The thickness-independent ultrasonic imaging method is shown to provide more accurate and quantitative materials characterization as compared to that for conventional ultrasonic c-scan and apparent velocity images. The method has been commercialized via a cooperative agreement between NASA Lewis Research Center and Sonix, Inc.

INTRODUCTION

Ultrasonic c-scan imaging is an effective nondestructive evaluation (NDE) technique used for materials analysis and quality control in the aerospace, electronics, and other industries [1]. C-scan imaging in its most conventional implementation is used to map variations in ultrasonic echo peak amplitude that occur when scanning across a material part [1]. In the *pulse-echo* c-scan configuration, echoes that are sometimes monitored or "gated" include those reflecting off the sample front and back surfaces [1]. For currently available commercial scan systems, the amplitude variations are generally scaled to values between zero and 255 [8-bit dynamic range] and displayed on video in terms of gray or color scale. In this manner, gray scale variations in the image are associated with amplitude, or attenuation, variations. An additional implementation of c-scan imaging involves mapping variations in the time of occurrence of an ultrasonic echo peak with respect to a reference time. These "time-of-flight" variations are scaled and displayed in a similar fashion to echo peak amplitude. Both amplitude and time-of-flight variations of a *back surface* echo can many times be related to variations in a volumetric microstructural property such as density which can affect physical material properties such as stiffness and thermal conductivity. However, a weakness of conventional ultrasonic c-scan imaging regarding both peak amplitude and time-of-flight modes is that gray scale variations in images for *back surface* reflections indicate part thickness variations as well as microstructural variations unless the part is uniformly thick.

BACKGROUND

By their very nature, both peak amplitude and time-of-flight measurements of back wall echoes are directly dependent on thickness. Back wall peak amplitude (A) is affected by thickness (d) according to the expression for pulse-echo attenuation where [1]

$$A = A_0 \exp(-\alpha[2d]) \qquad (1)$$

where A_0 is an initial reference amplitude, α is the material attenuation coefficient, and d is the thickness of the part. In practical application, the severity of the effect of thickness variation on peak amplitude depends on the frequency of ultrasound used, because the ultrasonic attenuation coefficient normally increases with increasing frequency. Generally it is desirable to use the highest frequency of ultrasound possible to maximize resolution of nonuniformity. As an example of the effect of thickness variations on peak amplitude, fig. 1 shows the effect of thickness on back wall ultrasonic signal amplitude obtained from PMR-15 polyimide samples for pulse-echo experiments conducted at NASA Lewis Research Center. Thickness of the samples ranged from approximately 2.3 to 5 mm. The experiments were conducted using both 5 and 20 MHz broadband transducers. Results are shown for the time-domain broadband back wall pulse. Figure 1 illustrates that the back wall echo amplitude increases with decreasing thickness at both frequencies, but much more significantly so at 20 MHz than at 5 MHz due to the considerably higher attenuation coefficient (α) at 20 MHz for PMR-15 material ($\alpha_{PMR\text{-}15}$ at 20 MHz = 2.761 ± 0.2 Neper/cm and $\alpha_{PMR\text{-}15}$ at 5 MHz = 1.184 ± 0.1 Neper/cm as obtained from experiments at NASA Lewis). Thus, peak amplitude c-scan images, including the effect of thickness variations on them, will be highly frequency-dependent.

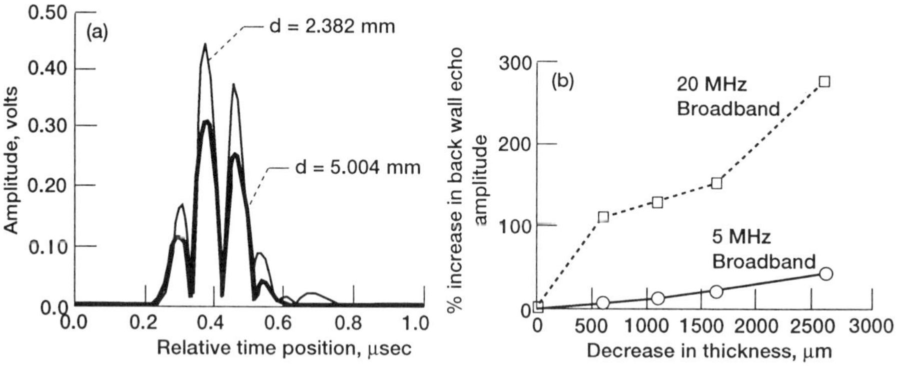

Figure 1.—Effect of thickness on ultrasonic echo peak amplitude for PMR-15 polymide sample. d = thickness. (a) Rectified back wall echo for two thicknesses using 5 MHz broadband transducer. (b) Percent increase in back wall echo peak amplitude versus decrease in thickness.

The situation is more straightforward for interpreting the effect of thickness variations on time-of-flight of ultrasonic pulses off the sample back wall. Back wall pulse-echo time-of-flight (2τ) is affected by thickness variation according to

$$2\tau = \frac{(2d)}{V} \tag{2}$$

where V is the velocity of ultrasound in the material. No frequency dependence exists as it does for attenuation (disregarding dispersion effects of frequency on material velocity (V)). Thickness effects on time-of-flight can also be interpreted by rearranging eq. (2) to calculate velocity according to

$$V = \frac{(2d)}{2\tau} \tag{3}$$

such that velocity is inversely proportional to time-of-flight. For velocity mapping from scan results, only one thickness value can be used practically in the velocity map calculation (eq. (3)). This thickness value is usually an average value obtained from several measurements at different sample locations. For scan locations where actual thickness is less than the value chosen for the calculation, time-of-flight will be less, and *apparent* velocity will be greater, than that if the scan location had the chosen value of thickness. The situation is opposite for scan locations where actual thickness is greater than the value chosen for the calculation. Hence, velocity and time-of-flight maps will be affected similarly (although inversely in terms of magnitude) by thickness variations, and velocity maps are used in this investigation to indicate time-of-flight variations.

Since the attenuation coefficient (α) and velocity (V) in a material will not be constant if microstructural variations are present, eqs. (1) to (3) indicate that maps of peak amplitude, time-of-flight, and velocity will show a combination of microstructural and thickness variations.

RELATED WORK

Several attempts to account for thickness variation effects in ultrasonic measurements were noted in the literature. Reference [2] used a two-transducer method whereby the transducers, located on opposite sides of the sample, were used in both pulse-echo and through-transmission mode to produce thickness-independent peak amplitude c-scans. Accurate thickness at each scan location was obtained via both transducers using the time-of-flights acquired from pulse-echo front-surface reflections off both surfaces of the sample, and using the known constant velocity in water and the known distance between the two transducers. Peak amplitude information was obtained in through-transmission mode. A peak amplitude c-scan free of thickness effects was then computed using eq. (1) where A, A_o, d, and α were the thickness-corrected peak amplitude, measured peak amplitude, measured thickness, and assumed constant material attenuation coefficient, respectively, at each scan location. Reference [3] also used a two-transducer method, but the transducers were employed solely in pulse-echo mode to produce thickness-independent ultrasonic velocity images. Thickness information was obtained in a similar manner to that described for ref. [2]. Additionally, time-of-flight from the first reflected pulse off the sample back surface was obtained by one of the transducers at each scan location. Knowing thickness (d) and time-of-flight (2τ) at each scan location, thickness-independent velocity (V) images were calculated according to eq. (3).

References [4 to 7] described a single-point ultrasonic velocity measurement method using a reflector plate located behind and separated from the sample, that does not require prior knowledge of sample thickness. The latter method, chosen for study in this investigation, was studied with success in prototypical scanning configurations in refs. [8 to 10], and incorporated into a commercial scan system in refs. [10] and [11]. Figure 2 shows a schematic of the immersion pulse-echo testing set-up required to use this method and the resulting ultrasonic waveforms. The mathematical basis for the method is derived in ref. [10] and results in ultrasonic velocity being calculated according to

$$V = c\left(\frac{\Delta t}{2\tau} + 1\right) \tag{4}$$

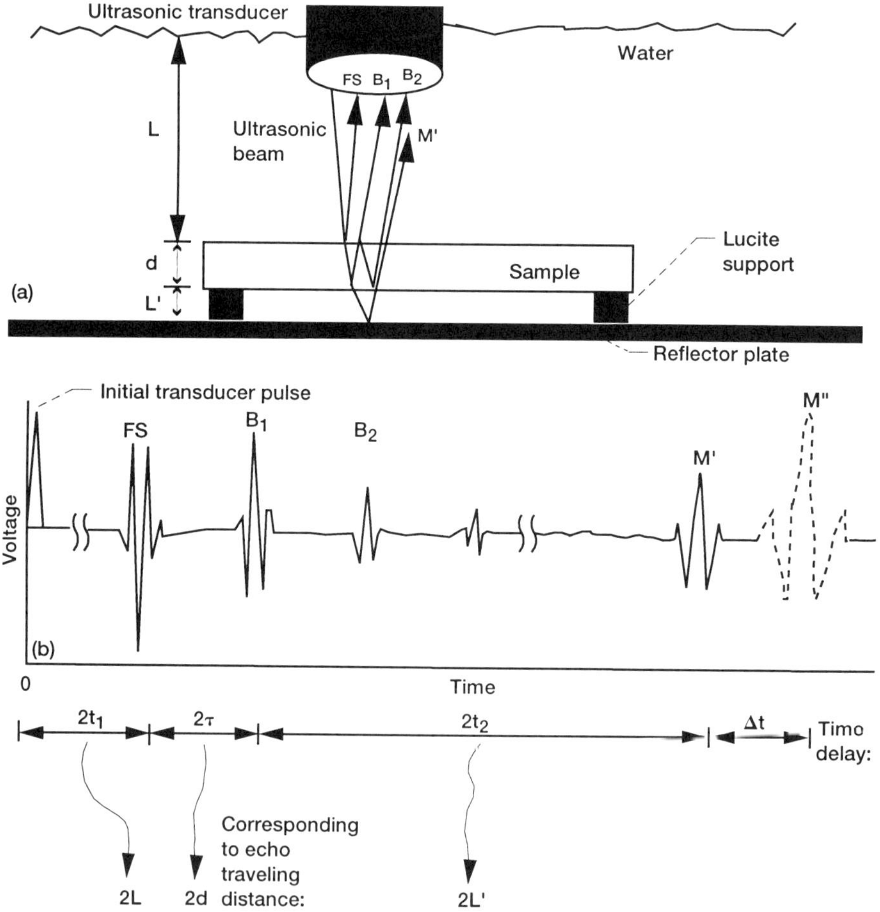

Figure 2.—Ultrasonic pulse-echo immersion testing. (a) Schematic of ultrasonic pulse-echo immersion testing. (b) Resulting waveforms.

where c is water velocity, 2τ is the pulse-echo time delay between between a front and back surface echo or between two successive back surface echoes, and Δt is the pulse-echo time difference between the first echo off the reflector plate front surface with and without the sample present, respectively. Water velocity (c) is determined from:

$$c(\text{mm}/\mu\text{sec}) = (1410 + 4.21T - 0.037T^2) \times 10^{-3} \tag{5}$$

where T is temperature in °C. Water velocity can also be determined by direct measurement using the time difference of ultrasonic wave travel in water between two transducer heights.

This thickness-independent ultrasonic imaging method does not require prior knowledge of sample thickness as shown in eq. (4), and, if engineered for scanning, the effect of thickness variation is eliminated in the resulting image. Precision and accuracy associated with this method are estimated at nearly 1% as described in ref. [10]. The methods of refs. [2 to 10] all require access to both sides of the sample, i.e. there is not a single-sided technique available for scanning that will result in thickness-independence.

SCOPE OF INVESTIGATION

In this investigation, the single-transducer-reflector plate method of refs. [4 to 10] and fig. 2 is applied to obtain thickness-independent ultrasonic images of monolithic silicon nitride ceramic plate and hollow tubular materials. Thickness-independent ultrasonic images for these materials are compared to c-scan peak amplitude images and apparent velocity images (the latter being obtained using eq. (3)). The single-transducer-reflector plate method was chosen over the two-transducer methods because of its simplicity and potential applicability to inspection of hollow tubular samples currently being fabricated for high-temperature structural duty. Some of these tubes recently inspected at NASA by ultrasonic methods have inner regions too small to contain a perpendicularly-positioned second transducer but large enough to contain a reflector plate.

MATERIALS

The silicon nitride plate material was in the form of a wedge of nominal thickness 3.3 mm and 36 mm length, and it had a 0.3 mm edge-to-edge thickness gradient (lengthwise) (approximately 10% of total thickness), 3 areas of higher-than-average (HTA) pore fraction (approximately 3 to 5% HTA as determined from optical analysis) and an edge-to-center pore fraction gradient of 2 to 3%. The thickness gradient was machined into the wedge. The silicon nitride tubular material was of nominal wall thickness 7.5 mm, 48 mm outside diameter, and 150 mm height. It was manufactured unintentionally with a 0.8 mm thickness variation (approximately 10% of total thickness) and currently unknown microstructural variation.

EXPERIMENTAL

The thickness-independent ultrasonic imaging method requires at least two scans to collect the necessary echoes required for calculation of time delays (fig. 2). The first of two scans is run with the sample in place, and echoes B1 (first back surface echo), B2 (second back surface echo), and M' (first echo off reflector plate front surface with sample present) are collected. Following this, the sample is removed and a second scan is run to collect M" (first echo off reflector plate front surface with sample removed). In this investigation, three scans were used to obtain all echoes as required using the version of this method that is commercially-available [11]. B1and B2 are obtained in scan 1, M' is obtained in scan 2, and M" is obtained in scan 3. A computerized cross-correlation method was used to calculate precise time delays [12].

The silicon nitride wedge was ultrasonically scanned with a 20 MHz (nominal center frequency) broadband unfocused transducer to obtain apparent velocity and thickness-independent velocity images, and additionally scanned with a 50 MHz broadband spherically-focused transducer to obtain conventional peak amplitude c-scan images. A machined tungsten reflector plate was positioned approximately 1 cm beneath the wedge. An 800 MHz analog-to-digital sampling rate was used during data acquisition. The scan grid was 45 by 90 points with a 0.5 mm scan increment for the apparent velocity and thickness-independent velocity scans, and 90 by 180 points with a 0.25 mm scan increment for the peak amplitude c-scan images. The scan set-up for the wedge is approximated by the schematic of fig. 2. Images were contrast-expanded to better reveal *global* material variation by filtering relatively high and/or low data values that occurred in very low numbers (stray data values). Further experimental details are given in ref. [10].

The silicon nitride tube was ultrasonically scanned with a 10 Mhz broadband spherically-focused transducer to obtain preliminary apparent velocity and thickness-independent image results. Photographs of the set-up for the tube scan are shown in fig. 3. A 250 MHz analog-to-digital sampling rate was used during data acquisition. The tube was scanned using a turntable assembly upon which the tube was placed. The transducer was positioned perpendicularly to the tube. The turntable was spun so that one scan line corresponded to a full rotation (360°) of the tube. Ultrasonic echoes were acquired at 1° increments. After the turntable completed a full rotation and thus returned to the scan starting position, the transducer was raised in height by 1 mm so that a new 360° line of data could be obtained. The scan proceeded in this manner such that 105 lines of data encompassing most of the tube height were obtained. The resulting scan grid was 360 by 105 points. A machined stainless steel reflector plate was suspended from the scanner bridge and positioned internal to the tube approximately 1 cm from the inside wall. In this manner, the reflector plate remained stationary throughout the scan. Data filtering for contrast expansion of tube ultrasonic images was used as before.

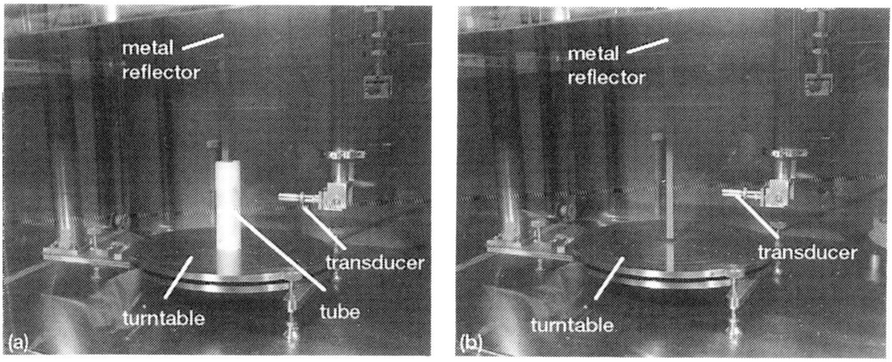

Figure 3.—Ultrasonic set-up using turntable scan for application to tubes. (a) Scan 1: sample present, collect B1, B2, M'. (b) Scan 2: sample removed, collect M".

RESULTS

Figures 4 and 5 show imaging results for the silicon nitride ceramic wedge with the 0.300 mm (10%) edge-to-edge thickness gradient (lengthwise), 3 areas of higher-than-average (HTA) pore fraction and an edge-to-center pore fraction gradient. Consider fig. 4. The peak amplitude c-scan image does not reveal the thickness or microstructural variations. This is likely due to the fact that the attenuation coefficient at ~50 MHz in silicon nitride (even with some porosity) is extremely low, such that a system with 8-bit digital dynamic range cannot resolve them. The *apparent* velocity image shows an ~10% velocity variation from top-to-bottom edge; some microstructural variation is evident but is mostly masked by the thickness variation. In this image, the HTA pore fraction area at the top of the disk is accentuated while the lower HTA pore fraction area is masked totally by the thickness gradient. In the thickness-independent image, the lower HTA pore fraction area is revealed, the top HTA area is de-emphasized, and an ~ 4% total velocity variation is seen. Additionally, the edge-to-center pore fraction gradient is disclosed. The

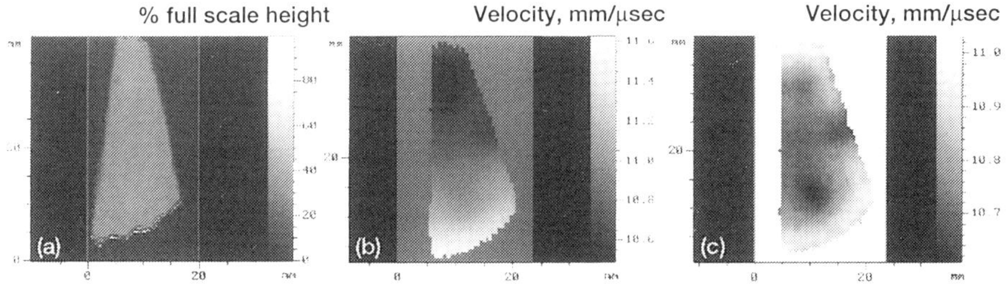

Figure 4.—Ultrasonic images of silicon nitride wedge having pore fraction variations and a 300 μm thickness gradient edge-to-edge (lengthwise). (a) C-scan peak amplitude image where back wall echo was gated, 50 MHz. (b) Apparent velocity image, 20 MHz. (c) Thickness-independent velocity image, 20 MHz.

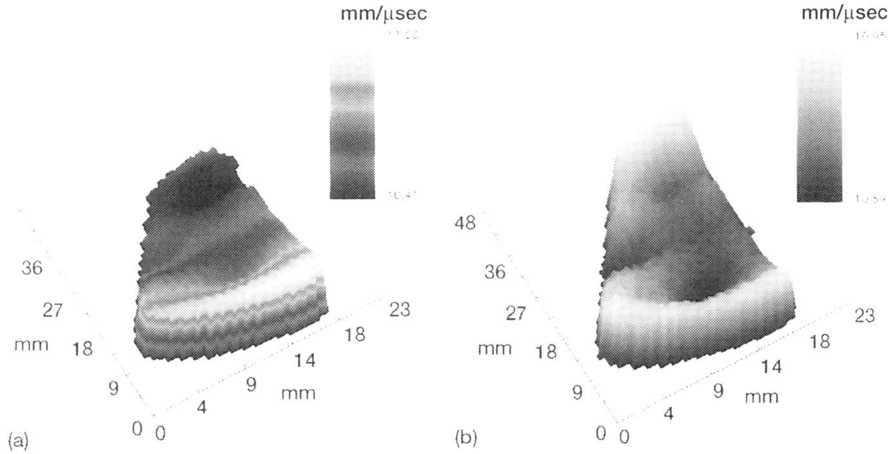

Figure 5.—Ultrasonic images of silicon nitride wedge, shown as 3-D renderings. (a) Apparent velocity. (b) Thickness-independent velocity.

latter image correctly portrays the microstructural condition in this material. Figure 5 shows three-dimensional renderings of the apparent velocity and thickness-independent velocity images which further highlight the differences between the results.

Figures 6 and 7 show preliminary imaging results for the silicon nitride tube of nominal wall thickness 7.5 mm and 150 mm height having a 0.8 mm (10%) thickness variation and currently unknown microstructural variation. Apparent velocity and thickness-independent velocity results are shown in flat, unwrapped form (fig. 6) and decaled onto a cylindrical / tubular model (fig. 7). Both views can be useful depending on the scan results. For these results, the tubular views (fig. 7) prove more useful since major image indications appear 180° apart (rotationally about the vertical axis) from each other, and one of the features is interrupted by the image end in the unwrapped view of the apparent velocity image of fig. 6(a). A funnel-like feature containing concentric open contours showing decreasing velocity from tube bottom-to-top is indicated in the apparent velocity image decal of fig. 7(a). This feature disappears for the most part in the thickness-independent image of fig. 7(b), indicating it was primarily due to thickness variation. Consider an orientation for the tube rotated 180° (about the vertical axis) from the views shown in figs. 7(a) and (b). For this orientation, the images of which are shown in figs. 7(c) and (d), an oval-like feature is

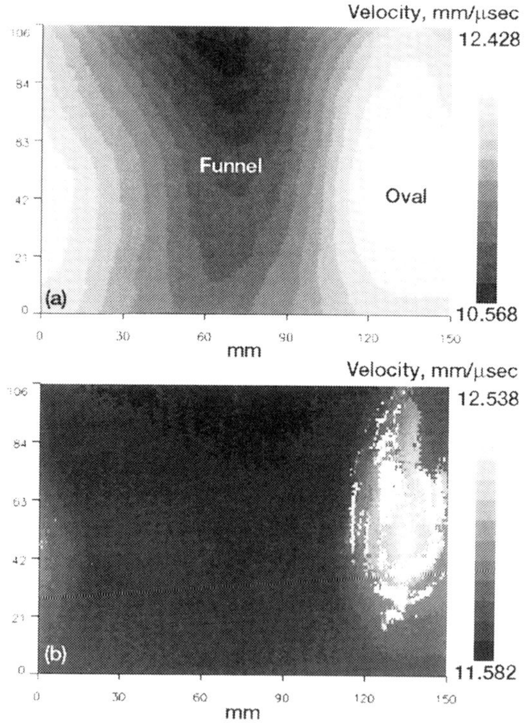

Figure 6.—"Unwrapped" ultrasonic Images of silicon nitride tube. (a) Apparent velocity. (b) Thickness-independent velocity.

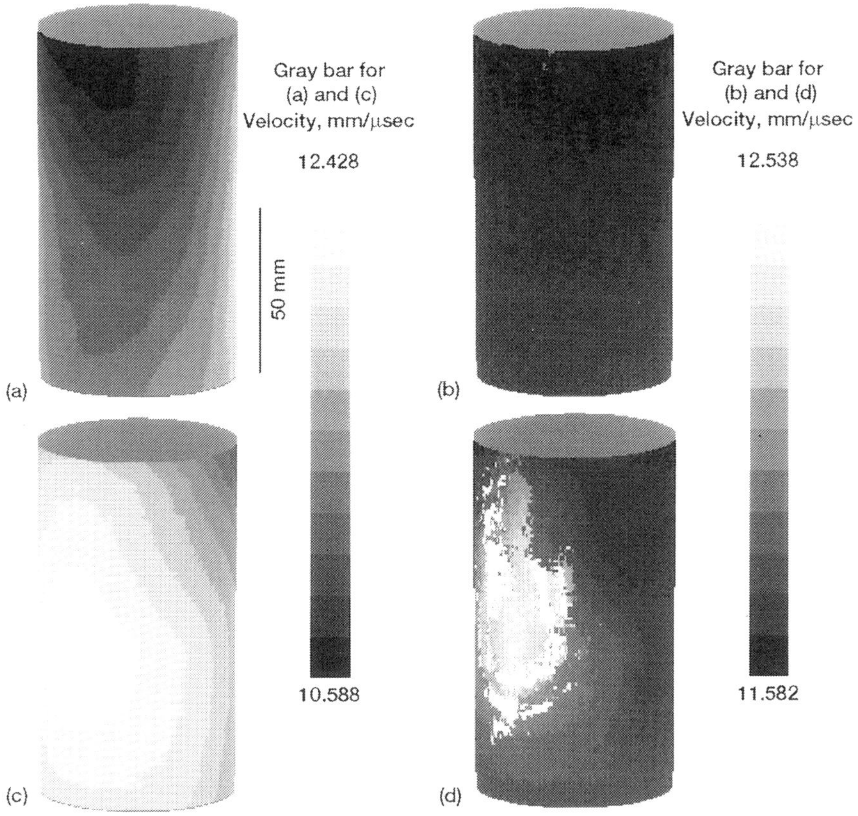

Figure 7.—Ultrasonic images of silicon nitride tube with ultrasonic info "decaled" onto cylinder model. (a) Apparent velocity, initial orientation. (b) Thickness-independent velocity image, initial orientation. (c) Apparent velocity, rotated 180° with respect to initial orientation. (d) Thickness-independent velocity image, rotated 180° with respect to initial orientation.

indicated in the apparent velocity image decal image of fig. 7(c). The oval-like feature contains concentric contours indicating different apparent velocity regions, but overall the feature shows higher-than-average apparent velocity as compared to the surrounding region. In the thickness-independent image of fig. 7(d), the overall oval-like feature remains, although the internal details have changed as compared to its portrayal in the apparent velocity image. The results of figs. 7(c) and (d) indicate that the features associated with the oval-like area may be the result of a combination of microstructural (possibly density/pore fraction) and thickness variation. Destructive testing with optical analysis is required to confirm these results. These preliminary results for the silicon nitride tube indicate the increased degree of inspection accuracy that the thickness-independent ultrasonic imaging method can provide for tube-type structures.

CONCLUSION

This article described a single-transducer ultrasonic imaging method based on measurement of ultrasonic velocity that eliminates the effect of thickness variation in the image. The method thus isolates ultrasonic variations due to material microstructure. Results were shown for silicon nitride ceramic plate and tubular materials. It was found that the peak amplitude c-scan image (gating the back wall echo) for the silicon nitride wedge was not visibly affected by the 300 μm (10%) thickness variation. Additionally, however, the peak amplitude c-scan did not reveal the microstructural variation in the wedge either. *Apparent* velocity images indicating time-of-flight variations showed significant superposition of thickness effects which masked the true picture of material microstructure for the silicon nitride wedge. For the wedge, the thickness-independent ultrasonic imaging method accurately quantified (in terms of velocity variation) the pore-fraction variations. For the tube, two major indications were revealed in the apparent velocity image results. Preliminary thickness-independent ultrasonic image results showed that one of these indications was due to thickness variation and that the other was likely a combination of microstructural and thickness variation. Further characterization needs to be performed on the tube sample to confirm these results. The use of the thickness-independent ultrasonic imaging method can result in significant cost savings, because the ultrasonic image, and consequently assessment of material and manufacturing quality, can be interpreted correctly without the need for machining to achieve precise thickness uniformity during nondestructive evaluation stages of material development.

REFERENCES

[1]. Nondestructive Testing Handbook, second edition, Volume 7 Ultrasonic Testing, eds. A.S. Birks, R.E. Green, and P. McIntire, American Society For Nondestructive Testing, 227–230, 387–390, 405, 225, (1991).

[2]. M. Bashyam, Thickness Compensation and Dynamic Range Improvement For Ultrasonic Imaging of Composite Materials. Proc. Of the 17th Annual Review of Progress in Quantitative Nondestructive Evaluation, La Jolla, CA, **10A**. Plenum Press, 1991,1035–1042, July 15–20, 1990.

[3]. J.J. Gruber, J.M. Smith, and R.H. Brockelman, Ultrasonic Velocity C-scans for Ceramic and Composite Material Characterization. Mater. Eval., **46**, [1], 90–96 (1988).

[4]. B.D. Sollish, Ultrasonic Velocity and Thickness Gage, United States Pat. No. 4,056,970, Nov. 8, 1977.

[5]. L. Pichè, Ultrasonic Velocity Measurement for the Determination of Density in Polyethylene. Polymer Engineering and Science, **24**, [17], 1354–1358, Mid-Dec. 1984.

[6]. I.Y. Kuo, B. Hete, and K.K. Shung, A novel method for the measurement of acoustic speed. J. Acoust. Soc. Am. **88**, [4], 1679–1682, Oct. 1992.

[7]. D.K. Hsu and M.S. Hughes, Simultaneous Ultrasonic Velocity and Sample Thickness Measurement and Application in Composites. J. Acoust. Soc. Am. **92**, [2], Pt. 1, 669–675, Aug. 1992.

[8]. V. Dayal, An Automated Simultaneous Measurement of Thickness and Wave Velocity by Ultrasound, Experimental Mechanics, **32**, [2], 197–202, Sept. 1992.

[9]. M.S. Hughes and D.K. Hsu, An Automated Algorithm for Simultaneously Producing Velocity and Thickness Images, Ultrasonics, **32** [1], 31–37, 1994.

[10]. D.J. Roth, Single Transducer Ultrasonic Imaging Method That Eliminates the Effect of Plate Thickness Variation in the Image. NASA TM–107184, 1996.

[11]. D.J. Roth, et. al., Commercial Implementation of Ultrasonic Velocity Imaging Methods via Cooperative Agreement Between NASA Lewis Research Center and Sonix, Inc. NASA TM–107138, 1996.

[12]. D.R. Hull, H.E. Kautz, and A. Vary, Measurement of Ultrasonic Velocity Using Phase-Slope and Cross-Correlation Methods. Mater. Eval., **43**, [11], 1455–1460, 1985.

NON-CONTACT ULTRASONIC CHARACTERIZATION OF CERAMICS AND COMPOSITES

Mahesh C. Bhardwaj
Ultran Laboratories, Inc.
1020 E. Boal Avenue
Boalsburg, PA 16827 USA

ABSTRACT

Conventional wisdom stipulates that ultrasonic analysis of ceramics and of other solids cannot be accomplished without direct transducer contact with test materials. Thus, non-contact ultrasound--i.e., through air coupling--is viewed as impossible. This impossibility is entirely due to the phenomenal acoustic impedance mismatch between the coupling air and the test media. After many years of struggle, this barrier has been broken through novel transducer designs, characterized by extremely high transduction in air and in other gaseous media. Now it is possible to propagate ultrasound in a non-contact mode through a wide variety of materials. We have successfully built and tested such transducers within the <250kHz to >5MHz frequency range on very thick to very thin, green and sintered ceramics under ambient atmospheric conditions.

This paper describes the quantitative and qualitative observations on a variety of ceramic and other materials by utilizing the newly developed non-contact methodology for density-velocity relationships and for defect detection. Also shared are some reflections about ultrasound from the author's experience.

INTRODUCTION

Analogous to any materials characterization method that utilizes a wave as the characterizing tool, ultrasound can also provide significant information about the medium through which it is propagated. The reflected and transmitted ultrasonic signals can be correlated and manipulated to detect internal discontinuities, to measure the physical properties and microstructure, and to generate the surface and internal images of a medium: Table-I. Besides being nondestructive, the advantages of ultrasound are numerous: it does

Table I. Significance of Ultrasonic Analysis, modified after Vary [1].

MEASUREMENT CATEGORY	MEASURED PARAMETERS	INFORMATION REVEALED
Time Domain	Velocities of Longitudinal, shear, and surface waves.	Direct correlation with density, porosity, defect detection, mechanical properties, residual & applied stresses, interfacial analysis, etc.
Frequency Domain	Frequency dependence of ultrasound absorption and attenuation. (Ultrasonic Spectroscopy)	Direct correlation with atomic structure, inter-granular relationships, phase detection, etc.
Image Domain	Time of flight, attenuation, or frequency domain monitoring of ultrasonic-signals by raster or multi-transducer scanning.	Planar or 3D images of surface and internal features of materials, including graphical representation of material heterogeneity in terms of microstructure, defects, and physical properties.

not require special sample preparation; it can be adapted to on-line manufacturing or testing environments; unlike x-ray and γ-ray methods, ultrasound is non-hazardous; the equipment involved is relatively easy to transport; and it can be used to test transparent or opaque materials. In fact, ultrasound can be used to characterize *any* medium through which it can be propagated. For example, with the exception of a vacuum, all three states of matter--solid, liquid, and gas--can be analyzed by ultrasound. However, the complexity of the test material geometry can be sometimes arduous to overcome. Also, the selection and design of suitable transducers can be extremely frustrating, particularly for the NDC of complex materials.

These impediments can be greatly minimized through the development of NDC educational and training curricula which are targeted towards the objectives of materials characterization [2]. It is a young field that needs to be understood, taught, and applied properly. Otherwise, there is a real danger of making poor choices, and of developing even poorer conclusions. Despite the dedicated efforts of a handful of educational and research institutions [3,4,5,6], it is this author's experience that this discipline--from the standpoint of materials science and engineering--is not yet fully developed. For example, the practical aspects of a wave-material interaction with respect to the crystallo-chemical makeup of the media through which ultrasound is propagated are not addressed in a manner that are congruent to other methods of materials characterization [2].

A one-to-one analogy and correlation with optics can create unreal perceptions about ultrasound, leading to erroneous expectations and conclusions. We forget that in ultrasound we do not have the luxury of light's rectilinear propagation and its small frequency and wavelength range: ~3,00 to ~7,00nm. In ultrasound we are dealing with an enormous frequency range--<50kHz to >1GHz--and an equally large range of wavelengths due to the diversity of test media through which it (ultrasound) can be propagated. When variations in apertures and attenuation are added, a complex spectrum of ultrasonic fields ensues. These seemingly academic subjects are of real value in defining the suitability of ultrasound and its limitations with respect to resolution, detectability, scattering, propagation, etc. Knowledge of these issues from a practical standpoint is vital.

Since its inception, ultrasonic NonDestructive Testing (NDT) has been preoccupied with the overt flaw detection in metals and alloys by either physically coupling transducers to test materials with oil and grease, or by submerging both the transducers and the test materials in water or like media. However, materials--such as powder compacts, green, liquid-sensitive and porous materials--cannot be reliably tested by conventional ultrasonics. On the other hand, the advantages of Non-Destructive Characterization (NDC) are obvious and of major significance at a time when the issues of quality and economics of the manufactured materials are extremely demanding. Ultrasound must first be developed in such a way that it is applicable to the wide range of materials at various stages of their processing without damaging or altering them during testing.

WORK RELATED TO NON-CONTACT ULTRASONICS

Recognizing the necessity of total materials quality (including the accuracy and reliability of the analyzed information), we have advanced ultrasonic NDC to new heights through the essential transducer and applications R&D with respect to the modern materials characterization objectives [7]. In order to facilitate the NDC of green, fragile, liquid-sensitive, and porous materials, we introduced the dry coupling longitudinal and shear wave transducers in 1983. These were quickly adapted to our earlier development of the short unipolar time envelope transducers. Dry coupling provided the means for test material protection, while the short pulse permitted accurate time of flight measurements. The combination of transducer acoustics with the coupling methodology facilitated quantitative NDC of a wide range of green and sintered ceramics and composites [8,9,10,11,12]. While the uses of the new transducers in aerospace, nuclear, materials, and bio-medical industries are numerous today, the proprietary nature of their applications prevents further disclosure.

Despite the arduous nature of non-contact ultrasonics, our struggle for its development did not cease with the advances in dry coupling technique. Transducers for air/gas propagation (<250kHz to ~20MHz) were a by-product of the dry coupling transducers. They found limited applications for high frequency receivers, surface profiling, proximity analysis, level sensing, and non-contact characterization of paper products. These earlier air/gas propagation transducers failed to generate high-strength quality transmitted signals required for the NDC of ceramics or metals.

In 1993 our struggle finally produced transduction in gaseous media to such an extent that--leaving aside non-contact ultrasound propagation in papers, rubbers, and low density materials--it was possible to accomplish this through materials such as dense ceramics, metals, and biological structures [13]. Based upon the development of non-contact ultrasound, work is also in progress for the critical medical diagnostic applications such as burnt skin and bed sore characterization [14]. Recently an excellent methodology for the measurement of elastic constants of polymer matrix composites has also been reported [15]. This investigation utilized capacitance air-coupled transducers. For a review on non-contact or air-coupled ultrasound, see reference [16].

COMPLEXITY OF NON-CONTACT ULTRASOUND

In principal, the non-contact ultrasonic *modus operandi* is analogous to the well-known contact and immersion techniques, but it varies dramatically in magnitude. For example, while the acoustic impedance mismatch between the transducer and the test materials is greatly minimized in contact ultrasound, it must be overcome in non-contact modality. The mismatch is of several orders of magnitude when one compares the acoustic impedances of air (4.0×10^2 kg/m^2.s) and of most industrial materials (30×10^6 kg/m^2.s). In addition, the phenomenal absorption of ultrasound by ambient air--particularly at frequencies greater than 1.8MHz--multiplies the complexity by magnitudes. Until the acoustic barrier between air and solids is broken, non-contact ultrasound cannot be realized. As will be seen in this paper, this is the barrier that has finally been broken.

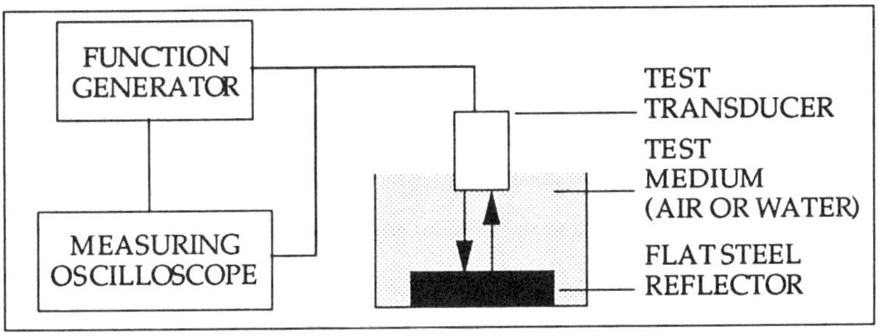

Fig. 1. Transducer characterization scheme for sensitivity comparison.

COMPARISON OF NON-CONTACT AND CONVENTIONAL NDT TRANSDUCERS

A simple, yet effective, unbiased transducer characterization scheme was developed to evaluate and to compare the sensitivity of non-contact transducers with those of conventional high-sensitivity NDT transducers. Transducers in pulse-echo reflection mode were excited by a single-burst 16V sine wave pulse from a function generator, Fig. 1. The reflected signal from a flat steel target--10mm away from the transducer in the test medium, received by the same transducer--was fed directly into the oscilloscope. The frequency of each test transducer was adjusted in the function generator to yield the maximum reflected signal amplitude. The sensitivity (loop gain) was determined according to:

$$\text{Sensitivity (dB)} = -20 \text{ Log } V_x/V_o \quad (1)$$

where V_x is the received signal amplitude in volts directly measured from the oscilloscope, and V_o is the voltage of the excitation pulse, in this case, 16V. Table II provides a one to one comparison for the sensitivities of non-contact and the conventional high sensitivity immersion NDT transducers, respectively, in ambient air and water. All transducers are planar, and typically, 25mm active area in diameter.

CURRENT LIMITATIONS OF NON-CONTACT ULTRASOUND

While the non-contact transducers have accomplished the task of propagation through virtually all materials, it is still not easy to apply them in the popular pulse-echo mode for the NDC of ceramics or metals. However, the transmission technique is perfectly applicable.

Table II. Comparison of the reflected sensitivities of non-contact transducers and the conventional NDT immersion transducers.

FREQUENCY (MHz)	TRANSDUCER SENSITIVITY (dB)	
	Non-Contact *(In Ambient Air)*	NDT Immersion *(In Water)*
0.25	-62	-36
0.5	-66	-32
1.0	-72	-30
2.0	-84	-28

Note that despite the exceptionally high transduction of current non-contact transducers (only about 40dB lesser than that in water, Table II), still high amplification of the received signals through test materials is required. This requirement increases as a function of increasing acoustic impedances of materials, further exacerbating the problems of interpreting transmitted signals. For example, besides the key signals corresponding to ultrasound propagation in the materials, one sees a number of unwanted signals related to material interfaces and edges, and from any other reflector (walls, equipment, operator, etc.) upon which ultrasound falls through air from the transducers. In time domain, the majority of such signals arrive much later than the signals of interest. It is observed that the first arrival always corresponds to the ultrasound transmitted signal through the test material. However, sometimes signals internally reflected through test materials can also be seen. The appearance of such signals, though of great value, depends upon the acoustic impedance and the frequency-dependent attenuation of the test material. Usually the internally reflected signals are observed when the test material impedances and the attenuation are relatively low. Figures 2 and 3 show the location of various signals when ultrasound is propagated through a material in the non-contact transmission mode under ambient atmospheric conditions. Reflected signals corresponding to air distances between the material surfaces and transmitting and receiving transducers can be removed from the

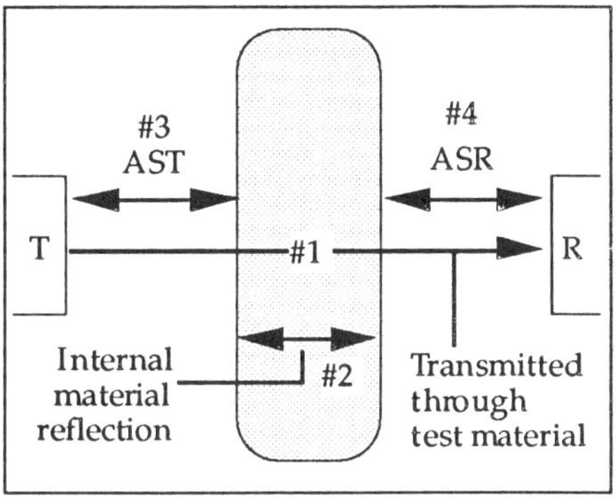

Fig. 2. Location of key ultrasonic signals in non-contact transmission mode.
#1--Straight through material.
#2--Internally reflected in material.
#3 & 4--Reflections from Air Separation Transmitter (AST) and Air Separation Receiver (ASR) from material surfaces.

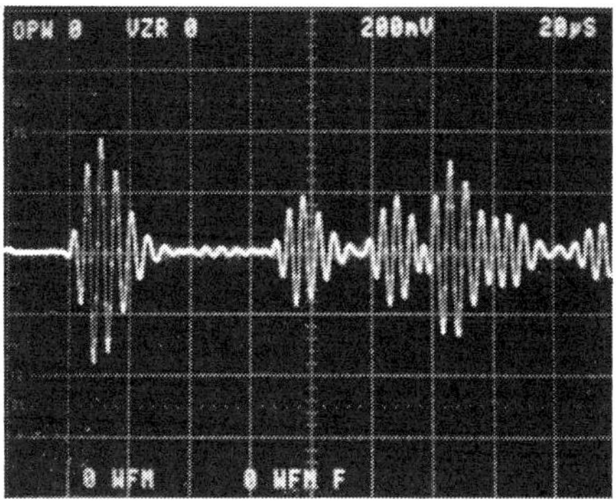

Fig. 3. Demonstration of key ultrasonic signals from an 80mm thick polystyrene. From left to right: Transmitted through material, internal reflection, AST (15mm) ASR (20mm).

region of interest in time domain by increasing the transducers' air separation distances.

VELOCITY MEASUREMENT

There are several ways to determine longitudinal wave velocity in the test materials by non-contact transmission mode. Here we describe a simple two-step technique, as shown in Fig. 4.

Measurement of Air Velocity

The first step is to align both transmitting and receiving transducers that are facing each other in air and are separated by a distance, d_a. The next step is to read the Time of Flight (ToF), t_a, corresponding to ultrasound travel in air directly from the oscilloscope, and then to determine the air velocity,

$$V_a = d_a/t_a \tag{2}$$

Measurement of ToF in the Test Material and its Velocity

After inserting the test material between the transmitting and receiving transducers, the complete ToF, T_c--corresponding to ultrasound travel in air as well as in the test material--is read directly

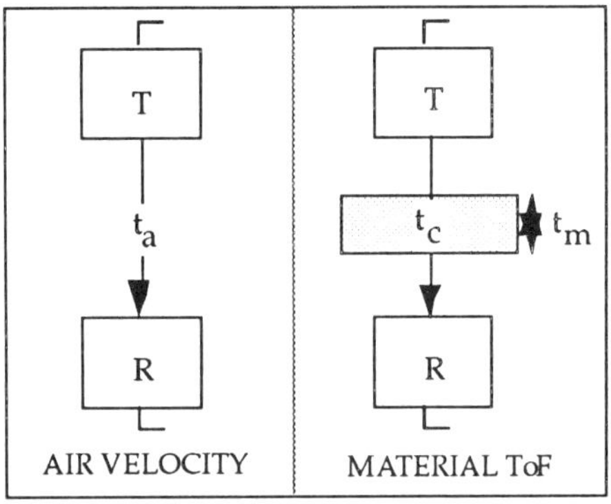

Fig. 4. Alignment of transducers and insertion of the test material for velocity measurements.

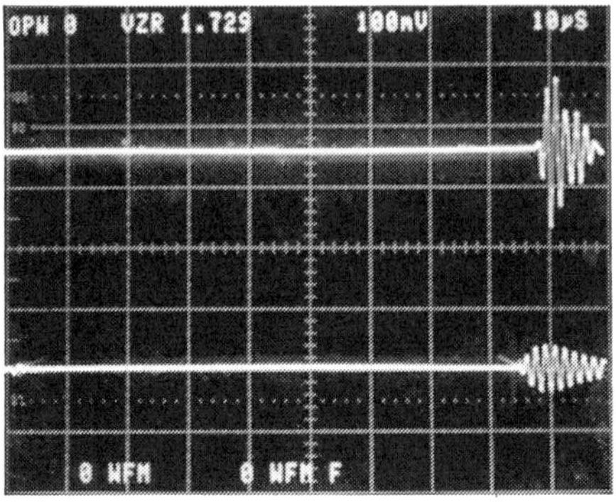

Fig. 5. Top trace: T and R in air 30mm apart. Bottom trace: With a 1.6mm green multi-layer ceramic.

from the oscilloscope. The ToF corresponding to the delay time in air is determined as:

$$t_o = (t_a - t_{am}) \tag{3}$$

where t_{am} is the ToF in air relative to the material thickness, d_m, i.e.,

$$t_{am} = d_m/V_a. \tag{4}$$

The ToF through the test material is thus

$$t_m = (t_c - t_o). \tag{5}$$

Therefore, the test material velocity, V_m can be determined as:

$$V_m = d_m/t_m \tag{6}$$

or

$$\boxed{V_m = \frac{d_m}{t_c - (t_a - t_{am})}} \tag{7}$$

As an example, Fig. 5 shows the transmitted signals corresponding to ultrasound travel in air and in a test material, in this case a green multi-layer ceramic. From these observations:

t_a = 86.13μs. V_a = 348.13m/s. t_c = 82.3μs. V_m = 2,100m/s.

Using equation 7, the velocity of this material is 2,100m/s. The velocity determined by the dry coupling technique of this sample is 2,240m/s. Assuming our measurements of the ToFs are accurate, it is reasonable to conclude that the velocity determined by the non-contact technique is more representative of the true value. Accuracy of the velocity calculation is a sensitive function of the accuracy of the measured ToF. For example, a variation of 10μm distance in air will introduce an error of ±20ns in the ToF measurement. For thin and for high velocity materials, this could make a considerable difference. Therefore, in order for the non-contact ultrasonic velocities to be reliable, it is imperative to make accurate and repeatable ToF measurements.

DEFECT DETECTION

Should ultrasound encounter a region in the test material characterized by defects such as internal cracks, pores, gross texture or compositional variations, its transmitted signal would be correspondingly different from that in the defect-free region. This is demonstrated in Fig. 6 on a 47mm thick polystyrene block.

The most frequently asked questions in ultrasonic NDC concern the resolution and detectability. The maximum resolution of two closely lying surfaces or discontinuities is the major goal in the pulse-echo or the direct reflection technique. While resolution is not the main issue in the direct transmission technique (applied in the present work), the ability to detect the smallest defect in a test medium is relevant. Detectability is directly proportional to the wavelength and to the size of the interrogating beam, particularly that of the receiving transducer. Table III provides an idea about the non-contact detectability limits in various materials as a function of the ultrasonic velocities.

VELOCITY-DENSITY RELATIONSHIPS

According to the velocity measurement scheme described in this paper, green and sintered ceramics (those varying in density or in other

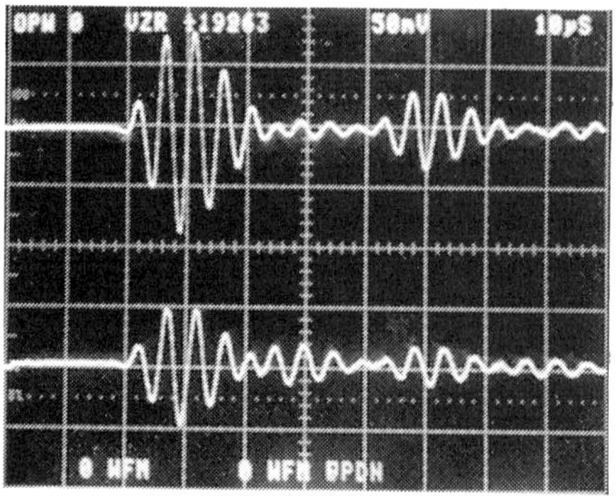

Fig. 6. An example of defect detection in a 47mm thick polystyrene block. Top Trace: The defect-free region. The first signal corresponds to transmission through the material, and the second to its internal reflection. Bottom trace: Reduced amplitudes with a 4mm cylindrical hole. Also observe the appearance of the hole indication between the transmitted and reflected signals.

Table III. Non-Contact ultrasonic detectability limits*

MATERIALS & VELOCITIES (m/s)	DETECTABILITY (mm), function of FREQUENCY 100kHz	200kHz	500kHz	1.0MHz	2.0MHz
1,500 Liquids, visco-elastic materials, tissues, green materials, etc.	1.6	0.8	0.3	0.15	0.08
3,000 Polymers, composites, bones, porous and granular materials.	3.3	1.6	0.6	0.3	0.15
6,000 Metals, ceramics, glasses, and composites.	6.6	3.3	1.2	0.6	0.3
12,000 Dense ceramics, oxides, nitrides, carbides, etc.	13.0	6.6	2.5	1.2	0.6

*NOTES

1. Detectability is the smallest detectable discontinuity in the test medium.
2. These calculations are based upon the interrogating beam size being equal to one wavelength. Note further that this requirement is not easily realized in non-contact applications.
3. The effects of attenuation and diffraction have not been considered in these calculations.

features) were investigated. The transmitting transducer was excited by a 400 volt pulse. Transmitted signals were received by a receiving transducer and amplified by a broadband amplifier. Depending upon the acoustic impedance and the frequency attenuation behavior of the test materials, transducer frequencies varied from 250kHz to 2.0MHz.

Figures 7 and 8, respectively, show rf A-scans of green ceramics for velocity measurements, and the velocity-density relationship for green Al_2O_3. Similarly, figures 9 and 10, respectively, show rf A-scans and the velocity-density relationship for sintered Al_2O_3.

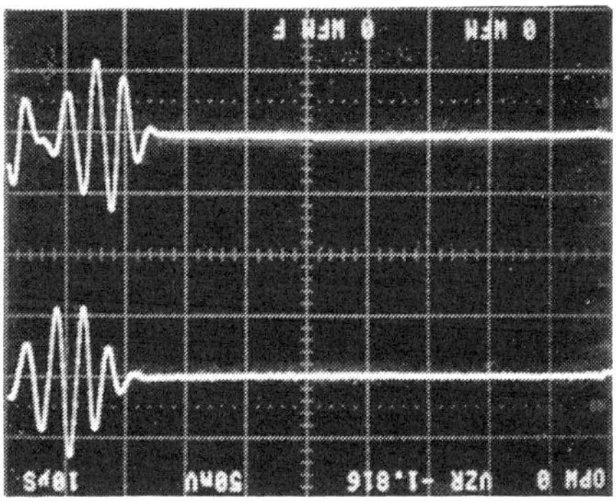

Fig. 7. Full transmitted signals of green ceramics. Top trace: Slip-cast porcelain. $Z = 1.9X10^6 kg/m^2.s$. V = 1,180m/s. Bottom trace: Pressed titania. $Z = 4.2X10^6 kg/m^2.s$, V = 1,900m/s.

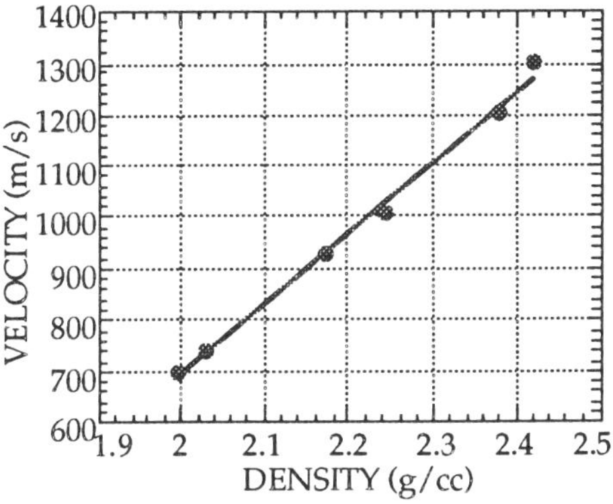

Fig. 8. Relationship between density and non-contact ultrasonic velocity for green Al_2O_3.

QUALITATIVE CHARACTERIZATION

We have applied the non-contact ultrasonic technique on a wide range of industrial materials and biological structures in order to demonstrate its potential and feasibility for various applications [11]. As stated earlier, if ultrasound can be propagated through a medium, then interpretation of transmitted signals holds the key to important information relative to quality and internal features of the medium. Here we provide the self-explanatory evidence of qualitative analysis for selected green and sintered ceramics, rocks, and composites. The observations reported here show not only feasibility of non-contact

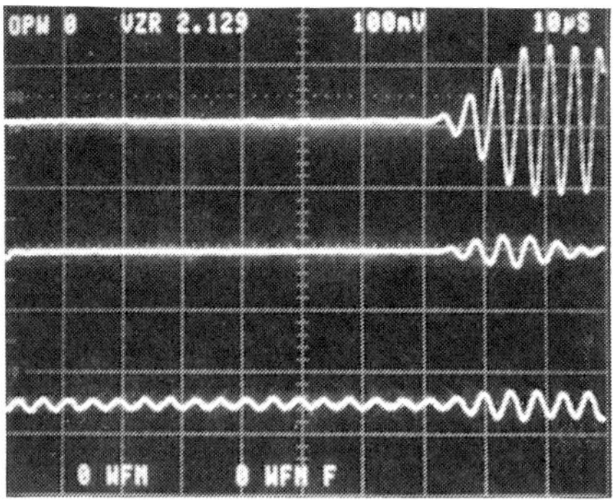

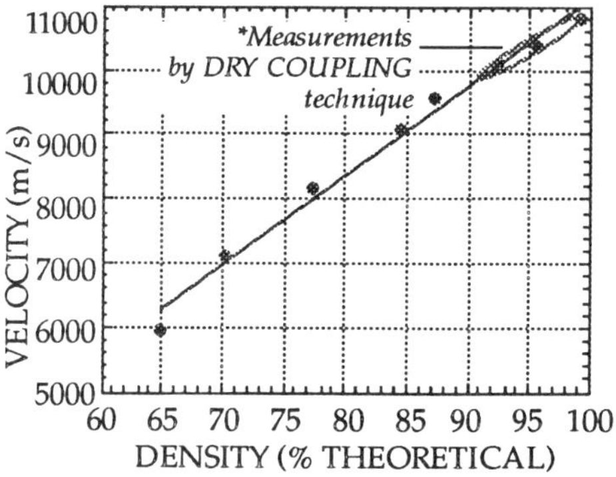

Fig. 9. Full transmitted signals of sintered Al_2O_3. Top trace: 65% dense. $Z = 15X10^6 kg/m^2.s$. V = 5,900m/s. Middle trace: 77% dense. $Z = 25X10^6 kg/m^2.s$. V = 8,100m/s. Bottom trace: 92% dense. $Z = 37X10^6 kg/m^2.s$. Velocity = ~10,000m/s

Fig. 10. Relationship between density and non-contact ultrasonic velocity for sintered Al_2O_3.

Non-contact technique could not be used reliably for very dense ceramics.

analysis of compositionally variant materials, but also of micro-structurally diverse materials of varying thicknesses.

Green Ceramics

Fig. 11 shows transmitted ultrasonic signals in the rf A-scan mode from a 76mm thick section of ZrO_2 brick. As expected, when ultrasound propagates through a defective region, the amplitude of the transmitted ultrasound is substantially decreased. Fig. 12 shows similar signals from a green ceramic multi-layer package.

Sintered Ceramics

Fig. 13 shows transmitted signals from two samples of a porous ZrO_2 refractory. Visually both look alike. However, when they were examined ultrasonically, one of them produced results typical of internal cracks. Fig. 14 is the same as Fig. 13, with the observations produced by dry coupling, presented here for comparison purposes.

Fig. 15 shows transmitted signals through various regions in a SiO_2 fiber block with porosity >50%. Fig. 16 shows similar signals at various points in a glazed ceramic tile.

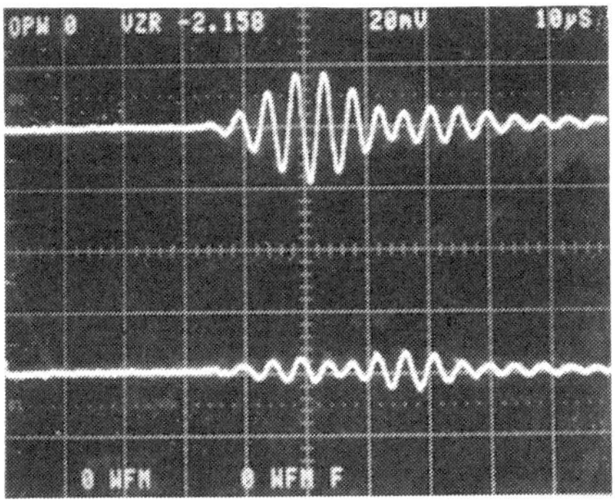

Fig. 11. Green ZrO_2 refractory - 76mm
AST: 20mm. ASR: 20mm.
Top trace: Defect-free region.
Bottom Trace: With 3mm cylindrical hole.

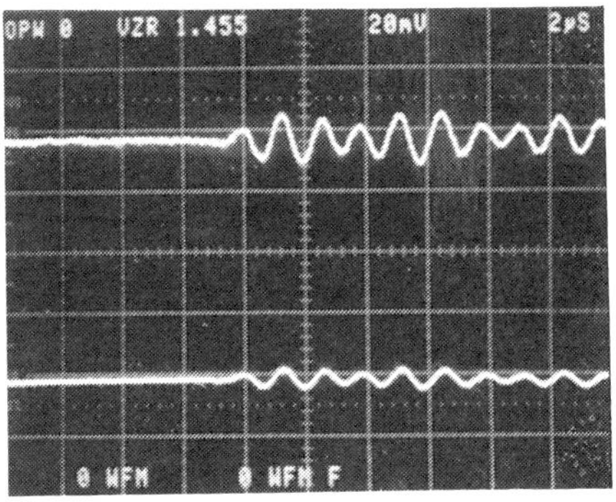

Fig. 12. Green ceramic multi-layer package 4.4mm. AST: 5mm. ASR: 5mm.
Top trace: Defect-free region.
Bottom trace: Presumably delamination or an internal crack.

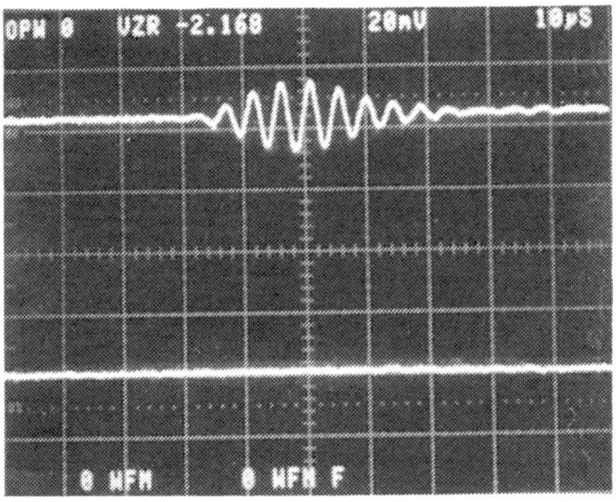

Fig. 13. Sintered ZrO_2 refractory - 100mm.
AST: 20mm. ASR: 20mm.
Top trace: Defect-free sample.
Bottom trace: Internally cracked sample.

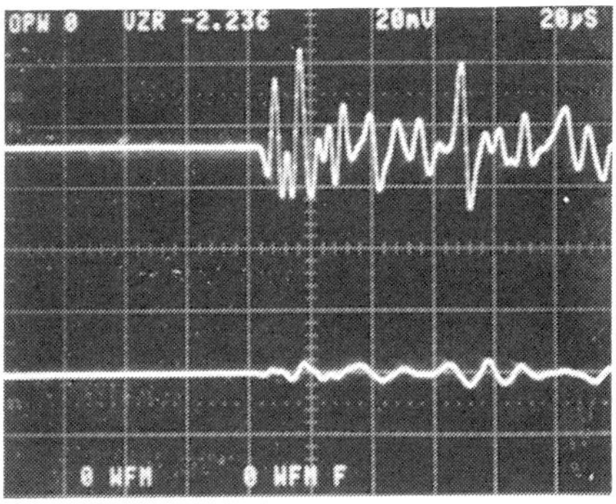

Fig. 14. Same as Fig. 13, but the analysis is performed by dry coupling for comparison.

Rocks

Fig. 17 shows transmitted signals through a core-drilled sample of sandstone at different locations.

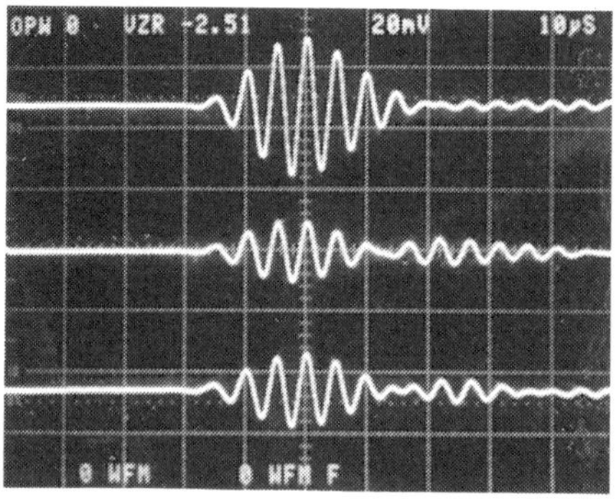

Fig. 15. Transmitted signals at various regions in a SiO_2 fiber block - 10mm. Porosity: >50%. AST: 10mm. ASR: 10mm. Significant variations in signal strength could be indicative of density variations or internal defects.

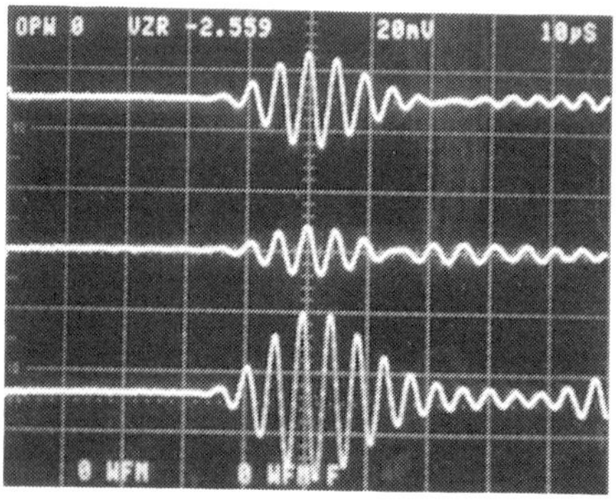

Fig. 16. Transmitted signals at various regions in a glazed ceramic tile - 6mm. AST: 10mm. ASR: 10mm. Significant variations in signal strength could be indicative of tile and/or glaze quality.

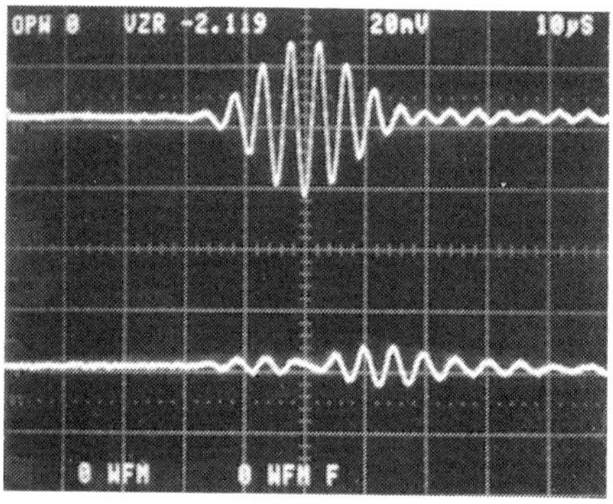

Fig. 17. Transmitted signals through different regions of a core-drilled sample of sandstone - 100mm.
Top trace: Defect-free region.
Bottom Trace: Presumably a cracked or a different texture region.

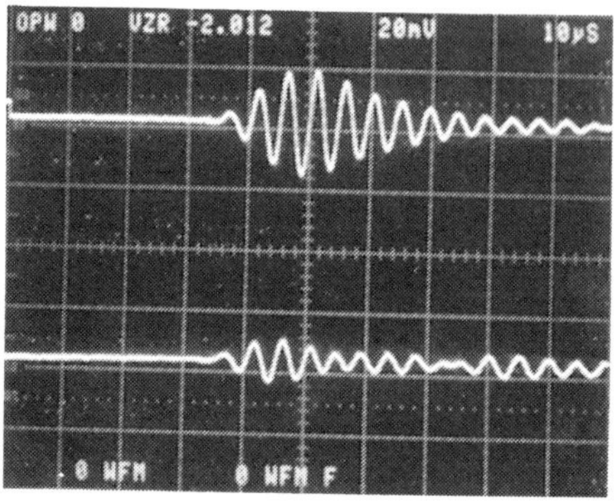

Fig. 18. Transmitted signals through 2D C-C disc brake - 20mm.
AST: 20mm. ASR: 20mm.
Top trace: Defect-free region
Bottom trace: Presumably an internal heterogeneity or a defective region.

Composites

2D Carbon-Carbon disc brakes for large aircraft are rather difficult to characterize nondestructively even by conventional ultrasonic means.

To our surprise, we discovered that ultrasound could be easily propagated through this material by non-contact means, Fig. 18.

Fig. 19 shows transmitted signals from a defect-free region and from an impact-damaged region in a 1.5mm 8-ply graphite fiber-reinforced plastic composite. Fig. 20 is a similar example for 6.3mm 12-ply glass fiber-reinforced plastic composite.

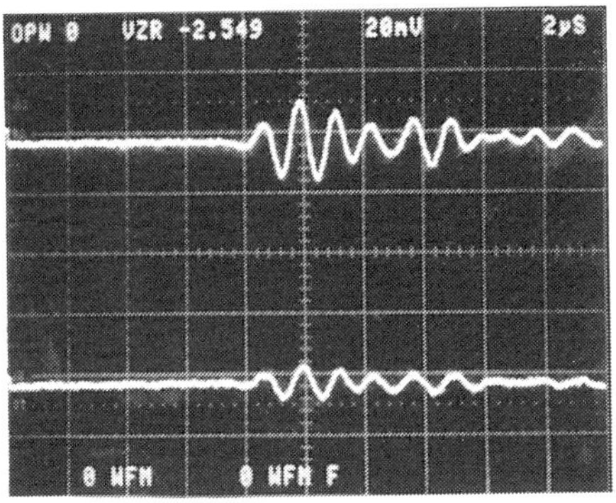

Fig. 19. Transmitted signals through a 1.5mm 8-ply graphite fiber reinforced plastic composite. AST: 5mm. ASR: 5mm. Top trace: Defect-free region Bottom trace: Impact-damaged region.

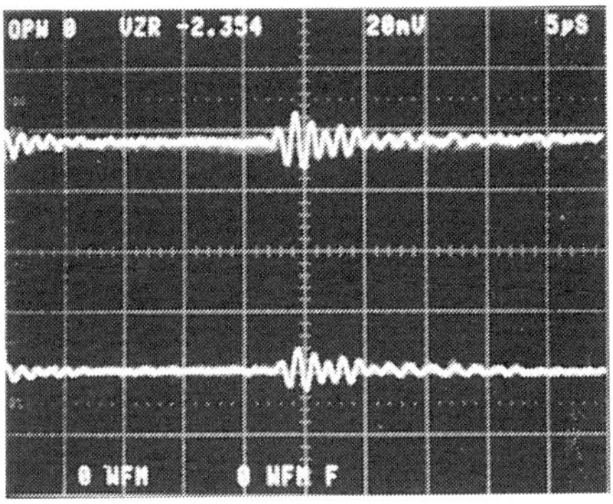

Fig. 20. Transmitted signals through a 6.3mm 12-ply glass fiber reinforced plastic composite. AST: 5mm. ASR: 5mm. Top trace: Defect-free region. Bottom trace: Impact-damaged region.

CONCLUSIONS

In this paper we have described the exciting practical development of non-contact ultrasonic methodology of immense value to the industry and to the society in general. However, this development should not be considered a panacea. This technique cannot replace the high frequency and high resolution ultrasonics necessary for microstructural evaluation and for imaging and microscopy of dense and high acoustic impedance materials. On the other hand, as shown in this paper, non-contact ultrasound can be relatively easily and reliably used for green and for porous materials characterization. In the author's opinion, in many cases, such applications have not been reliably done by conventional non-destructive ultrasonics.

Non-contact ultrasonic applications are naturally limited by the laws of physics. For example, under ambient conditions we are presumably limited to ~5MHz frequency for thin green and sintered ceramics and for thick materials to ~500kHz frequency. Similarly, for gross defect detection and for density-velocity estimations in dense and high velocity materials, the current non-contact methodology is adequate. For such materials, it should not be trusted for small defect detection and for precise velocity-density relationships.

Some of these problems can be minimized by subjecting transducers and test materials to environments of high air pressures, or by using other gases, particularly those that are less attenuative (such as hydrogen or helium). While investigating high velocity materials (dense oxides, carbides, nitrides, borides and diamond and diamond-like materials) by the non-contact ultrasonic technique, one should also be fully conscious of deleterious diffraction effects since the applicable non-contact frequencies are still very low. All these factors are deemed significant and should be evaluated with respect to test material composition and microstructure as well as with the objectives of NDC.

ACKNOWLEDGMENTS

The work reported here is supported by our desire to continue advancing ultrasound for both the personal satisfaction of our creative personnel and for significant applications in our increasingly complex world. For this, the author is thankful to all the staff members of Ultran, particularly to M.E. Langron and H.A. Holderman.

REFERENCES

1. A. Vary, *Mechanics of NonDestructive Testing,* W.W. Stinchcomb, ed., Plenum Press, New York (1980).
2. Committee on Characterization of Materials, *"Characterization of Materials,"* National Academy of Sciences & National Academy of Engineering, Washington, DC, Publication MAB-229-M, March 1967.
3. A.B. Bhatia, *"Ultrasonic Absorption, An Introduction to the Theory of Sound Absorption and dispersion in Gases, Liquids, and Solids,"* Dover Publications, Inc., New York (1967).
4. R. Truell, C. Elbaum, and B. Chick, *"Ultrasonic Methods in Solid State Physics,"* Academic Press, New York (1969).
5. G. Kino, *"Acoustic Waves--Devices, Imaging, & Analog Signal Processing,"* Prentice-Hall, Inc., Englewood Cliffs, NJ (1987).
6. H.N.G. Wadley, P.A. Parrish, B.B. Rath, and S.M. Wolf, editors, *"Intelligent Processing of Materials and Advanced Sensors,"* The Metallurgical Society, Warrendale, PA (1987).

7. Bhardwaj, M.C., "Evolution, Practical Concepts and Examples of Ultrasonic NDC," Ceramic Monographs, Supplements to *Interceram* 41 (1992) [7/8] #4.5 and 42(1993) [1] #4.5 - Handbook of Ceramics, Verlag Schmidt GmbH, Frieburg, Germany.
8. M.C. Bhardwaj, "High Resolution Ultrasonic Non-destructive Characterization," *Cer. Bull.* **69**, 9 (1990).
9. M.C. Bhardwaj and K. Trippett, "Nondestructive Characterization of Green and Sintered Ceramics," Proceedings of the *First International Symposium of Engineering Ceramics,* S. Kimura and K. Niihara, Editors, Koda, Aichi-Prefecture, Japan, October 21-23, 1991.
10. M.C. Bhardwaj and A. Bhalla, "Ultrasonic Characterization of Ceramic Superconductors," *J. Mat. Sci. Lett.*, **10** (1991).
11. N. Kulkarni, B. Moudgil, and M. Bhardwaj, "Ultrasonic Characterization of Green and Sintered Ceramics: I, Time Domain," *Cer. Bull,* **73,** 6 (1994).
12. N. Kulkarni, B. Moudgil, B. and M. Bhardwaj, "Ultrasonic Characterization of Green and Sintered Ceramics: II, Frequency Domain," *Cer. Bull,* **73,** 7 (1994).
13. M.C. Bhardwaj, "Evolution of Non-Contact Ultrasonic Analysis and its Applications for Hidden Objects Detection," *Innovations in Materials Res.*, **1,** 2 (1996).
14. J.P. Jones, M. Bhardwaj, V. Vanderkam, and B. Achauer, "Non-Contact Ultrasonic Imaging for the Evaluation of Burn-Depth and other Biomedical Applications," a paper presented at the 23rd International Symposium of Acoustic Imaging, Boston, MA, April 13-16, 1997. To be published in *Acoustical Imaging,* Vol. 23, Plenum Press, NY (1997).
15. B. Hosten, D.A. Hutchins, and D.W. Schindel, "Measurement of Elastic Constants in Composite Materials Using Air-Coupled Ultrasonic Bulk Waves," *J. Acoust. Soc. Am.*, Vol. 99, #4, (1996).
16. C. Wykes, "Advances in Air-Coupled Ultrasonic Transducers," *Nondestr. Tes. Eval.*, **12** (1995).

X-RAY NDE OF FINE CERAMICS WITH IMAGE SUBTRACTION METHOD

Yasushi Ikeda, Yasutoshi Mizuta and Katsuhiro Onda
Japan Fine Ceramics Center
2-4-1, Mutsuno, Atsuta-ku, Nagoya 456, JAPAN

ABSTRACT

In order to detect very small defects in ceramic products with high speed and high reliability, a new micro-focus X-ray testing method has been developed. By introducing an image subtraction process in radioscopy (image intensifier), the X-ray images of defects were extracted from various background noises of the imaging system, and the contrast enhancement to the defect images was greatly improved. A pore smaller than 100 μm in silicon nitride of a few of 10 mm in thickness has been detected with the method. Much better detectability of these defects than conventional radioscopy methods could be obtained, which is also considerably superior to usuall film methods. The required testing time for the new method is about 1 minute, which is very smaller than those for conventional film method.

INTRODUCTION

Ceramic materials have usually rather fragile properties, so that very small defects can be the origin of brittle fracture[1,2]. Therefore, it is very important to inspect such small defects in ceramic products. For this purpose, a micro-focus X-ray testing system has been used, and imaging with more than several magnifications is usually employed for detection. The film method, however, is very slow because the X-ray intensity of micro-focus X-ray system is rather weak. Therefore, a high speed X-ray TV system is expected to be used[3]. The X-ray TV system, however, has usually both rather low spatial resolution and low contrast resolution. Therefore this TV method could not be applied to the inspection of fine ceramics so far. A new method having high detectability and high speed for testing is required.

For this purpose, we developed a new radioscopic method using an image subtraction process. The new method is very simple and very effective for NDT of ceramics.

PRINCIPLE OF THE NEW RADIOSCOPY METHOD

To date, conventional X-ray TV systems have poor image quality and low image resolution, which can not be directly applied for the detection of very small defects in fine ceramics. This poor spatial image resolution, however, can be improved by using a micro-focus X-ray source and taking high enough magnification on imaging. We are using a micro-focus X-ray generator (ANDREX, MX-4) with an average focusing size of 10μm, a sample manipulator and an image intensifier system (TOSHIBA, 9in./ TV tube camera) followed with an image processing system. The image processor (JAPAN AVIONICS, IMAGE Σ-3) has the functions of both integration and subtraction with an image field of 640 x 480 pixels and an 8 bit A/D digitizer and a D/A converter. The images are stored with digital format.

The principle of image magnification[4] for the micro-focus X-ray system is shown in Fig.1. The magnification factor M is defined as the ratio of the distance between the focal point and the imaging device (FID) to that between the focal point and the object (FOD). By using micro-focus X-ray, the penumbra Ug for the image, which is given by (M-1)d, can be attained as very small. For example, if the size of focal point,d, is 10μm, and the imaging magnification is taken as 10, the magnification brings only 90 μm to the penumbra. Then, if a radioscopy system has the spatial resolution of 4 line pares per mm (150 μm lines can be visible separately), the penumbra can be hidden within the spatial resolution, and cause no blur in the image. On the other hand, since the spatial resolution of 4 line pares per mm of the radioscopy system (an image intensifier with a following TV camera) is corresponding to the resolution of 40 line pares per mm at the specimen under the magnification of 10, a spatial resolution of about 15 μm is attained for the object. In our usual testing, pore defects of approximately 50 to 100 μm in size are the target of detection. Then an imaging magnification of 10 or 20 can yield the enough spatial resolution, theoretically.

On the other hand, the more difficult cause of poor detectability is due to the poorer contrast resolution for X-ray images. As the image contrast is intrinsically given by the X-ray absorption properties of the specimen, it is not so easy to improve the problem. Moreover, the distribution of scattering X-rays and the non-uniformity of the sensitivity of the pixels of the image intensifier (I.I.) with TV system hide the image information and degrade the image contrast significantly. Therefore, these degradation factors must be minimized from the X-ray imaging. For the purpose of the present study, a new method of radioscopy has been devised, which uses an image subtraction process combined with moving objects. The outline and the principle of the method are given in Fig.2[5,6].

In this subtraction method, first an integral image of the specimen to be tested is obtained, which can eliminate random noises of the I.I. with a TV system, and a fairly smooth and noise-less image is obtained. The integral imaging process, however, is not enough for further image contrast enhancement. Because a concern with the integral nature of the image process is that very small contrast differences are hidden by the nonuniform sen sitivity of detector pixels, shading effects with the optical system, and scattered x-rays. Therefore, to make whole contrast enhancement does not identify the image of such small defects. To exclude these unfavorable effects and extract only the true image of the defects, the image subtraction process must be applied.

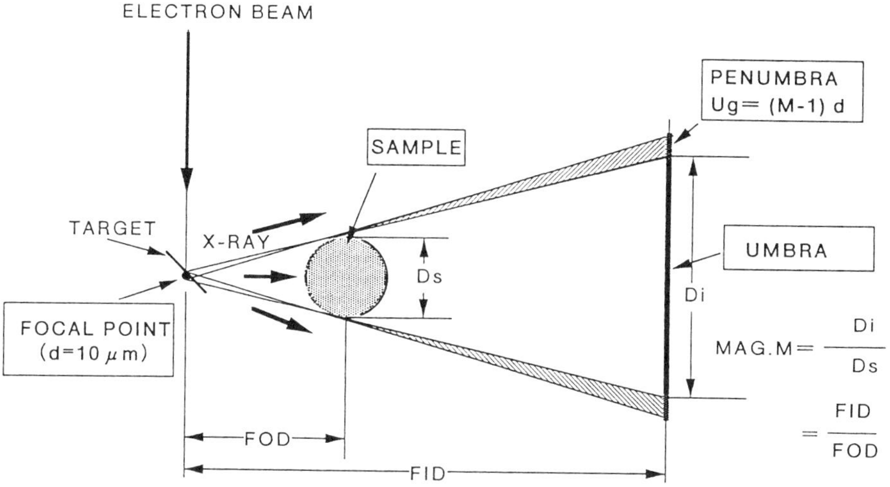

Fig.1 Principle of micro-focus X-ray imaging

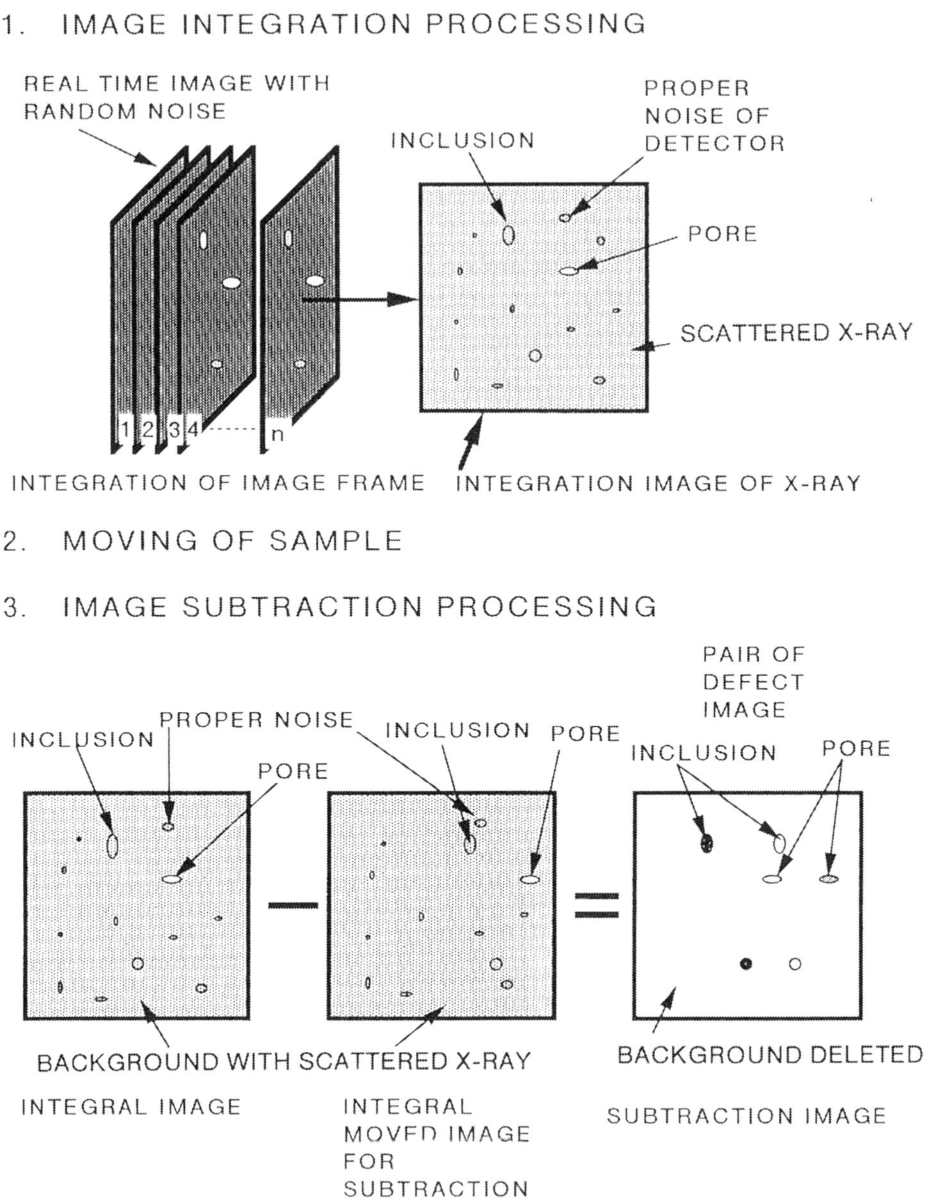

Fig.2 Principle of the new image subtraction of X-ray radioscopy

In usual image subtraction, some reference images were prepared previously. Sometimes such preparation of reference image is difficult because of the shape of specimens and unknown composition of materials for NDT of ceramics. Therefore, we devised a new method of subtraction, which is very simple and very effective for image subtraction without any references. In our method, one of the relative positions of the components of the test system, that is, the sample, the detector and the X-ray source is slightly changed. Next, subtraction of image has been carried out. With this moving operation, only the moving point in the test specimen has been extracted. Noisy points and shading in the image due to the X-ray TV system and scattering X-rays could be excluded significantly and a very flat intensity for the image field could be obtained. Then, the final contrast enhancement of the image could be very effectively executed on the extracted image of defects. In the final X-ray image, one pair of images, i.e., white with integration and black with subtraction for each defect, are superposed in the same image, being apart from each other with the given distance of moving. This pairing of the image can allow much easier and more reliable detection of defects because of the information about a given defect increases more than double.

SOME RESULTS FOR SILICON NITRIDE CERAMICS

Figure 3 shows the X-ray images of a silicon nitride plate taken by conventional image integration methd (Fig.3a) and by the new image subtraction method (Fig.3b). The test sample is a silicon nitride plate of 5 mm in thickness, on which surface a thin silicon nitride film of 40 μm in thickness is attached. The thin film is like a penetrameter and has a penetrated hole, which diameter is 40 μm. In Fig.3a, the image of the hole is very faint and almost in visible. However, in Fig.3b, the pair image of the hole can be clearly visible. The thin film has also a crack, which image is also much clearer in Fig.3b. In Fig.3b, we can see another defect, that is, a condensation of some additives for the ceramic, which absorbs more X-rays than silicon nitride. Therefore, the order of black and white images in the pair image is reverse to that of the hole. From the experiments we carried out, we found that the sensitivity of the method attained to better than 0.1%, which is significantly superior to conventional film methods.

Figure 4 shows an X-ray image of additive concentrations in a silicon nitride ceramic turbine part. In this case, the concentration area of an additive component of about 100 μm in size very clearly.

Figure 5 shows pores included in a thick silicon nitride part, which thickness is more than 30 mm at the test position. In the figure, the detected pores have the diameters of about 300 to 500 μm. Conventional film method could not be applicable to test this sample because of the large thickness of it.

(a) Integral method

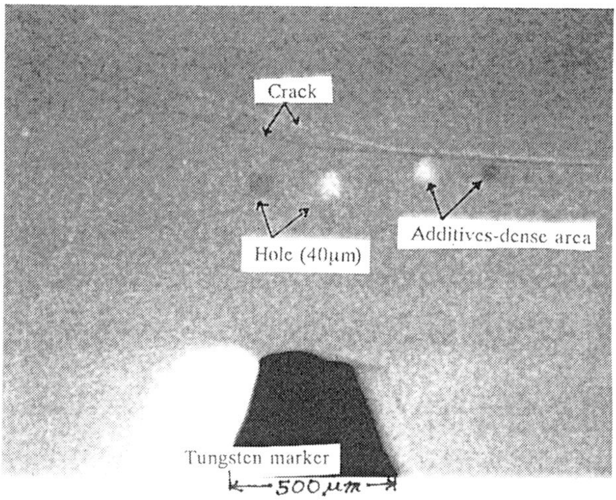

(b) Subtraction method

Fig. 3 The X-ray images of a silicon nitride plate of 5 mm in thickness attached with a thin silicon nitride film of 40 μm in thickness including a penetrated hole and a crack

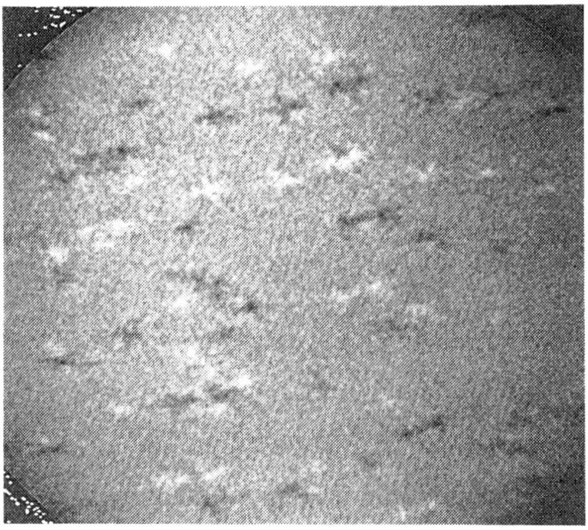

Fig. 4 X-ray image of additive concentrations in a silicon nitride ceramic part.

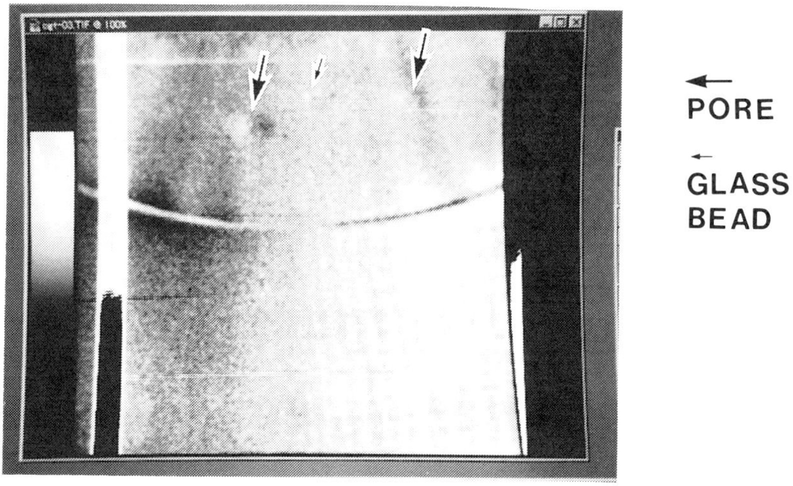

Fig. 5 X-ray image of pores included in a thick silicon nitride part, which thickness is more than 30 mm at the test position.

SUMMARIES

The image subtraction radioscopy showed very simple and very sensitive method for detection of defects in ceramics by using micro-focus X-ray. It is very useful to detect small defects in ceramics and the sensitivity is even more superior to conventional film method with very small required test time.

REFERENCES

1) Lange F.F.(1974) "Origin and use of fracture mechanics" Fracture mechanics of ceramics, vol. **1,** Bradt R.C., Hasselman D.P.H. & Lange F.F.(eds), Plenum Press, New York, pp3-16.

2) Goebbels K. (1986) "Nondestructive evaluation of ceramic components for engines" Ceramic materials and components for engines, Bunk K.& Hausner H. (eds), Deutshe Keram. Ges., pp779-787.

3) Amin K.E.and Leo T.P. (1987) "Radiographic Detectability Limits For Seeded Defects In Both Green and Densified Silicon Nitride" NDT of High Performance Ceramics, Vary & Schneider(eds.) The Am. Ceram. Soc., pp211-232.

4) Cotter D.J. and Koenigsberg W.D. (1987) "Microfocus Radiography of High Performance Silicon Nitride Ceramics" ibid, pp233-253.

5) Ikeda Y. Mizuta Y. and Tobita H " X-ray image subtraction method for detection of small defects in ceramics" Proc. 8-th Asia-Paciffic Conf. NDT, 1995, pp349-355.

6) Ikeda Y. and Mizuta Y."A new method of X-ray Fluoroscopic Testing for Small Defects in Fine Ceramics with Image Subtraction" Proc. 1st. US-Japan Sym. on Advanced NDT, 1996, pp435-440.

SONIC DETECTION OF MANUFACTURING FLAWS IN CERAMIC COMPONENTS

David A. Pape
Division of Mechanical Engineering
Alfred University
Alfred, New York 14802-1232

ABSTRACT

A new approach for detecting manufacturing flaws based upon the sonic vibrational spectra of the ceramic component is presented. The method involves constructing a nominal vibrational spectrum using experimental modal analysis techniques. A "sonic signature" for a particular piece is obtained by testing a number of like components. Once the modal norms are established, subsequent parts having frequencies outside of acceptable ranges are rejected as defective. Subsequent destructive testing has shown the validity of the sonic approach. This new method is powerful for applications where information regarding the existence of a defect is desired, and can be applied on line as a quality control check in a manufacturing facility.

INTRODUCTION

A disadvantage of most NDE methods currently in use is that the component must be investigated in a piece-meal fashion. That is, the investigation proceeds in a point by point manner until the entire structure has been scanned for defects. For instance, ultrasonic NDE methods use high frequency mechanical vibration energy to detect and locate structural discontinuities and flaws. An electric pulse is generated in a test instrument and transmitted to a transducer, which is attached to the test structure. The transducer converts the pulse to small amplitude, high frequency vibrations which are then transmitted into the test object where the ultrasonic energy is attenuated, scattered, reflected or resonated to indicate the presence of flaws or defects within the material. A coupling medium must be used, and the test object must either be immersed in water or the transducers coupled with oil or grease. Test frequencies range from 40 kHz to 50 MHz, with 2.25 and 5 MHz being most commonly used for flaw detection [1].

Although ultrasonic testing may allow flaws in a component to be precisely located, it can be difficult to use for structures with complicated geometries, and a skilled operator must perform the testing. Furthermore, in instances where only the existence (rather than the location) of the flaw is desired, the method may be needlessly complex.

Experimental modal analysis techniques have been used for many years as an alternative to ultrasonic testing for locating flaws and damage in various structures [2]. Modal analysis is the process of determining the dynamic behavior of a structure or component, assuming that the dynamic response of a structure is the sum of a discrete set of independent modes. A mode is described by it's modal parameters, namely natural frequency, damping ratio, and mode shape. A continuous body will, in theory, have infinitely many modes of vibration. In practice, only a finite number of these modes are needed to accurately describe the structure. In addition, the number of modes that can be measured is limited by the dynamic range of the instruments and transducers as well.

A fundamental assumption of modal testing is that a mode of vibration can be excited at any point on a structure, except at nodes of vibration where it has no motion [3]. Thus, the natural frequency and damping are global properties which can be obtained from a single measurement.

A structure's modal properties (natural frequencies, damping ratios, and mode shapes) are dependent on its physical properties (mass, stiffness). These, in turn, depend upon the geometry and condition of the component. Flaws such as cracks in a component generally cause a decrease in stiffness and an increase in damping [4]. Thus, any changes or defects in the structure will show up in the modal properties, which points to this method for global flaw detection.

The advantages of modal testing as a NDE method have been recognized by many authors. Richardson and Mannan [5] list several potential benefits:

- Modal properties can be measured on any structure that can be made to resonate. Therefore, any structure that is constructed such that acoustic energy can readily travel within its boundaries will resonate.

- Modes are sensitive indicators of the physical characteristics (mass, stiffness, damping) of a structure.

- Faults can be detected at unmeasured points, due to the global nature of modes. Therefore, measured changes can be used to detect faults in areas of the structure that are inaccessible for direct measurement.

- Modal testing is non-destructive. Modal parameters can be identified from frequency response functions that are measured at low excitation force levels, with little risk of damaging the structure during testing.

A variety of flaw detection techniques using the modal analysis approach have been proposed. Penny et. al. [6] use a statistical method on the measured natural frequency data of a portal frame. Zimmerman et. al [7] uses the frequency response functions directly to locate damage on a truss. Defects in a cantilever beam are investigated by Fox [8] using mode shapes and modal displacements. The modal analysis approach has also been used for ceramic components. Kan and Bedwi [9] tested specimens of silicon carbide, alumina, and silicon nitride with artificially induced cracks. Pape [10] tested a series of twenty alumina station post insulators, and was able to detect one with a crack using only natural frequency data. In this paper a method which is closely related to the modal analysis method is proposed.

SONIC SIGNATURE METHOD

The primary objective of this work is to develop a simple nondestructive test that can be used to determine whether or not a component is flawed. It is not intended as a method to find *the location* of any flaws or damage. Furthermore, the test must be simple enough to use in a production environment. A method meeting these criterion is described below.

The proposed method is closely related to the modal analysis techniques described above, with two distinct differences. First, rather than mount an accelerometer to the component to acquire vibration data, a microphone located in close proximity to the test component is used to obtain the response signal. The modal properties are then calculated with the impact force as input and the resulting sound pressure level as output. In contrast to an accelerometer, the microphone does not need to be physically mounted to be component, which would be beneficial in a high volume situations such as manufacturing. We can use the microphone because we do not need information on the motion of a specific point of the component. Rather, we are looking for global properties.

All of the work described in the preceding section involves comparing the modal properties of a component prior to damage to the modal properties *of that same component* after damage has occurred. Thus, the second key difference for the proposed method is that instead of performing before and after testing on a single component, we propose to test a large number of like components to assemble a statistical basis for the modal properties. These norms will then represent a "sonic signature" for the component and flawed components may be detected based on any significant differences from the norms. The key assumption here is that changes in modal properties due to damage are greater than differences due to variability of the product as well as the experimental uncertainties.

The term "sonic signature" rather than "acoustic signature " is used herein so as not to confuse the method with another distinctly different technique, namely acoustic emission. With acoustic emission, straining of the object produces bursts of energy, which can then be detected by ultrasonic transducers mounted on the object.

We further note that the term "ultrasonic" is generally used for sound with a frequency above 20,000 Hz, that is, above the range of human hearing. However, as described above, ultrasonic testing is typically performed with frequencies orders of magnitude higher than this level. The sonic signature method could be performed in the 0-100,000Hz range, using standard microphones and signal processing instrumentation. While technically there is some overlap in the frequency range of the two methods, they are actually quite distinct.

Modal Analysis Theory

The dynamic response of a component depends upon it's mass, stiffness, and damping. Assuming a linear, single degree of freedom system with viscous damping, these physical properties can be used to write an equation of motion by applying Newton's second law:

$$M\frac{d^2x(t)}{dt^2} + C\frac{dx(t)}{dt} + Kx(t) = f(t) \qquad (1)$$

with

$x(t)$	position, as a function of time, of the degree of freedom
M	mass of the body
C	viscous damping coefficient
K	linear stiffness coefficient
$f(t)$	an arbitrary time dependent force

Equation (1) may be solved if the initial conditions (position and velocity) and the applied force are known. For the special case where $f(t)$ and C are zero, we find that the system oscillates as a sinusoid, $x(t) = Xe^{i\omega t}$, with a natural frequency given by $\omega = \sqrt{K/M}$. If the damping term is included (with $f(t)$ remaining zero), we have the case of damped, free vibration, and the system oscillates as:

$$x(t) = Xe^{-\zeta\omega t}e^{i\varpi t} \qquad (2)$$

where

$\zeta = C/2\sqrt{KM}$ is the damping ratio
$\varpi = \omega\sqrt{1-\zeta^2}$ is the damped natural frequency.

If we consider now a force of the form $f(t) = Fe^{i\omega t}$ the equation of motion can be written as [Ewins]

$$(K - \omega^2 M + i\omega C)Xe^{i\omega t} = Fe^{i\omega t} \qquad (3)$$

from which we can extract the frequency response function $H(\omega)$:

$$H(\omega) = \frac{X}{F} = \frac{1}{(K-\omega^2 M)+i(\omega C)} \qquad (4)$$

Note that the magnitude of the FRF is:

$$|H(\omega)| = \frac{1}{\sqrt{(K-\omega^2 M)+i(\omega C)}} \qquad (5)$$

which is plotted in Figure 1. The peak corresponds to the natural frequency. A key point is that this FRF depends only on the physical parameters of the system, and is independent of the excitation force [3]. This is preferable to a simple Fourier transform, which would show different magnitudes depending on the level of input force.

Figure 1: Frequency Response Function

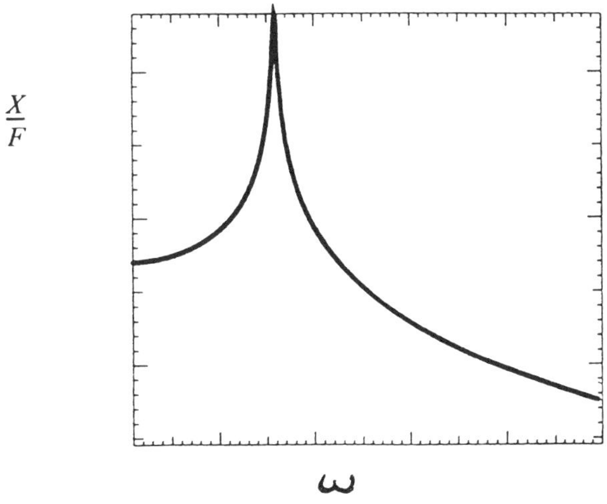

The vibration of a body and the resulting sound are intimately connected. When a body is subjected to a broadband force (e.g. an impact) numerous structural modes are excited. An equivalent viewpoint is that compression, bending, shear, and torsional waves have been set up in the body. It can be shown that structural bending waves contribute directly to sound radiation [11]. In this work we utilize the airborne sound waves caused by mechanical vibration (impact) of the body

The wave equation for airborne sound in three dimensions is

$$\frac{\partial^2 p}{\partial t^2} = c^2 \nabla^2 p$$

where

p is the sound pressure (Pa)
c is the propagation velocity of airborne sound waves (m/s)
t is time (s)
∇ is the laplacian, $\nabla^2 = \frac{\partial^2}{\partial x^2} + \frac{\partial^2}{\partial y^2} + \frac{\partial^2}{\partial z^2}$

Since sound pressure is the difference between instantaneous absolute pressure and ambient pressure, we often use the mean-square sound pressure, given by

$$p_{rms}^2 = \frac{1}{T} \int_0^T p^2 dt$$

with T the averaging time. Because of the wide range of audible sound (20 μPa to 200 Pa), sound measurements are commonly converted to the decibel (dB) scale. This is accomplished with the following definition:

$$L_p = 10 \log \left(\frac{p^2{rms}}{p^2{ref}} \right) = 20 \log \left(\frac{p_{rms}}{p_{ref}} \right)$$

where

L_p is the sound pressure level (dB)
p_{ref} is the reference pressure (20 μPa)

In the following, frequency response functions constructed using an input force and the output sound pressure level are used for flaw detection.

EXAMPLES OF FLAW DETECTION

This section presents the techniques used to obtain the modal information required for flaw detection. The basis for this damage detection method is the availability of accurate modal data including both the input to the component (excitation force) as well as the resulting output (sound pressure). This approach enables the calculation of a frequency response function (FRF) for the component.

Experimental Setup

The experimental setup used in these tests is shown in Figure 2. The test object was placed on a thick foam pad to isolate it from the table below. Excitation was provided by striking the object with an impact hammer (PCB model 086C03) with integral force transducer. The hammer was fitted with a hard steel tip which produced a broadband excitation from 0 to 7,000 Hz. The response was measured with an electrocet microphone (Radio Shack model 33-1063) placed approximately 3cm from the test object. This microphone has a useful frequency range of 50-10,000 Hz. The force input and sound pressure output signals were recorded and analyzed using a spectrum analyzer (Data Physics model DP420). This analyzer provides analog to digital conversion of the force and sound pressure, and calculates the frequency response function in real time. A frequency range of 0 to 5,000 Hz was used for all of the tests. No special sound isolation was provided for the testing; the tests were carried out in a normal lab, with typical background noise from HVAC and other sources.

Figure 2: Experimental Setup

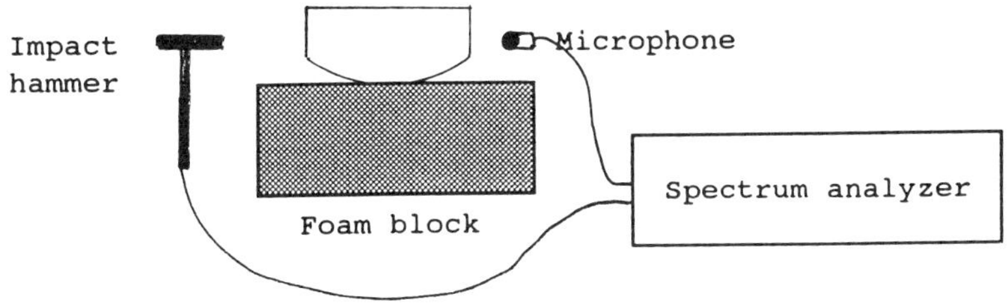

Experimental Results

In order to verify that this sonic signature method works, we must first verify that the sonic method is able to detect flawed components. A common ceramic item, a small bowl, is used for the initial testing. Finally, a large number of pin type insulators are tested to show the applicability of this method in manufacturing situations. In both cases frequency domain testing is used to obtain the sonic signatures.

Bowls: For the initial investigation four common whiteware bowls were tested. These bowls are oval shaped, 17 cm long by 13 cm wide and 5 cm deep. They are tested in an unglazed condition. One bowl has a hairline surface crack on the inside surface.

A typical frequency response function for this component is shown in Figure 3. As this figure shows, the first two natural frequencies fall within the 0-5000 Hz test range. The FRF for the cracked bowl is shown in Figure 4. Natural frequency results for the three good bowls, as well as the one containing the crack, are given in Table 1.

Table 1: Natural Frequency Data for Bowls

	Frequency, Hz	
Bowl	Mode 1	Mode 2
1	1,892	3,896
2	1,923	3,937
3	1,902	3,916
4	1,688 1,749	3,520 3,651
mean (1-3)	1,906	3,916

Although there is insufficient data to obtain valid statistical data, the three undamaged bowls all show frequencies within one percent of the mean, while the bowl with the hairline crack shows obvious and measurable differences. Additionally, the cracked bowl shows additional modes, possibly due to reflection at the crack boundaries.

Figure 3: Frequency Response Function, Unflawed Bowl

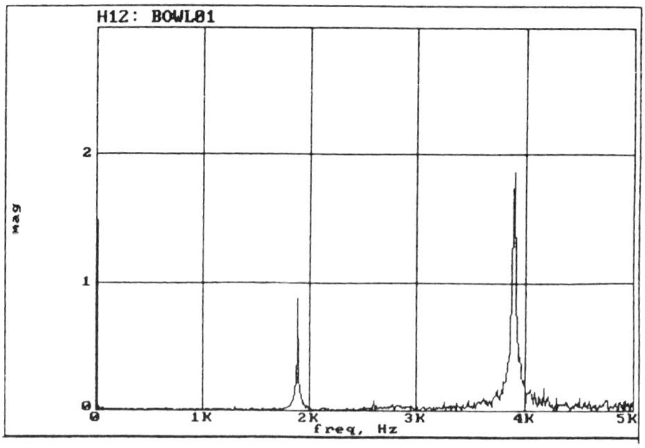

Figure 4: Frequency Response Function, Bowl with Crack

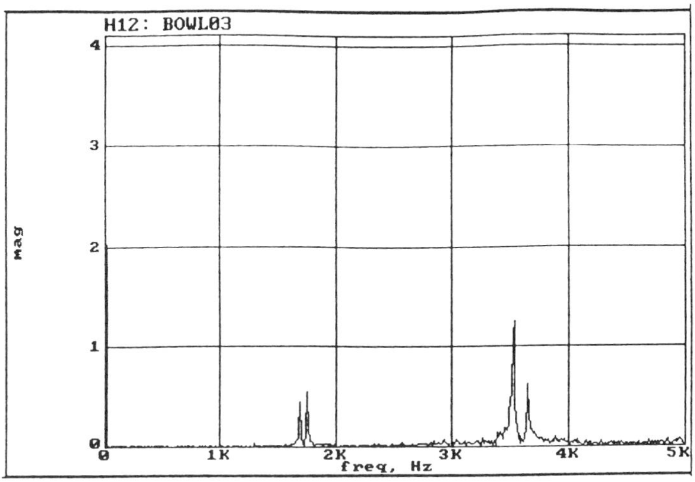

Insulators: Because in this work we assume no a priori knowledge of the modal data of a particular piece, we must rely on comparison to a nominal set of data obtained from previous testing of a large number of components. In order to develop a relatively large statistical basis for the sonic signature method, 47 pin type 19 kV high voltage insulators, 14 cm high by 21.5 cm diameter, were tested. For this component, only one natural frequency was found in the 0-5000 Hz range of the test. After preliminary statistical data was compiled for these components, two outlying points were found by applying Chauvenet's criteria[12]. The statistical data for the remaining 45 insulators is given in Table 2. A histogram showing this frequency distribution is shown in Figure 6.

Table 2: Natural Frequency Data for 45 Unflawed Insulators.

	Frequency (Hz)
mean	3828 Hz
standard deviation	51.6 Hz
minimum	3744 Hz
maximum	3967 Hz

Figure 6: Natural frequency Data for Unflawed Insulators

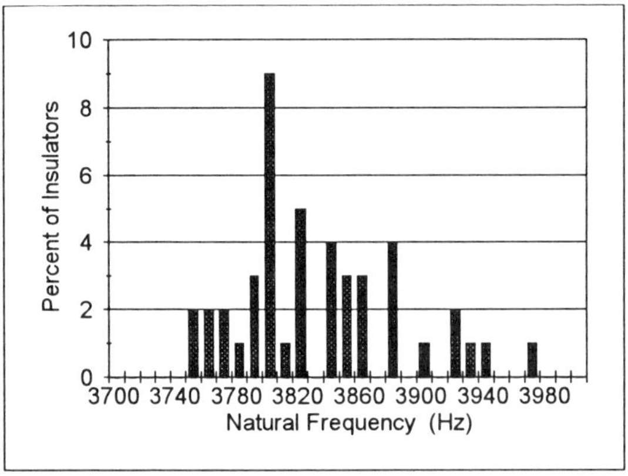

Finally, measurements were made with 27 insulators known to have small dunt cracks. These cracks are typically invisible to the naked eye, and often in an inaccessible location. Statistical data for these insulators is shown in Table 3, and plotted in Figure 7. It is apparent from both the table and the plot that the cracked insulators shown a much wider scatter and the histogram shows a non-gaussian distribution which points to defects as opposed to random experimental error or manufacturing differences.

Table 3: Natural Frequency Data for 27 Flawed Insulators.

	Frequency (Hz)
mean	3465 Hz
standard deviation	198 Hz
minimum	2838 Hz
maximum	3713 Hz

It is interesting to note that there is no overlap in the two data sets for these insulators. That is, the minimum frequency for the unflawed insulator (3744 Hz) is greater than the maximum frequency for the cracked insulators (3713 Hz). In fact, of the cracked insulators, none fall within 2.0 standard deviations (95.4 % confidence range) of the mean for the good insulators and only one falls within 2.57 standard deviations (99% confidence range).

Figure 7: Frequency Distribution for Cracked Insulators

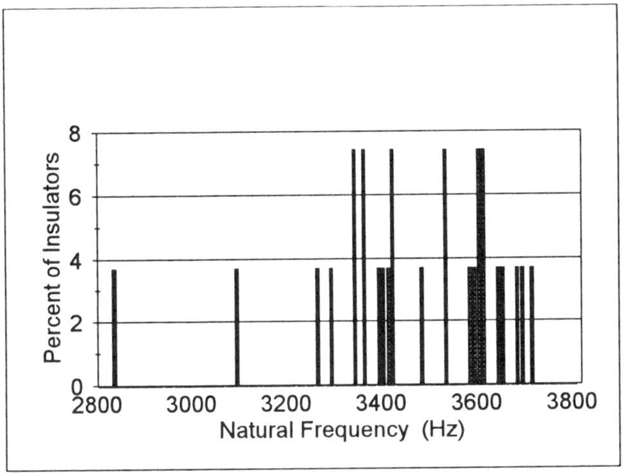

CONCLUSIONS

The sonic signature of two common ceramic components is constructed using the frequency response function created by impacting the component and measuring the resulting sound. By assembling a statistical base from a number of unflawed components, defects can be detected based upon differences from the mean values. Defective components were successfully found in both cases tested. Although a relatively low frequency range was used for these tests (0-5000 Hz), the capability of current transducers and instrumentation is in the range of 0 to 100,000Hz. This higher frequency sonic testing may allow even smaller defects visible only in higher modes to be found. In addition, gradual changes in the manufacturing process may cause a subtle shift in the "sonic signature" may also be detected. Automation of this test could make it practical to test each and every component before shipping.

REFERENCES

1. Baumeister, T., Marks' Standard Handbook for Mechanical Engineers, McGraw-Hill, Eighth Ed., 1978.

2. Cawley, P., and Adams, R.D., "The Location of Defects in Structures from Measurements of Natural Frequencies", Journal of Strain Analysis, Vol 14, no. 2, 1979, pp49-57.

3. Ewins, D., Modal Testing: Theory and Practice, Research Studies Press, 1984.

4. Richardson, M., and Mannan, M.A., "Correlating Minute Structural Faults with Changes in Modal Parameters", Proceedings of the 11th International Modal Analysis Conference, 1993, pp. 893-898.

5. Richardson, M., and Mannan, M.A., "Remote Detection and Location of Structural Faults using Modal Parameters", Proceedings of the 10th International Modal Analysis Conference, 1992, pp. 502-507.

6. Penny, J.E.T., Wilson, D.A.L., and Friswell, M.I., "Damage Location in Structures using Vibration Data", Proceedings of the 10th International Modal Analysis Conference, 1992, pp. 861-867.

7. Zimmerman, D.C., Simmermacher, T., and Kaouk, M., "Structural Damage Detection using Frequency Response Functions", Proceedings of the 13th International Modal Analysis Conference, 1995, pp.179-184.

8. Fox, C.H.J., "The Location of Defects in Structures: A Comparison of the Use of Natural Frequency and Mode Shape Data", Proceedings of the 10th International Modal Analysis Conference, 1992, pp. 522-528.

9. Kan, C.D., and Bedwi, N., "Nondestructive Evaluation of Ceramics Using System Identification Technique", Proceedings of the 12th International Modal Analysis Conference, 1994, pp.1173-1178.

10. Pape, D..A., "A Modal Analysis Approach to Flaw Detection in Ceramic Insulators", Proceedings of the 10th International Modal Analysis Conference, 1992, pp. 35-40.

11. Norton, M.P., Fundamentals of Noise and Vibration Analysis for Engineers, Cambridge University Press, 1989.

12. Holman, J. P., Experimental Methods for Engineers, 6th Ed., McGraw-Hill, 1994

ACOUSTIC EMISSION STUDY OF MICROCRACKING IN WHITEWARES

Darren Stohr, Vernon Burdick, and James Varner

Whiteware Research Center
NYS College of Ceramics at Alfred University
Alfred, NY 14802

Whiteware ceramics contain a glassy phase and several crystalline phases. Differences in contraction characteristics and, in some systems, phase transitions produce residual stresses during cooling after sintering that, in turn, produce microcracks. In this study, acoustic emission was monitored during cooling of whiteware specimens to detect microcracks as they formed. Acoustic signals were transmitted from the specimen to the transducer using an acoustic waveguide in contact with the specimen. Several commercial whiteware compositions and three different cooling schedules were investigated. Little acoustic activity occurs until the temperature falls below about 800° C. Influences of composition and cooling schedules on acoustic activity, i.e., on microcracking, will be presented. The effectiveness of different ways of presenting and analyzing acoustic emission data will be discussed (e.g., hits, energy, frequency spectra, etc.). Initial results of post-test microscopy will also be shown.

INTRODUCTION

Whitewares consist of crystalline phases, such as mullite and quartz, imbedded in a glassy matrix. These crystalline phases expand and contract at different rates during heating and cooling. Some of the crystalline phases also undergo a phase transformation which is accompanied by a large volume change, eg., β-quartz to α-quartz. These phase transitions and mismatches in thermal expansion coefficients cause stresses to develop in the body. When the stress becomes large enough, microcracking can occur. Microcracking reduces the mechanical strength and, in electrical porcelain affects electrical properties as well. In order to improve the performance of whitewares, reducing the amount of microcracking is desirable. In this study, the effect of the cooling rate and body composition are being examined using acoustic emission (AE).

EXPERIMENTAL PROCEDURE

Extruded samples of silica and alumina porcelains form corporate members of the Whiteware Research Center were shaped into 6.5 cm long cylinders with a 1.8 cm diameter. One side of the cylinder was leveled to a width of 1.6 cm so that the sample could be positioned in the furnace without moving. The furnace was designed to be capable of rapid heating and cooling rates. Each sample was fired to 1300° C at a rate of 10 K/min and held for 30 minutes. The samples were cooled at three rates, 1 K/min, 5 K/min, and 10 K/min.

The samples rest on an alumina waveguide with a piece of thin platinum foil placed between the sample and the waveguide. This foil was used to prevent reactions between the body and waveguide. The waveguide extends out of the furnace to a high-temperature transducer, where the AE signal passes through a 60 dB pre-amplifier. The AE collection unit is a Physical Acoustics Corporation model DSP-32/16.

Temperature was measured using a type K thermocouple, and recorded by the AE system. An amplifier was used to condition the thermocouple output. The thermocouple was inserted into the furnace at about 1040° C during the cooling cycle, making sure it is nearly touching the sample, and data measurements started.

The post-fired microstructure of the as-fired surfaces was examined using scanning electron microscopy.

RESULTS AND DISCUSSION

The first parameter investigated was the number of hits, or packets of phonons, received as a function of temperature. The hits are separated into small 3.75 K bins. The hits are recorded as a hit sum per bin, which is an indication of the acoustic activity taking place in that temperature interval. There is little acoustic activity above 800° C in the silica bodies. Moderate activity is present from 800° C to about 500° C. There is a large increase in activity below about 500° C. The higher the cooling rate, the lower the onset of the increased activity as seen in figures 1, 2, and 3. Composition does not affect the onset of the increased activity. Figures 4, 5, and 6 show the activity of an alumina body.

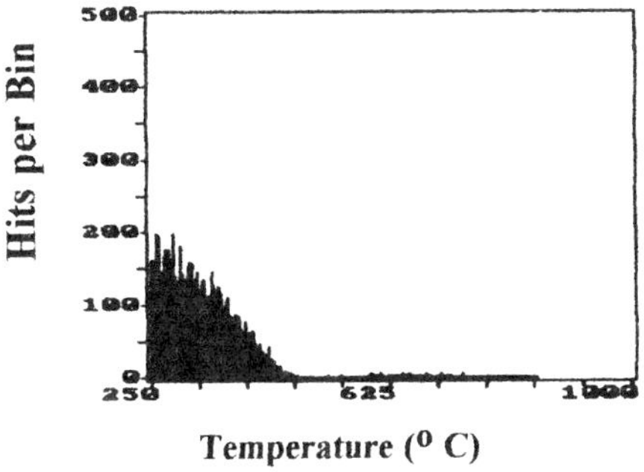

Figure 1: Hits per Bin vs Temperature of Silica Body Cooled at 10K/min

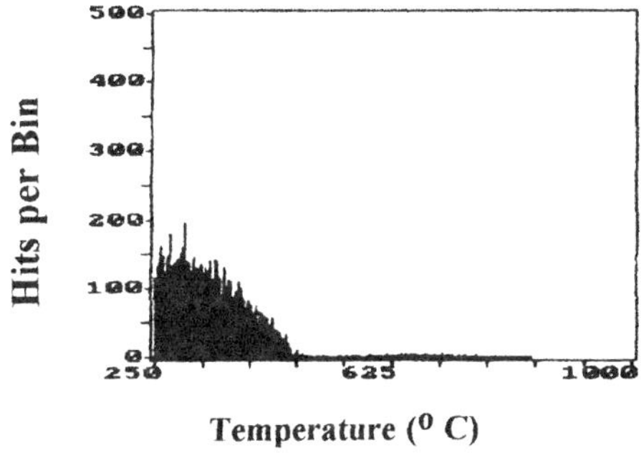

Figure 2: Hits per Bin vs Temperature of Silica Body Cooled at 5K/min

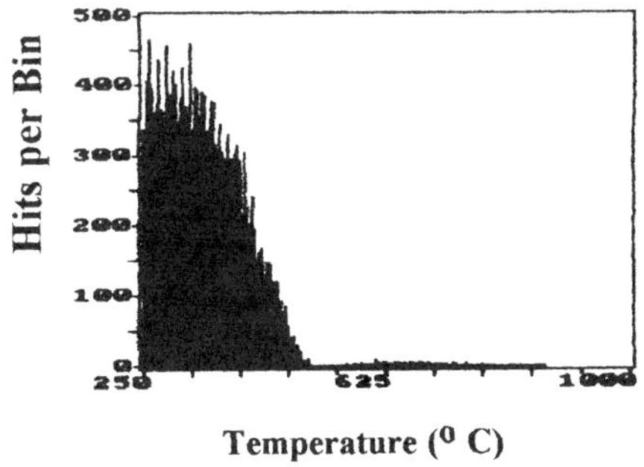

Figure 3: Hits per Bin vs Temperature of Silica Body Cooled at 1K/min

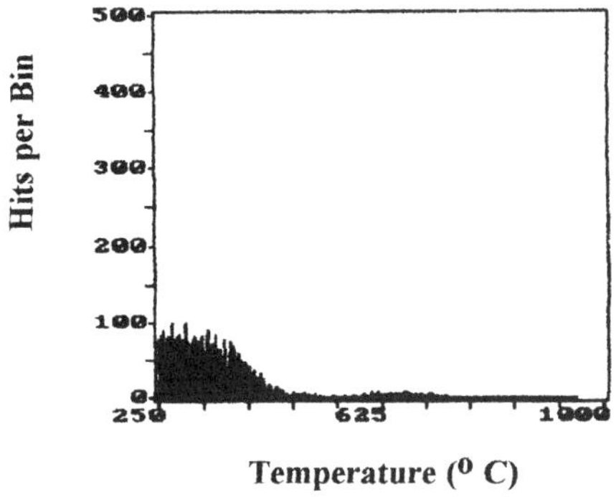

Figure 4: Hits per Bin vs Temperature of Alumina Body Cooled at 10K/min

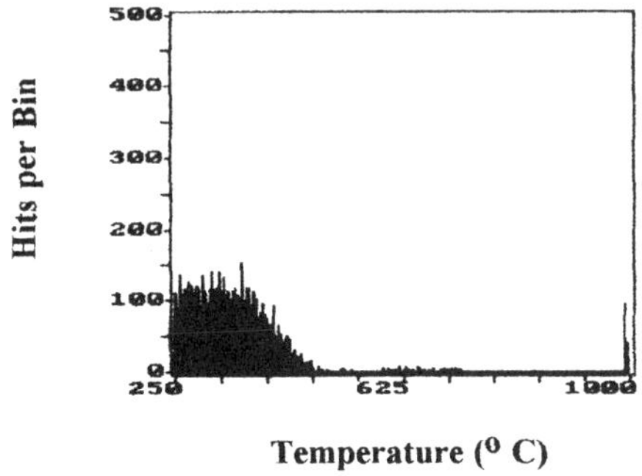

Figure 5: Hits per Bin vs Temperature of Alumina Body Cooled at 5K/min

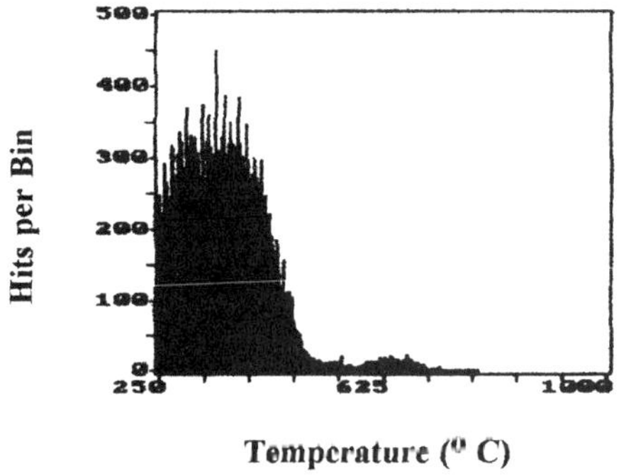

Figure 6: Hits per Bin vs Temperature of Alumina Body Cooled at 1K/min

These results are consistent with a previous study.[1]

The higher activity below 500° C corresponds to microcracking. Microcracking occurs both around the quartz grains in the glassy phase and in the grains themselves as the thermal stresses increase with decreasing temperature. The volumetric phase transition from β-quartz to α-quartz should also produce an AE signal. The quartz transition occurs at 573° C under equilibrium conditions. The temperature of the sample lags behind the air temperature, and there is a further lag within the sample itself. The inside of the sample is at a higher temperature than the outside. This is magnified by a faster cooling rate. This temperature lag can explain the lower temperature for the onset of AE activity for the faster cooling rates.

The phase transformation is fast, thus the temperature is very consistent. It would be expected that the activity would increase just below 573° C if the phase transition itself was being observed. The lag of the sample temperature would probably not be great enough to shift the increased activity to such a degree. The acoustic activity must then be attributed to the formation and extension of microcracks.

The activity between 500° C and 800° C appears to have two possible sources. The first is microcrack formation and extension. The microcracking would probably occur in the glassy phase around large crystalline grains and possibly within the grains themselves. The second possible explanation is related to the Pt foil used to separate the sample from the waveguide. There are thermal mismatches between the sample and the foil and between the foil and the waveguide. The thermal mismatches might cause a quick release of energy as the Pt foil slides. However, there should not be many such events, so this effect can be ignored.

The average energy per bin was also examined, figures 7, 8, and 9. The energy is a relative quantity. It is the amplitude over a threshold value set by the user integrated with respect to time. In this system, a high-energy event is the formation of a microcrack, while a low-energy event could be a smaller microcrack or the extension of an existing microcrack. It is assumed that the attenuation of the energy signal is about the same regardless of the location of the event in the sample. This is assumed, since the signal must travel through the Pt foil, down the waveguide, and into the transducer. There are two interfaces that would reduce the phonon signal plus the long waveguide makes this assumption reasonable.

The low-activity zone (above about 500° C) is characterized by high average energy. This indicates that the activity is due to the formation of microcracks. The average energy per bin is low for the high-activity zone (below about 500° C). There are some high-energy events at these lower temperatures,

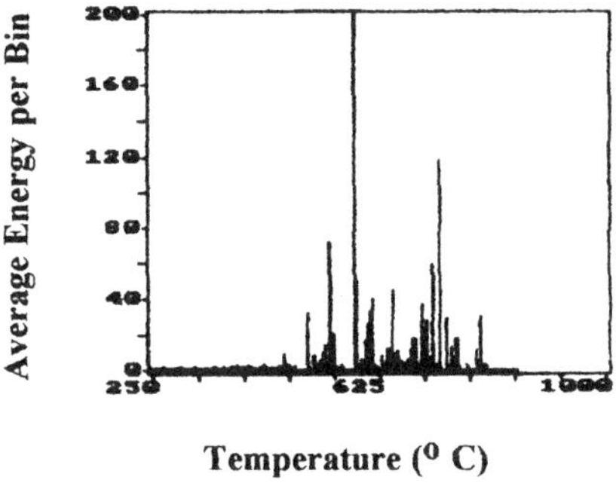

Figure 7: Average Energy per Bin vs Temperature of Silica Body Cooled at 10K/min

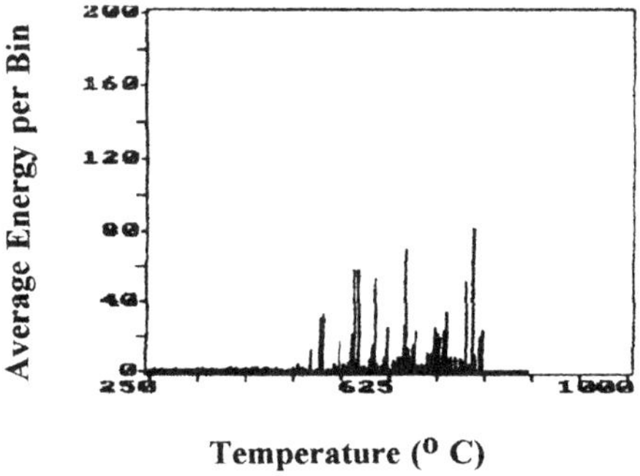

Figure 8: Average Energy per Bin vs Temperature of Silica Body Cooled at 5K/min

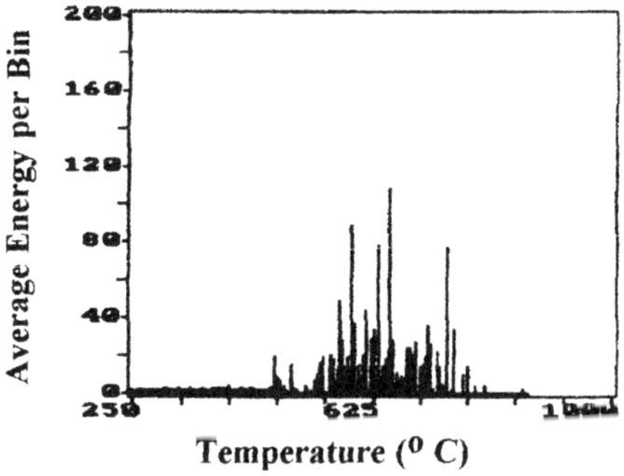

Figure 9: Average Energy per Bin vs Temperature of Silica Body Cooled at 1K/min

but the number of low-energy events dominates. This indicates that either smaller cracks are formed or there is extension of existing flaws.

The waveform of the signal can also be examined. Due to the large number of hits recorded, only a small sample of the results was examined. The waveforms support the average-energy results. The maximum amplitude of the signal decreases with decreasing temperature, as shown in figures 10, 11, and 12. The time of the waveform, i.e., the event duration, appears to be shorter with decreasing temperature, although this is not conclusive, due to the limitation of the waveform output.

Another method for examining the AE data is by plotting cumulative hits vs amplitude[2], figures 13 and 14. The slope of the log-log plot is found. A higher percentage of high-energy events yields a lower slope, and a lower percentage of high-energy events yields a higher slope. A low slope indicates that a larger number of microcracks are being formed. A high slope indicates that there is a greater proportion of extending of microcracks. There appears to be little difference in the slopes of the fast-cooled and slow-cooled specimens. However, these are preliminary results, which are not conclusive. The slope does not appear to be a single value. Instead, there are at least two different slopes present. The number of samples investigated in this was small at this point due to the difficulty in managing the data. The large number of events makes it practically difficult for all the AE parameters to be recorded on each cooling run without making the data files too cumbersome. Work is continuing in this area.

For the silica bodies, the number of hits is larger for the slower cooling rates. It is not clear if this is a result of the longer time available to record information, or if there is more microcracking occurring. If the results of the evaluation of cumulative hits vs amplitude is correct, it appears that the longer time available to record the data is responsible. More work is needed in these areas, and microstructure evidence must be used to support this conclusion.

The number of hits for the alumina body appears to be similar to that of the silica body except for the fastest cooling rate. The faster cooling rates show an apparent decrease in AE activity. This indicates that the addition of alumina is beneficial for faster cooling rates but not necessarily for the slower cooling rates.

Initial results from the examination of the microstructure appear promising. Microcracks can be detected on the as-fired surface, i.e., on surfaces free from post-test mechanical handling. Cracks can be seen in and around grains and in the glassy phase, figures 15 and 16. Other sample preparation techniques, such as polishing and chemical etching, need to be examined, because the sample surfaces are rough.

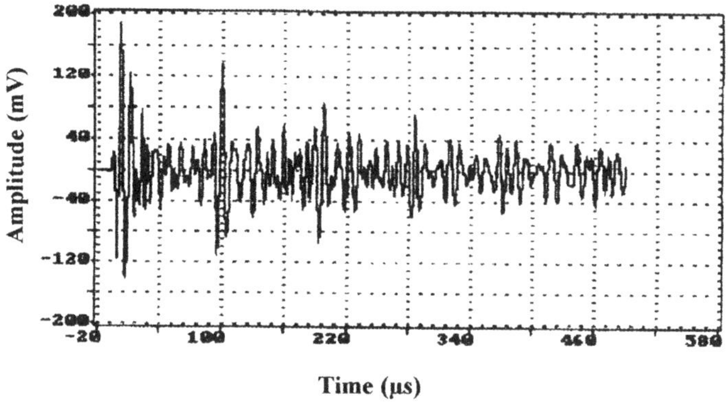

Figure 10: Waveform of Silica Body at 680° C with a Cooling Rate of 10K/min

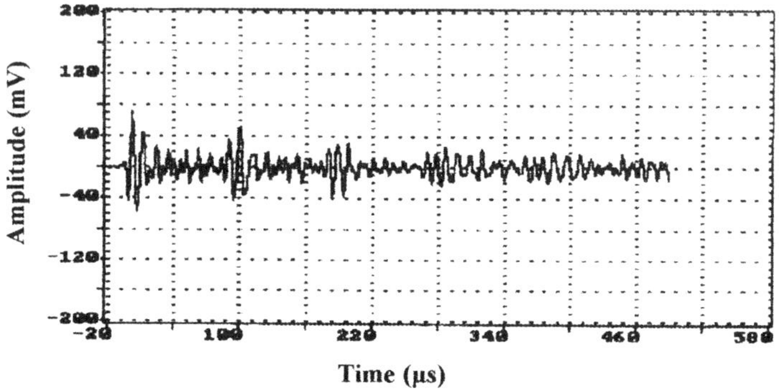

Figure 11: Waveform of Silica Body at 529° C with a Cooling Rate of 10K/min

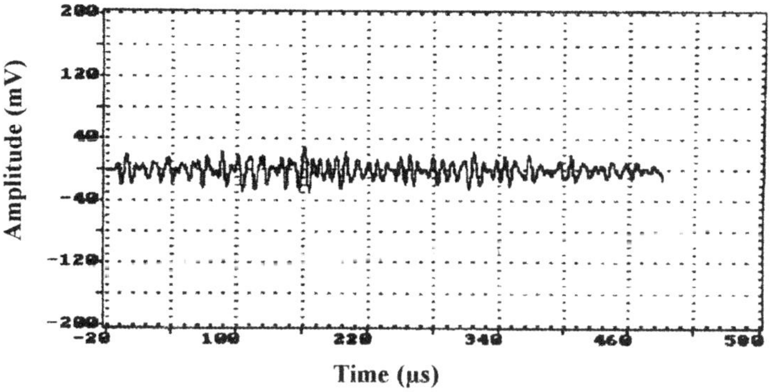

Figure 12: Waveform of Silica Body at 392° C with a Cooling Rate of 10K/min

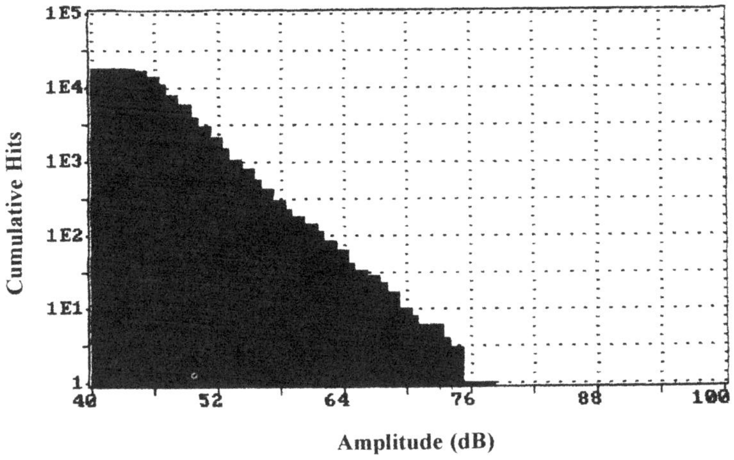

Figure 13: Log Cumulative Hits vs Amplitude (dB) of Silica Body Cooled at 10K/min

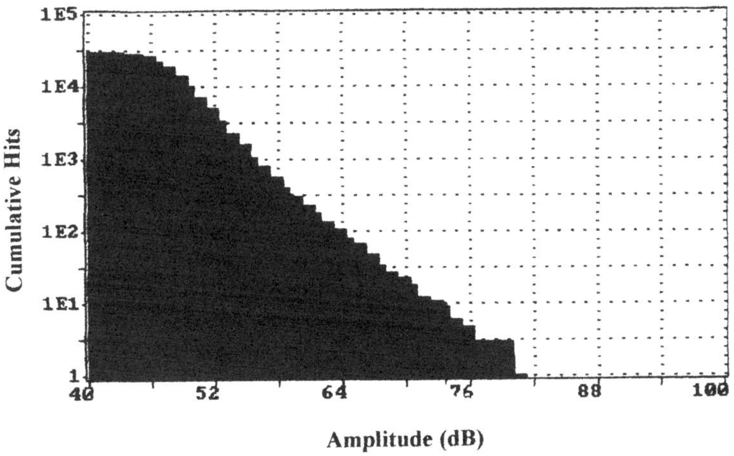

Figure 14: Log Cumulative Hits vs Amplitude (dB) of Silica Body Cooled at 1K/min

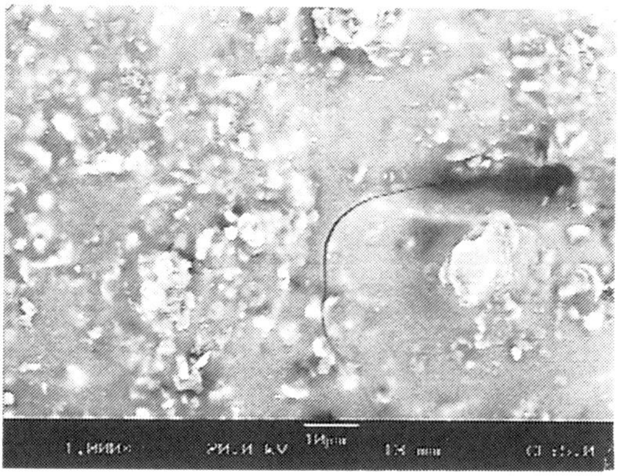

Figure 15: Tangental Microcrack Around Grain

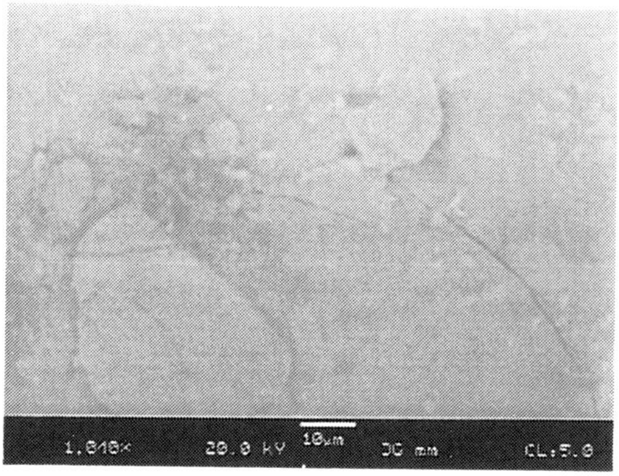

Figure 16: Microcracks Through Grain and In Glassy Phase

CONCLUSION

The effect of cooling rate and composition on microcracking in whitewares was examined by monitoring and analyzing acoustic emission activity during cooling. There is low activity above 500° C. It appears that the microcrack formation is occurring, because there are high energies associated with these events. The high activity below 500° C is due to the formation of microcracks, extension of existing microcracks and possibly the β-quartz to α-quartz transition. There is little or no activity above 800° C.

The slower cooling rates have a higher number of events occurring, but it appears this is due to time available for data collection. The addition of alumina to the whiteware body appears to be more effective in reducing microcracking at the faster cooling rates than at the slower rates.

REFERENCES

[1]J. Herrmann, "Microcracking During Post-Firing Cooling Process," Internal Research Report, NYS College of Ceramics at Alfred, April 1994.

[2]A.A. Pollock, "Acoustic Emission Amplitude Distributions," *International advances in Nondestructive Testing*, Vol. 7 215-239 (1981).

[3]A.A. Griffith, "The Phenomena of Rupture and Flow in Solids," *Philosophical Transactions,* Series A, Vol. 221, (1920) p 163-198.

[4] D.B. Binns, "Some Physical Properties of Two-Phase Crystal-Glass Solids," *Science of Ceramics, Vol. 1,*G.H. Stewart, ed., Academic Press, London, (1962) p 315-334.

[5] D. Weyl, "Influence of Internal Strains on Texture and Mechanical Strength of Porcelains," *Ber. Deut. Keram. Ges.*, **36** [10] (1959) p 319-324.

[6]J.Selsing, "Internal Stresses in Ceramics," *Journal of the American Ceramic Society - Discussions and Notes*, **44** [8], (1961) p 419.

[7]R.W. Davidge and T. J. Green, "The Strength of Two-Phase Ceramic/Glass Materials," *Journal of Materials Science*, **3** (1968) p 629-634.

[8]F.F. Lange, "Criteria for Crack Extension and Arrest in Residual, Localized Stress Fields Associated with Second Phase Particles," *Fracture Mechanics of Ceramics, Vol. 2*, (1974) p 599-609.

EFFECT OF THE DEGREE OF DOMAIN REVERSAL ON ACOUSTIC EMISSION IN 1MOL%MnO_2-DOPED $(Pb_{0.85}La_{0.15})TiO_3$ FERROELECTRIC CERAMICS

D.G. Choi and S.K. Choi*
Department of Materials Science and Engineering
Korea Advanced Institute of Science and Technology
373-1 Kusong-dong, Yusong-gu
Taejon, 305-701, Korea

ABSTRACT

Acoustic emission from the domain reversal was investigated using 1mol%MnO_2-doped $(Pb_{0.85}La_{0.15})TiO_3$ ferroelectric ceramics with various domain sizes. Domain size was changed by the control of grain size. The degree of 90° domain reversal was quantitatively evaluated by using X-ray diffraction technique and acoustic emission during poling process was measured. All the samples showed an abrupt increase of 90° domain reversal in the early stage of the poling and its saturation above approximately 10min. It was also found that the larger the domain, the larger the 90° domain reversal. As the poling time and the domain size increased, the peak amplitude distribution of acoustic emission shifted toward higher peak amplitude region. This result could be explained by the fact that the amount of elastic energy released during the domain reversal is related to the degree of domain reversal and the domain structure.

INTRODUCTION

Grains of ferroelectric ceramics consist of a large number of domains which are formed in order to accommodate both elastic strain and depolarization field produced when the cubic paraelectric phase transforms into the noncubic ferroelectric phase[1,2]. In general, ferroelectric domains with tetragonal symmetry are grouped into two categories: 90° and 180° domains[1,2]. They

---

* Member, American Ceramic Society.

are switched favorably towards the field direction under a high d.c. electric field, which is called "poling". The poling process is essential to create piezoelectric and pyroelectric properties in ferroelectric ceramics. Because the domain rearrangement during the poling process has a direct influence on the macroscopic properties of ferroelectric ceramics, numerous efforts have been made to obtain a better understanding of the static and the dynamic properties of ferroelectric domains.

Several techniques, including X-ray diffraction (XRD)[3-5], strain gauge[6], and acoustic emission (AE)[7-11], have been applied to the study of domain dynamics. The XRD and the strain gauge techniques can only be used to make macroscopical observations of the degree of 90° domain reversal by measuring the change of X-ray diffraction peaks or strain, respectively. AE is defined as the stress waves which are generated during the elastic energy release associated with a deformation process in a material[12]. It is known that the AE technique is a very powerful tool in the interpretation of dislocation dynamics of ductile materials, or cracking behavior of brittle materials[12]. In the application of the AE technique to the ferroelectric materials, several studies have also revealed that AE is generated from domain reversal[7-11], paraelectric-to-ferroelectric phase transition at Curie temperature[10], and microcracking[10,13]. But, progressive results have not yet been achieved in the understanding of domain reversal related AE.

In this study, the characteristics of AE from the domain reversal in the ferroelectric ceramics during the poling process was investigated. $(Pb_{0.85}La_{0.15})TiO_3$ + $0.01MnO_2$ ferroelectric ceramics were prepared to have various domain sizes which were changed by the control of the grain size. The degree of 90° domain reversal and nonuniform strain with poling time were quantitatively evaluated by X-ray diffraction technique, and the domain structure was observed by scanning electron microscopy (SEM). The peak amplitude distribution of AE was discussed in terms of the amount of elastic energy released during the domain reversal.

EXPERIMENTAL

The chemical composition of $(Pb_{0.85}La_{0.15})TiO_3$ was weighed in stoichiometric from high-purity oxide powders, PbO, La_2O_3, and TiO_2. The oxide powders were mixed in a ball mill for 24h and calcined at 900 ℃ for 3h. The calcined powder was ground with the addition of 1mol% MnO_2 powder and 1wt% polyvinyl alcohol as a binder in a ball mill for 20h and dried in an oven. Samples were pressed into a 19mm disks and sintered at 1200 ℃ and 1250 ℃ for 1 to 30h. The average grain size was in the range of 1.6~10.1 μm. According to the X-ray diffraction analysis, all of them were confirmed to have a perovskite phase.

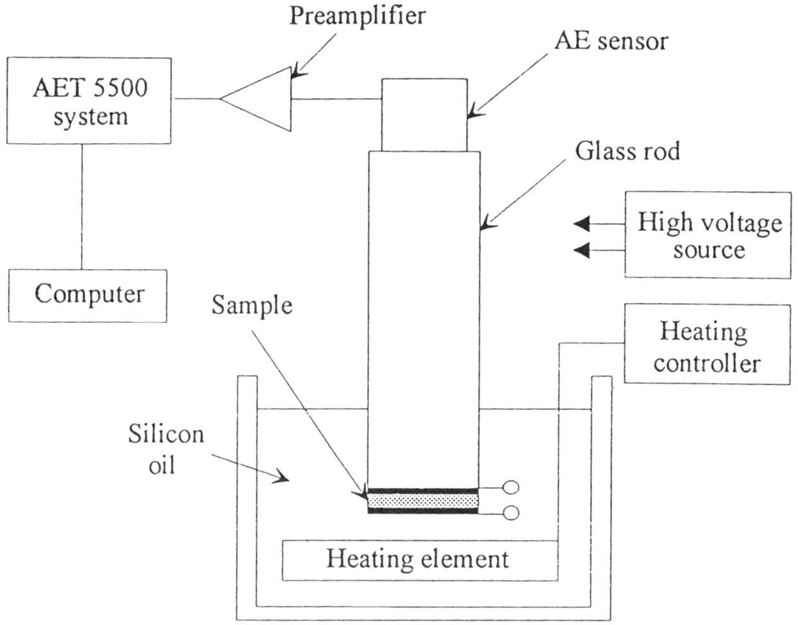

Fig. 1. A schematic representation of the experimental set-up for the AE measurement.

Their relative density was approximately equal to be 96 ± 1%. The samples were polished to 0.45mm and electroded on both sides by gold sputtering.

All the samples were poled by applying an electric field of 3kV/mm in a silicon oil bath at 105 ℃. AE measurements on the ferroelectric ceramics under the poling process were carried out using the experimental set-up schematically represented in Fig. 1. The AE was monitored by an AET 5500 system(Hartford Stream Inspection Technologies, U.S.A.). The AE signals were detected by an AE sensor with a resonant frequency of 175kHz via a glass rod as a waveguide and filtered by a band filter with a range of 125 to 1,000kHz. The overall gain was 70dB, and the threshold voltage was chosen as 0.1V.

Domain size and nonuniform strain were determined by the XRD technique. XRD measurements were performed on an X-ray diffractometer with the following operating conditions: CuK_α radiation, tube voltage 40kV, and tube current 40mA. The diffraction peaks of (111) and (222) planes which had no convoluted peaks in the neighborhood were chosen and the integral widths of $K_{\alpha 1}$ diffraction peaks, separated from the averaged K_α lines, were obtained. A coarse grained TiO_2 ceramic which was sintered at 1200 ℃ for 3h and of which

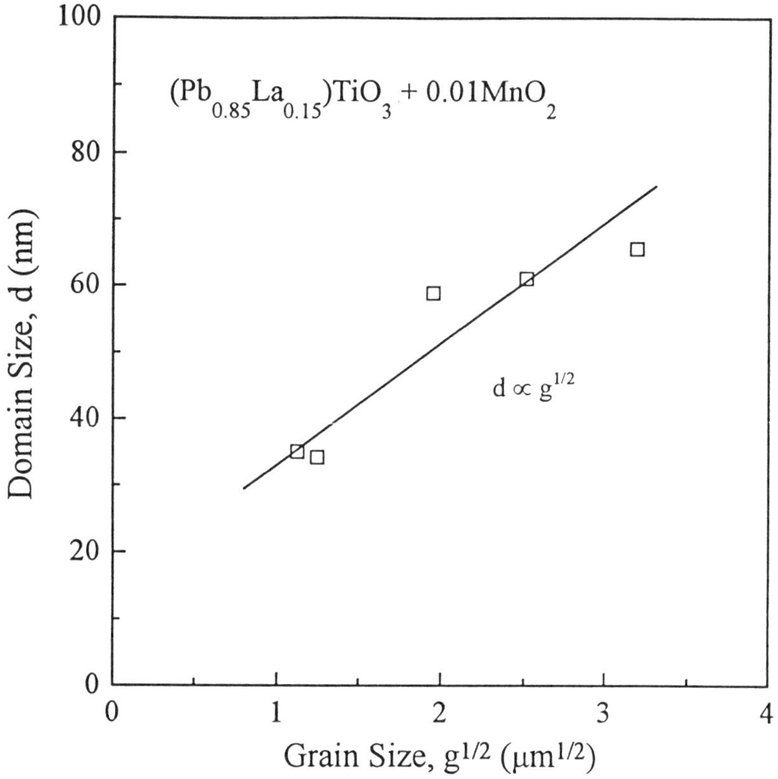

Fig. 2. Variation of domain size as a function of the square root of grain size of $(Pb_{0.85}La_{0.15})TiO_3 + 0.01MnO_2$ ceramics.

grain size was about 10 μ m was used as a standard sample for the calibration of peak broadening caused by the equipment. By using a Gauss squared method[14], the domain size and the nonuniform strain were calculated, respectively. The changes of the intensities of the (002) and (200) diffraction peaks with poling time were measured, and the degree of 90° domain reversal was calculated by using the equation that Mendiola and Pardo[3] suggested. The domain structure was characterized by SEM.

RESULTS AND DISCUSSION

Figure 2 shows the variation of the domain size as a function of the square root of grain size in the range of 1.6~10.1 μ m. The domain size is directly propotional to the square root of grain size. Arlt[15] theoretically calculated the dependence of the domain size(d) on the grain size(g), roughly expressed by

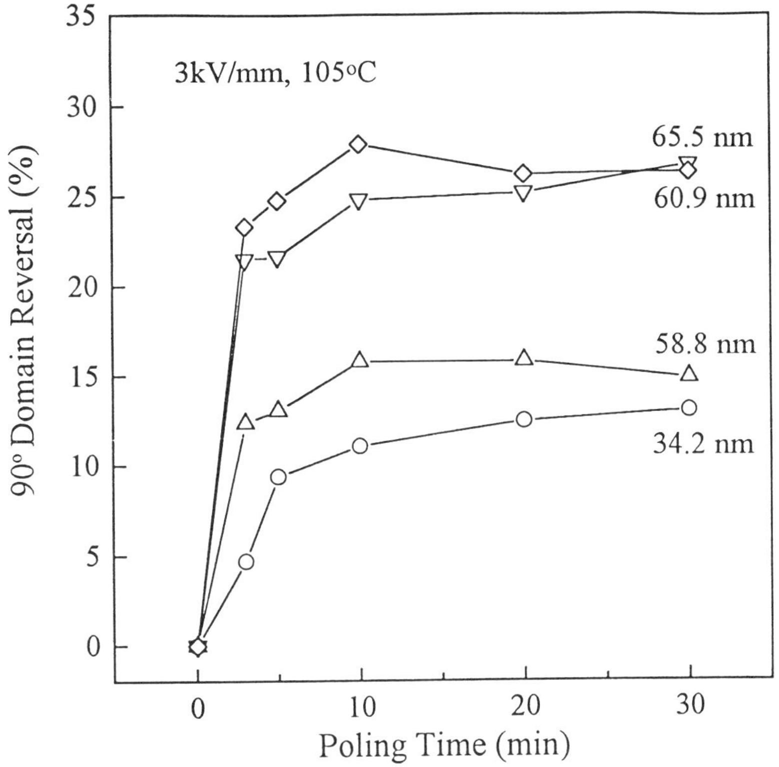

Fig. 3. Variation of 90° domain reversal of $(Pb_{0.85}La_{0.15})TiO_3$ + $0.01MnO_2$ ceramics with various domain size as a function of poling time.

$d \propto g^{1/2}$, and also confirmed that relationship in fine-grained $BaTiO_3$ ceramics of which the grain size is in the range of 1~10 μm with light microscopy and SEM.

Figure 3 shows the variation of 90° domain reversals of the samples with different domain sizes as a function of the poling time. It seems that the reversal behaviors of 90° domains of all the samples are similar. All of them show an abrupt increase of the 90° domain reversal in the early stage of the poling process and saturation above approximately 10min. On the other hand, there is a clear difference in the degree of 90° domain reversal even though all the samples were poled at the same condition. Figure 3 indicates that the amount of 90° domain reversal increases, as the domain size of sample increases. Kang et al.[16] have suggested that the increase of the pyroelectric coefficient of $(Pb_{0.9}La_{0.1})TiO_3$ ceramics with the grain size results from the enhancement of

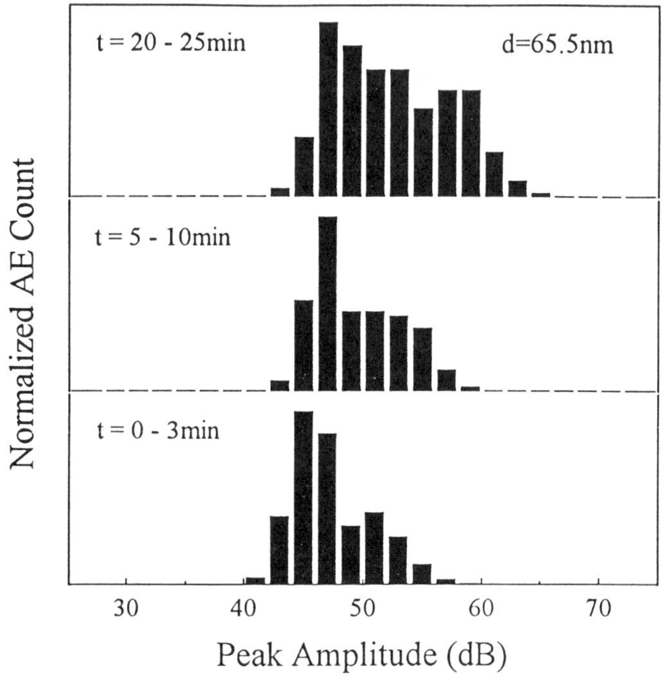

Fig. 4. Poling time dependence of the peak amplitude distribution of AE measured from $(Pb_{0.85}La_{0.15})TiO_3$ + $0.01MnO_2$ ceramic of which the average domain size is 65.5nm.

90° domain reversal.

The previous study[17] revealed that most of the AE from the $(Pb_{0.85}La_{0.15})TiO_3$ + $0.01MnO_2$ ceramics during the poling process occurred due to the 90° domain reversal and to a lesser extent 180° domain reversal. Therefore, in order to investigate the characteristics of AE with the degree of 90° domain reversal, the peak amplitude distribution of AE with the poling time has been analyzed for the sample with an average domain size of 65.5nm, which shows the most active 90° domain reversal, and is shown in Fig. 4. The peak amplitude is given in decibels above 10 μV voltage at the preamplifier input. The distributions are normalized for each maximum AE count. According to the observed AE generation behavior, the AE generation constantly occurs during all the poling, even though the degree of 90° domain reversal is saturated above 10min (Fig. 3). This is explained by the alternative switching behavior of the 90° domains[17] which means that the 90° domains reverse toward field directions or their original directions above a certain poling time.

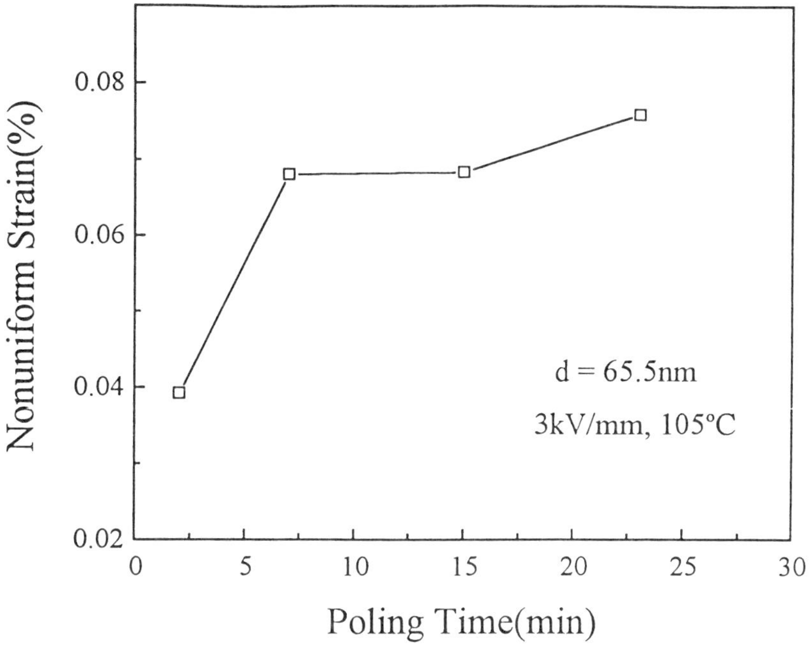

Fig. 5. Variation of nonuniform strain with poling time for $(Pb_{0.85}La_{0.15})TiO_3$ + $0.01MnO_2$ ceramic of which the average domain size is 65.5nm.

A trend seen in Fig. 4 is that the distribution shifts toward a higher peak amplitude level, as the poling time increases, that is, the degree of 90 ° domain reversal increases. Pan and Cao[11] have supposed that the peak amplitude(V_o) of AE observed from the transducer is proportional to the square root of the energy(ΔE_g) released during a given deformation process,

$$V_o = \varphi (\Delta E_g)^{1/2} \text{------------------------------} (1)$$

where φ is a proportionality factor, representing both the material properties and the preamplifier characteristics. Therefore, Figure 4 indicates that the amount of elastic energy released during the domain reversal becomes larger, as the degree of the domain reversal increases.

Iwasaki et al.[8] suggested that the the domains of ferroelectric ceramics have the characteristics threshold field levels to move the domain wall or to reverse their polarization vector. It supports the distribution of the threshold level for domain reversal that the observed peak amplitudes are distributed over a range of

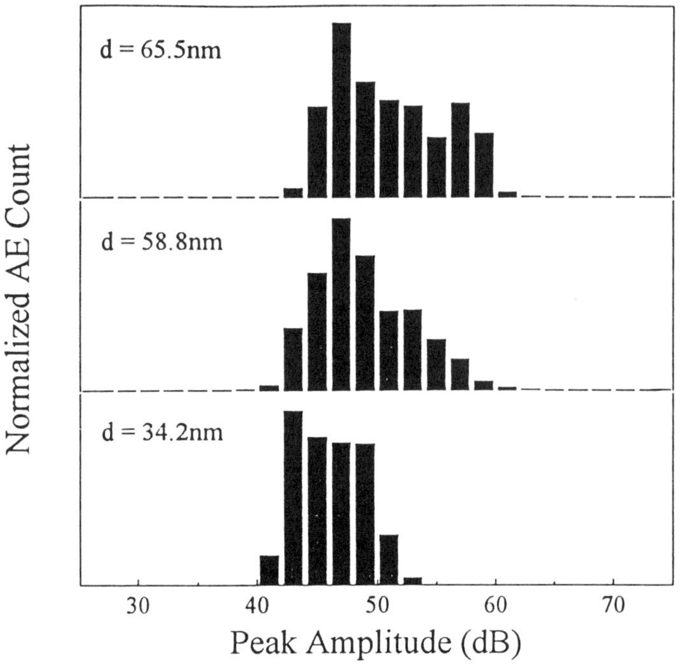

Fig. 6. Domain size dependence of the peak amplitude distribution of AE measured from $(Pb_{0.85}La_{0.15})TiO_3$ + $0.01MnO_2$ ceramics during the poling for 30min.

40 to 60dB(Fig. 4). When an electric field is applied to a sample, the supplied energy is first stored in the sample in the form of strain energy. When the local strain energy in a given domain exceeds the threshold level of the domain, the domain reversal occurs. But, the 90 ° domain reversal in a grain clamped by its neighboring grains in all three dimensions induces a large mismatch strain due to the tetragonality deviated from one. Figure 5 shows that the nonuniform strain increases abruptly in the early stage of poling in which the 90 ° domain reversal increases (Fig.3) and saturates above around 7min. As a result, as the degree of 90 ° domain reversal increases, the threshold level will be redistributed toward a higher energy level. This indicates that a larger amount of strain energy have to be accumulated to reverse its neighboring domains. When this happens, the resulting AE will have a higher peak amplitude. Therefore, the variation of the peak amplitude distribution with the poling time in Fig. 4 can be interpreted to be due to the difference of the degree of domain reversal. In particular, a remarkable shift of distribution toward a higher peak amplitude region is seen in the poling time of 20~25min at which the domain reversal sustains its saturation state. This

suggests that the constant application of an electric field without a contribution to the domain reversal toward the field direction induces a very large elastic stress in the material[17].

Figure 6 shows the domain size dependence of the peak amplitude distribution of AE measured from the samples during the poling process for 30min. As the domain becomes larger, the distribution shifts toward higher peak amplitude region. It is natural to consider that the reversal of the large domain requires the accumulation of a larger amount of strain energy than that of small domain. Therefore, AE originated from the reversal of large domains will occur with a high peak amplitude.

However, it is known that the domain structure in ferroelectric ceramics is dependent on the grain size[15]. It is expected that the domain structure will affect the domain reversal behavior. The domain structures of the ceramics of which the average domain sizes are 34.2nm(a) and 65.5nm(b), respectively, are shown in Fig. 7. As shown in Fig. 2, the former ceramic shows fine grains and the latter ceramic exhibits coarse grains. The 90° domains are well developed regardless of the grain size. Fine-grained ceramics consisting of small domains mainly show a simple lamellar domain structure. In contrast, coarse-grained ceramics consist of complex domain structures, including lamellar, banded, and irregular domain structure. The simple lamellar domain structure contains only grain boundary mismatch strains whereas the banded domain structure contains domain boundary mismatch strain as well as grain boundary mismatch strain[15]. In addition, it is thought that the irregular domain structure contains more complex and heterogeneous strains. Consequently, it is believed that the domain reversal in the banded or in the irregular domain structure will be accompanied by a larger amount of elastic energy release than that in the simple banded domain structure. In Fig. 7, the sample with an average domain of 34.2nm, mainly containing the simple lamellar domain structure, shows the most narrow distribution width in the low amplitude region. This is an evidence that the peak amplitude distribution of AE from the domain reversal is related to the domain structure.

CONCLUSION

The characteristics of AE from the domain reversal in the ferroelectric ceramics during the poling process were investigated. For this study, $(Pb_{0.85}La_{0.15})TiO_3 + 0.01MnO_2$ ferroelectric ceramics with various domain sizes which were changed by the control of the grain size were prepared. The results obtained in this study are summarized as follows.

1. All the samples showed an abrupt increase of the 90° domain reversal in the early stage of the poling process and its saturation above approximately

Fig. 7. SEM micrographes showing the domain structure of ceramics of which the average domain sizes are (a) 34.2nm in fine grains and (b) 65.5nm in the coarse grains, respectively.

10min. It was also found that the larger the domain size, the more the 90° domain reversal.

2. As the poling time and the domain size increase, the peak amplitude distribution of AE shifted toward a higher peak amplitude region. This result could be interpreted by the fact that the amount of the elastic energy released from the domain reversal was affected by the degree of 90° domain reversal and the domain structure.

REFERENCES

1. Y. Xu, "Perovskite-type Ferroelectrics ; Part I"; pp.101-215 in Ferroelectric Materials and Their Applications, Elsevier Science Publishers B.V., Netherlands , 1991.

2. D. Berlincourt and H.H.A. Hrueger, "Domain Processes in Lead Titanate Zirconate and Barium Titanate Ceramics", *Journal of Applied Physics*, **30** 1804-08 (1959).

3. J. Mandiola and L. Pardo, "A XRD Study of 90° Domains in Tetragonal PLZT under Poling", *Ferroelectrics*, **54** 199-202 (1984).

4. S. Li, C.Y. Huang, A.S. Bhalla, and L.E. Cross, "90° Domain Reversal in $Pb(Zr_xTi_{1-x})O_3$ Ceramics," *Ferroelectrics Letter*, **16** 7-19 (1993).

5. C.C. Li, "The Measurement of Space Distribution of Domains in Polarized PZT Ceramics", *Ferroelectrics*, **37** 623-26 (1981).

6. E.C. Subbarao, M.C. McQuarrie, and W.R. Buessem, "Domain Effects in Polycrystalline Barium Titanate", *Journal of Applied Physics*, **28** 1194-1200 (1957).

7. P.Buchman, "On Acoustic Emission from Ferroelectric Crystals", *Solid-State Electronics*, **15** 142-44 (1972).

8. H. Iwasaki and M. Izumi, "Acoustic Emission and Piezoactivity during Poling in PZT-like Ferroelectric Ceramics", *Ferroelectrics*, **37** 563-66 (1981).

9. I.J. Mohamad, E.F. Lambson, A.J. Miller, and G.A. Saunders, "A Threshold Electric Field for Acoustic Emission from Ferroelectric $Pb_5Ge_3O_{11}$", *Physics Letter*, **71A**(1), 115-18 (1979).

10. K. Uchino, T. Hirose, and A.M. Varaprasad, "Acoustic Emission in Piezoelectric/Electrostrictive Actuators", *Japanese Journal of Applied Physics*, **26** (Supplememt 26-2), 167-70 (1987).

11. W. Pan and H. Cao, "Acoustic Emission during Polarization Switching in Ferroelectric Ceramics", *Ferroelectrics*, **129** 119-26 (1992).

12. R.G. Liptai, D.O. Harris, and C.A. Tatro, "An Introduction to Acoustic Emission" ; pp.3-40 in Acoustic Emission, Edited by R.G. Liptai, D.O. Harris, and C.A. Tatro, ASTM STP 505, Philadelphia, 1972.

13. V. Spikanta and E.C. Subbarao, "Acoustic Emission in Ferroelectric Lead

Titanate Ceramics : Origin and Recombination of Microcracks", *Acta Metallurigica et Materialia*, **40**(5), 1091-100 (1992).

[14.] B. Morosin, E.J. Graeber, and R.A. Grtaham, "Line Broadening Studies on Highly Defective Al_2O_3 Produced by High Pressure Shock Loading", *Advanced X-Ray Analysis*, **27** 369-78 (1983).

[15.] G. Arlt, "Review Twinning in Ferroelectric and Ferroelastic Ceramics : Stress Relief", *Journal of Materials Science*, **25** 2655-66, (1990).

[16]. B.S. Kang, D.G. Choi, and S.K. Choi, "Effects of Grain Size on Pyroelectric and Dielectric Properties of $Pb_{0.9}La_{0.1}TiO_3$ Ceramics", *Journal of Materials Science*, submitted.

[17.] D.G. Choi and S.K. Choi, "Dynamical Behaviour of Domains during Poling by Acoustic Emission Measurements in La-modified PbTiO3 Ferroelectric Ceramics", *Journal of Materials Science*, **32** 421-25 (1997).

LIGHT SCATTERING OF PARTICLES IN GLASSES

M.H. Goller[+] , L.D. Pye
New York State College of Ceramics
Alfred University
Alfred, NY 14802

ABSTRACT

Small metal particles in glasses are well known to show wavelength selective, size dependent absorption effects in the wavelength range for visible light. This results in a red color for gold and yellow hues for silver particles. This study uses an algorithm based on the Mie theory to calculate and separate the scattering and absorption components of the observed UV-VIS spectra. Comparison with experimental data from silver containing glasses are made. The calculations show that the overall observed extinction has a significant wavelength dependence which can be attributed to the scattering contribution. The dielectric constants used in the model calculation are critical for the development of the UV-VIS spectra.

INTRODUCTION

Light scattering techniques are a well established tool for the determination of particle sizes in gasses and liquids. The applied methods are generally based on the light scattering theory developed by Mie[1], which is in turn directly derived from Maxwell's equations. The key problem in the mathematical description is to properly define all boundary conditions for the interaction of the electromagnetic field with the particle suspended in the medium. This includes the nature and geometry of the particle as well as the optical properties of the particle.

The system in this work is defined as precious metal spheres that are incorporated in an aluminosilicate glass. These glasses are commercially used as filters. The group of photosensitive glasses, which are based on gold or silver doping, are also included in this group.

[+] Now with: Corning Incorporated, Corning, NY 14831

The intention of this work was to establish a rapid size determination tool for the metal spheres in the glass, which, besides the interesting optical properties, is of interest for subsequent steps such as crystallization of the glass.

THEORETICAL BACKGROUND

This work concentrated on the adaptation of the Mie theory to calculate the UV-VIS spectra for the systems gold particle/glass and silver particle/glass. Two key boundary conditions are required to allow the use of Mie's calculations:

- the scattering particles are isolated from each other, i.e. only single scattering events are considered.
- the particle size is small compared to the wavelength range investigated.

Over the years it has been shown that the precious metal particles in the glass are indeed separated and far enough apart such that the condition is fulfilled. The second condition results in the definition of a size parameter x/λ, which has to be smaller than one.

Another advantage of this system is that the metal particles are almost perfect spheres, which allows simplification of the calculations.

The general formulation of the calculation was adapted from the derivation of the angular scattering of a sphere in a matrix described by Bohren and Huffman[2]. The final result of the derivation is the calculation of the Mie coefficients, which are then used in the expression of the coefficients or cross sections for extinction, scattering and absorption. The derivation can be modified for various geometries. In the present case, a spherical coordinate system is suitable. A large number of publications show the appropriateness of this assumption for the glass/metal system ([3],[4]). After establishing a solution of the wave equation and the determination of the generating wave functions, Bohren and Huffman then express the electric and magnetic field vectors. Two systems need to be considered: the field vectors in the sphere and the field vectors in the surrounding matrix. The analysis of the electric and magnetic fields at the interface sphere/matrix leads to four scattering coefficients. Two of these coefficients, a_n and b_n are used in the description of the external field, which in turn is used for the calculation of the observed UV-VIS spectra.

The connection with the mathematical treatment and the experimental observation is obtained by the cross sections of the different optical phenomena occurring at the sphere. These are absorption, scattering and extinction. The extinction property is actually the sum of the absorption and scattering property. The properties that can be described are:

- Energy transfer rate in the specific process: W_a = absorption, W_{ext} = extinction, W_{sca} = scattering
- Cross section of the process, where the cross section is defined as the energy transfer rate divided by the incident irradiance I_i: $C = \frac{W}{I_i}$ for all processes.

The solution of the energy transfer from the matrix into the sphere and from the sphere back into the matrix allows the calculation of the scattering coefficient, which includes the above mentioned coefficients a_n and b_n.

$$C_{sca} = \frac{W_{sca}}{I_i} = \frac{2\pi}{\mathrm{k}^2}\sum_{n=1}^{\infty}(2n+1)\left(|a_n|^2 + |b_n|^2\right) \tag{1}$$

The relation for the extinction cross section C_{ext}:

$$C_{ext} = \frac{W_{ext}}{I_i} = \frac{2\pi}{\mathrm{k}^2}\sum_{n=1}^{\infty}(2n+1)\operatorname{Re}\{a_n + b_n\} \tag{2}$$

The absorption cross section is calculated from the difference of the extinction cross section and the absorption cross section:

$$C_{abs} = C_{ext} - C_{sca} \tag{3}$$

The actual calculation of these cross sections involve summations, which need to be terminated after a certain number of terms. The size parameter x/λ is used to determine this cut off number. Large spheres require a larger number of terms to be carried in the summation.

The calculation of the Bessel function was solved by Bohren and Huffman by using a logarithmic derivative

$$D_n(\mathrm{k}r) = \frac{d}{d(\mathrm{k}r)}\ln\psi_n(\mathrm{k}r) \tag{4}$$

with Ψ being the wave function, k the wave vector and r the radius of the sphere. The scattering coefficients are then expressed in terms of this derivative:

$$a_n = \frac{\left[D_n(mx)/m + n/x\right]\psi_n(x) - \psi_{n-1}(x)}{\left[D_n(mx)/m + n/x\right]\xi_n(x) - \xi_{n-1}(x)}$$

$$b_n = \frac{\left[mD_n(mx)/m + n/x\right]\psi_n(x) - \psi_{n-1}(x)}{\left[mD_n(mx)/m + n/x\right]\xi_n(x) - \xi_{n-1}(x)} \tag{5}$$

In these coefficients, m is a defined relative refractive index between matrix and sphere.

The evaluation of the Bessel functions is done using downward or upward recurrence to calculate the values for $D_n(mx)$ in eq. 5. The selection of the recurrence process is determined by the Bessel function that is used in the wave function. $D_n(mx)$ terms that incorporate the wave function Ψ_n are evaluated by downward recurrence, terms that incorporate the wave function ξ_n are evaluated using upward recurrence.

The evaluation was performed using a FORTRAN program. The proper operation of the program was tested using angular scattering data by Bohren and Huffman. After that the calculation was modified for wavelength dependent analysis.

EXPERIMENTAL PROCEDURE

The experimental UV-VIS spectra for silver were measured using a commercial Li-aluminosilicate based photosensitive glass. The glass was irradiated with an ORIEL 1000W lamp for 30min and then heat treated at 525°C for varying times. The resulting spectra are shown in fig. 1. The spectra for gold were obtained from an experimental Mg-aluminosilicate glass. This glass was not photosensitive. The gold particles were precipitated at 660°C. The heat treatment temperature for both glasses is above the glass transition temperature. The spectra of the two glasses both show a characteristic absorption band. The resonance frequency for silver is found at ~450nm, the resonance frequency for gold is

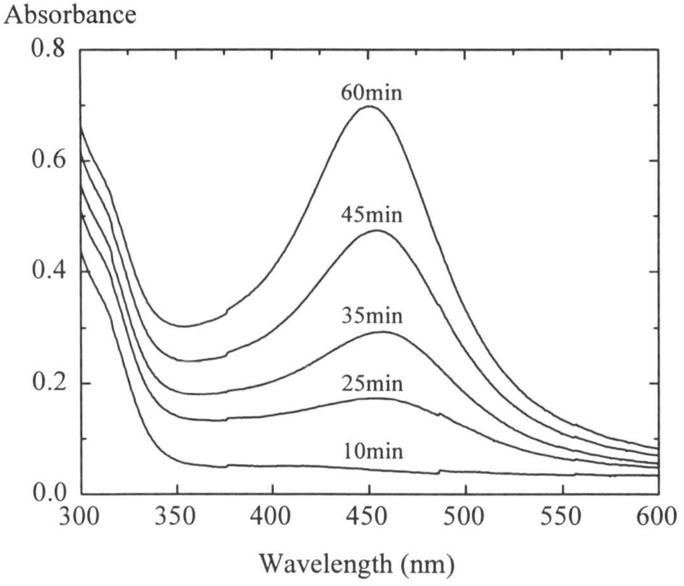

Figure 1: Experimental spectra for Ag-based photosensitive glass. Irradiated 30min, heat treated at 525°C

located at ~540nm.

The calculated UV-VIS spectra depend strongly on the applied optical constants n and k for the system. Various reported values for n and k were examined. The best results with the highest consistency were dielectric constants for silver and gold reported by Quinten[5]. These data were obtained by precipitating metal particles in a sol. The particle size of the silver spheres was measured using a transmission electron microscope. This value was then used to calculate the optical constants from experimental spectra. The wavelength dependence of the refractive index n and the absorption coefficient k for a silver/sol system are shown in fig. 2. The calculated UV-VIS spectra (fig. 3) for a silver sphere in a silicate glass with a refractive index of n=1.5 can be separated into scattering and into the absorption contribution.

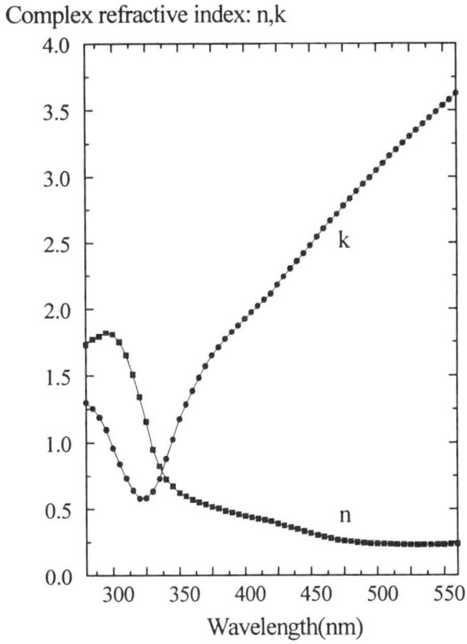

Figure 2: n-k data for silver in sol

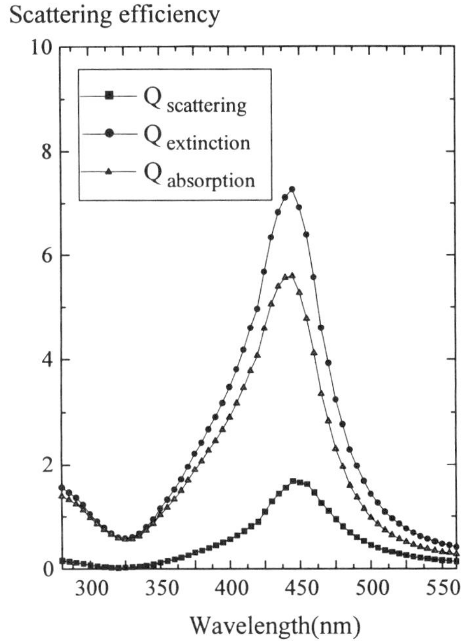

Figure 3: Calc. absorption, scattering and extinction spectra for silver sphere in glass, r = 20nm

The data for gold show the same behavior. Both, experimental spectra and calculated spectra show the resonance in the same wavelength range. The n-k-data for the gold system were also taken from Quinten's work.

RESULTS AND DISCUSSION

The calculated spectra shown in fig. 3 are typical for spectra of very small metal particles. The overall extinction is essentially caused by the absorption of the radiation in the particle. The scattering contribution is small. The scattering coefficient increases steadily with increasing particle size and finally dominates the spectrum. The size dependence for the extinction, absorption and scattering are illustrated in figs. 4-6.

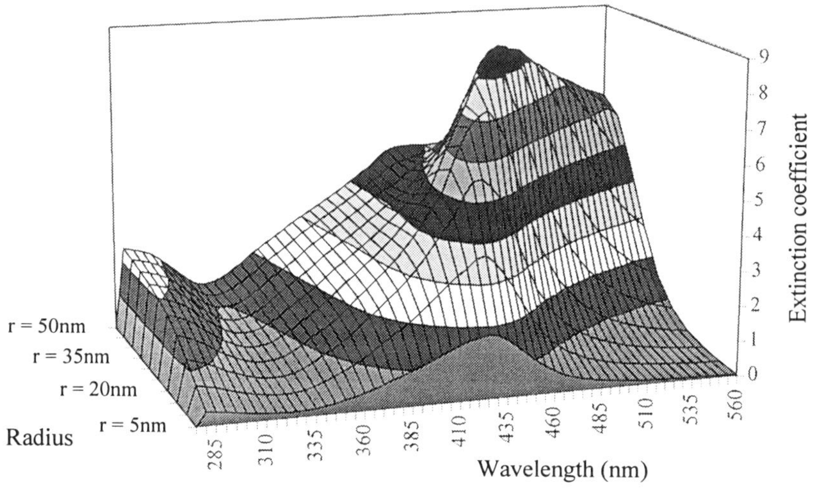

Figure 4: Silver sphere in glass, calculated extinction coefficient

The UV-VIS spectra of metal particles in glass are very sensitive to the optical properties and the modeling technique used in the analysis or simulation. The case study used experimental data for silver and gold spheres obtained from a related, but still different system: metal particles in a sol. A comparison of the experimental data for silver-in-glass with simulated spectra shows a general agreement with the spectra development, but it also shows a significant discrepancy between the two spectra series.

The resonance type peak form is achieved for both systems. The peak position agrees very well for small particle sizes. For both, experimental data and simulation, the peak is located for silver at ~440nm, while for gold the peak can be found at ~540nm. Both experiment and simulation show a comparable peak shape for small particle sizes. This shows that the general description of the problem is correct.

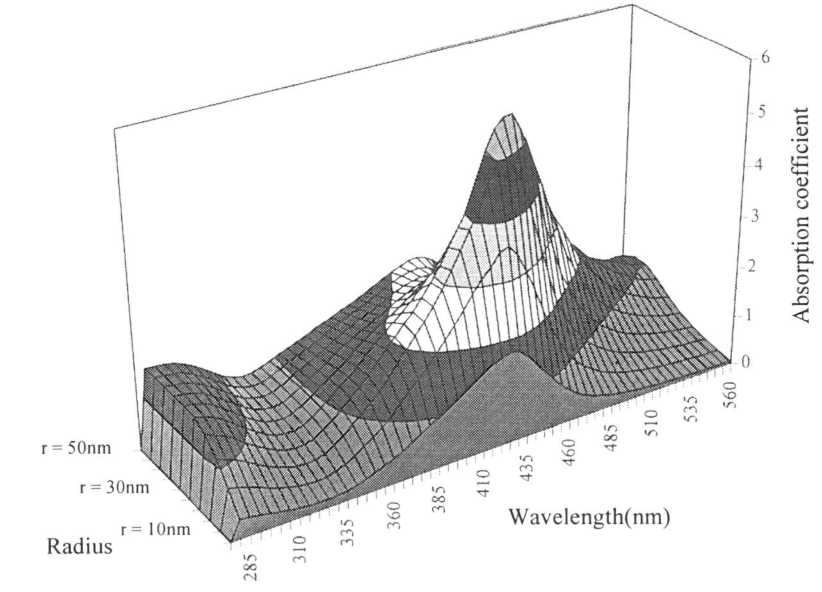

Figure 5: Silver sphere in glass, calculated absorption coefficient

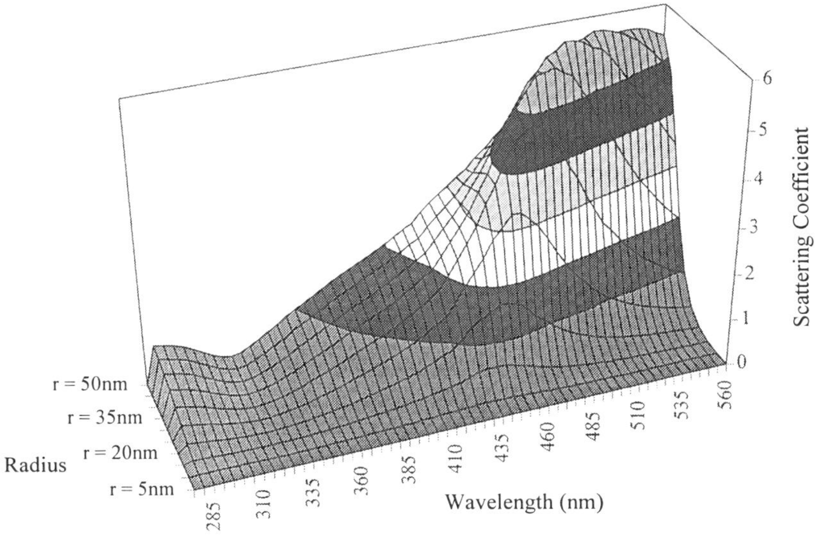

Figure 6: Silver sphere in glass, calculated scattering coefficient $Q_{scattering}$

The main discrepancy lies in the strong peak shift that is observed for the larger particle sizes in the simulation. This peak shift is a result of the scattering contribution. This difference in the peak development is present in both glass/particle systems, silver and gold.

This peak shift is not seen in the experimental data. The development of silver particles was extended through longer heat treatment to yield very large particles, but the peak shift is still not observed in the significance seen in the calculation (fig. 7).

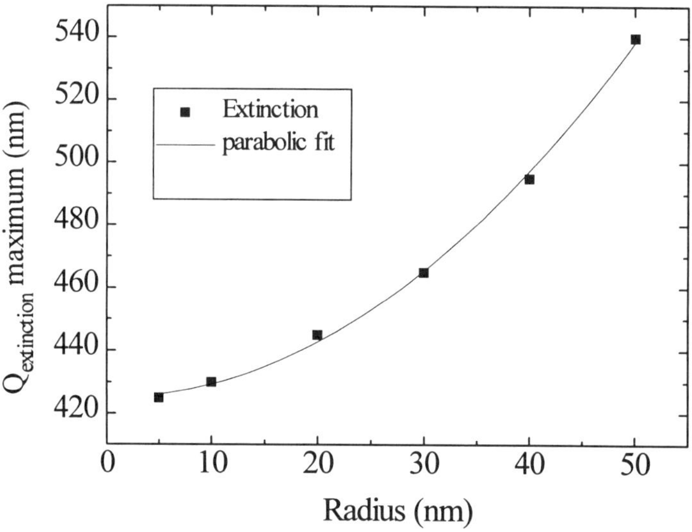

Figure 7: Size dependence of the calculated extinction

The initial goal of this work was to use a key parameter in the spectral development, such as the peak position as an indicator for the size of the particles in the glass matrix. Other authors[4] use the peak width of the resonance peak as an indicator. However, these models do not model the increased absorption on the short wavelength side of the peak, which is due to interband transition in the metal particles. The absorption peak (fig. 5) does show a clear trend of an increasingly narrower peak width in the initial stages of the calculation. However, the interference with the scattering contribution does not allow a simple fitting procedure.

As an additional complication, the peak width in the experimental glass is generally broader than that of the modeled spectra. Two explanations can be suggested for this behavior: incorrect dielectric constants or a particle size distribution that is non-negligible.

CONCLUSIONS

UV-VIS spectroscopy and calculated spectra are compared with the intent to obtain a fast size determination method for small metal spheres in a glass matrix. The overall agreement between calculation and experimental data is observed. However, a strong peak shift tendency in the calculated spectra is not repeated in the experiment. One possible explanation of this difference between the model and the experiment is the use of optical constants that are not appropriate for the system. The used data are obtained from a metal particle/sol system. The investigated system, however, is a metal particle/glass system. There are several possible factors that result in a subtle but for the calculation substantial variation in the optical constants. Imaginable factors are surface effects, stress effects, differences in the interface in the interface between the sphere and matrix material.

The only proper method to eliminate these systematic influences is to determine optical constants that describe the investigates system accurately. That implies that the method used by Quinten of backcalculating the optical constants for a given particle size needs to be applied for this system.

The calculation method as such however is correct and should be pursued and fine tuned to be used as an additional tool for the particle size determination of small metal particles in glass.

REFERENCES

1. G. Mie, "Beiträge zur Optik trüber Medien, speziell kolloider Metallösungen," *Ann. Phys., IV. Folge*, **25** [3] 377-445 (1908).

2. C. Bohren, D. Huffman, Absorption and Scattering of Light by Small Particles. Wiley, New York, 1983.

3. R.H. Doremus, "Optical Properties of Small Gold Particles," J. Chem. Phys., 40 [8] 2389-2396 (1964).

4. M. Mennig, K.-J. Berg, "Determination of Size, Shape and Concentration of Spheroidal Silver Colloids Embedded in Glass by VIS-Spectroscopy," Mat. Sci. and Eng. B9 421-424 (1991).

5. M. Quinten, "Analysis of Optical Properties of Aggregated Colloidal Systems," Dissertation, University of Saarland, Saarbrücken, 1989.

ELASTIC MODULUS CALCULATIONS FROM LOAD/DISPLACEMENT CURVES USING SPHERICAL AND POINTED INDENTERS

L. Riester, M. K. Ferber, and K. Breder
High Temperature Materials Laboratory
Oak Ridge National Laboratory,
Oak Ridge, TN. 37831-6069
R. J. Bridge
University of Connecticut
Storrs, Conn. 06269-3222

ABSTRACT

Hard materials with known mechanical properties were probed with spherical and Berkovich indenters using a Nanoindenter(TM). A simple analysis of the loading portion of the indentation experiment was found to give reasonable estimates of the elastic modulus for a variety of brittle materials.

INTRODUCTION

When indentation methods are used to measure the elastic properties of a hard material, there is wide disagreement as to whether it is preferable to use pointed, spherical, or flat bottomed indenters.[1-4] The advantages to using pointed diamond indenters are that their shapes conform more readily to the natural facets of diamond, and they can be calibrated to account for small imperfections in manufacturing.[1] On the other hand, pointed indenters may be less suited for measuring the elastic properties of multilayered coatings and thin films. The advantage of a spherical indenter over the sharp tips is that plasticity is delayed to greater indentation depths which makes it well suited to measure elastic properties. The disadvantages of using a spherical diamond indenter are that it is extremely difficult to manufacture a diamond indenter that conforms to a perfect spherical shape. This makes calibration to offset imperfections very controversial. Swain et al. have used spherical indenters successfully and have managed to calculate aberrations due to imperfections in the shape.[3, 4] Based on these theoretical premises and experimental methods, this study attempts to measure the elastic modulus of a variety of materials with known mechanical properties using two spherical and one Berkovich indenter.

BACKGROUND

The simplest indentation experiment which can be performed is that using a spherical indenter. The indentation can remain completely elastic up to a relatively high indentation load, at which point cracking or plastic deformation may occur. The theory of two elastic bodies in contact was originally described by Hertz[5], hence, Hertzian indentation. The elastic stress field beneath the indenter is symmetrical and well described, and a full solution for the three dimensional stress field has been reported.[6, 7] The most common use of the Hertzian indentation has probably been the Brinell hardness test, in which the plastic deformation after spherical indentation provides a measure of the material hardness. This test has long traditions, primarily as an empirical test, but has been set on solid theoretical foundations by Hill et al.[2] The Hertzian test has also been used to determine the fracture toughness,[8-11] again utilizing the well defined symmetrical stress field and the development of ring and cone cracks. Recently, Swain et al.[4] have combined the new depth-sensing hardness testing equipment (mechanical microprobe) with spherical indenters, and implemented elastic/plastic indentation mechanics to determine mechanical properties of thin films. However, very little work utilizing the elastic part of the Hertzian indentation cycle seems to be available in the literature. This is a somewhat puzzling fact since it is a very straightforward experiment with solid theoretical base as described above. Yoffe[12] explored the use of Hertzian indentation for large deformations, i.e., the very practical situation of indenting a highly elastic (rubber) solid with spheres, and corrections to the Hertzian solutions for large deformations were given. Zeng et al.[13] used Hertzian indentation to determine the elastic modulus of several brittle materials by utilizing the well known relationship between the contact area and the elastic constants for the materials. This approach required a measurement of the contact area, and a method for doing this was devised. With the use of the depth sensing indentation devices, another option for measuring the elastic modulus becomes available, utilizing the relationship that exists between elastic indentation depth and elastic constants. This is the approach taken in this paper.

Hertzian theory readily provides the following relationships between the contact radius and elastic constants:

$$a = \left(\frac{3PR}{4} \frac{1}{E^*} \right)^{\frac{1}{3}} \tag{1}$$

where a is the elastic contact radius, P the indentation load, R the radius of the indenter and E* the "composite Modulus" given by:

$$\frac{1}{E^*} = \frac{1 - \nu_1^2}{E_1} + \frac{1 - \nu_2^2}{E_2} \tag{2}$$

Here ν and E are Poisson's ratio and Young's modulus, respectively, and the subscripts 1 and 2 refer to the indenter and test material, respectively. The maximum elastic deformation δ is likewise given as:

$$\delta = \left(\frac{3P}{4R^{\frac{1}{2}}} \frac{1}{E^*} \right)^{\frac{2}{3}} \tag{3}$$

and in the case of small elastic deformation the relationship between the contact area and contact depth is a function of the indentation sphere radius only:

$$a = (R\delta)^{\frac{1}{2}} \quad (4)$$

Hence, using a depth sensing indentation system with a spherical indenter of known mechanical properties, the elastic modulus of the test material can be obtained assuming the Poisson's ratio is known. For highly elastic materials indented with a stiff indenter, Yoffe[12] found reasonable agreements assuming infinite stiffness of the indenter.

EXPERIMENTAL

Indentations were made with a Nanoindenter I (Nano Instruments) on chips of glass, silicon and sapphire and on metallographically polished specimens (0.5 μm) of fused silica, polycrystalline aluminum oxide, silicon nitride, sintered α-silicon carbide and titanium alloy (for microscopy only). Two spherical diamond indenters (20 μm and 2 μm radius) were purchased from Synton/Imetra. The Berkovich indenter was supplied by Nano Instruments. Single cycle indents were made using load control with loads of 24, 48, 72 and 100 mN; a record of load, displacement and time from each indent was stored. The loading segment from each load vs. displacement curve was extracted. Ten indents from each material which corresponded to the lowest load were averaged. The portion of the loading curve which was linear on a plot of load vs. $\delta^{3/2}$ (5mN or less) was considered for the calculations of the modulus since only that portion could be assumed to constitute the region of elastic contact.

Scanning electron microscopy (Hitachi S-800) and a Scanning Force Microprobe (Topometrix) techniques were used to examine impressions produced by the various indenters in a soft titanium alloy. To avoid errors introduced by the SFM, indents of varying sizes were sampled and scanned in several directions. Both forward and reverse scans were examined. Measurements of the depth and width of the indentations were made using the Topometrix software to determine the radius of the spherical indenters and for direct comparison to the load/displacement curve which provided data on the actual plasticity/elastic recovery of the indentation process.

RESULTS AND DISCUSSION

Nanoindentation

By looking for inflection points on a plot of load vs. δ^2 in the loading curve of the load-displacement cycle, it was relatively easy to determine the regions where the indentations were completely elastic or where some plastic deformation had started to occur. A plot of load vs. $\delta^{3/2}$, on the other hand, was linear even for loads in the plastic region of the indentations, unexpectedly so for the Berkovich indenter (see Fig. 1). This allowed for a more direct comparison to be made between the three indenters. Since a plot of load vs. $\delta^{3/2}$ was linear, the Berkovich indenter was assumed to be spherical in the region of elastic contact even though sphericity was not actually observed in the microscopic images (Figs. 2 and 3). From the fits of indentation vs. elastic deformation, the elastic modulus was determined using Eq. 3.

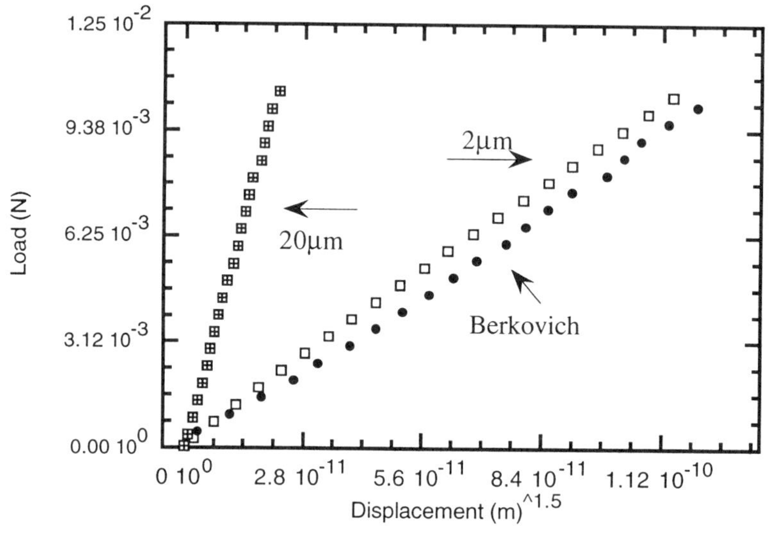

Fig. 1. Load vs. elastic deflection for the three indenters.

As a first assumption, a Poisson's ratio of 0.17 was assumed for all materials since there was some uncertainty about the values to be used for some of the materials. Table I lists the calculated values. The Elastic Modulus and Poisson's ratio for the indenter (diamond) were set to 1140 GPa and 0.07 respectively. The radii of the indenters were as specified by the manufacturer.

From the results in Table I it is apparent that the results are only partially reasonable, and that the 2 μm indenter underestimates the moduli while the 20 μm indenter overestimates the moduli. It is further apparent that the assumption of an infinitely stiff indenter (Yoffe) is relatively good for the low modulus materials (fused silica and glass).

Following Swain[4], the possible reasons for the differences in the moduli were explored, and the most reasonable assumption was that the actual radii of the indenters were not those quoted by the manufacturer. To investigate this possibility, the radius of the indenter was treated as a variable in the calculations of standard materials with the result that elastic moduli were forced to conform to literature values. Table II illustrates the effect on modulus measurements if the shape of all indenters is assumed to be spherical and if the radius is varied to obtain standard values. Results from the pointed Berkovich indenter are included for comparison.

Table I. Elastic moduli calculated using the nominal values of the indenter radii.

Material	E-Modulus* 20 μm Indenter [GPa]	E-Modulus+ 20 μm Indenter [GPa]	E-Modulus* 2 μm Indenter [GPa]	E-Modulus+ 2 μm Indenter [GPa]	E-Modulus# Berkovich Indenter [GPa]
Silicon Nitride	397.9	293.10			348.00
Sapphire	497	343.60	285.10	227.00	486.90
Alumina	501.1	345.60	260.90	211.40	490.90
Glass	91.7	84.70	42.00	40.50	81.20
Fused Silica	86	79.80	50.20	48.00	71.40
Silicon	193.2	164.60	106.00	96.80	180.20
Sintered SiC	422.5	306.30	284.80	226.80	488.80

* Hertz Theory
+ According to Yoffe,[12] assuming infinitely stiff indenter.
\# According to Oliver and Pharr method.[1]

Table II. Elastic Moduli calculated using the best fit values of the indenter radii.

Material	E-Modulus* 27μm Rad. Indenter (GPa)	E-Modulus* 1μm Rad. Indenter (GPa)	E-Modulus* 0.5μm Rad. Berkovich (GPa)	E-Modulus# Berkovich from unload. (GPa)
Silicon Nitride	326.21		273.71	348.00
Sapphire	402.71	451.03	379.33	486.90
Alumina	405.84	408.69	367.27	490.90
Glass	77.99	60.38	80.02	81.20
Fused Silica	73.19	72.28	83.84	71.40
Silicon	162.32	156.12	144.27	180.20
Sintered SiC	345.35	450.50	422.52	488.80

*Hertz Theory
According to Oliver and Pharr method[1]

Microscopy

Microscopy was undertaken to obtain measurement which conformed reasonably well to the actual dimensions of the radii of the indenters. A soft titanium alloy was selected for this purpose because it has only minimal elastic recovery when indented which makes it easy to study residual impressions. Electron Microscopic images of impressions of the three different indenters into a titanium alloy are shown in Fig. 2. It was clearly seen that the 2 μm spherical indenter did not conform to a shape which qualifies as spherical even at low loads. Rather it seemed that the spherical tip was ground from a four-sided pyramid. A side view of the indentation site also showed material pile-up. The 20 μm spherical and the Berkovich indenters appeared to have the expected shapes.

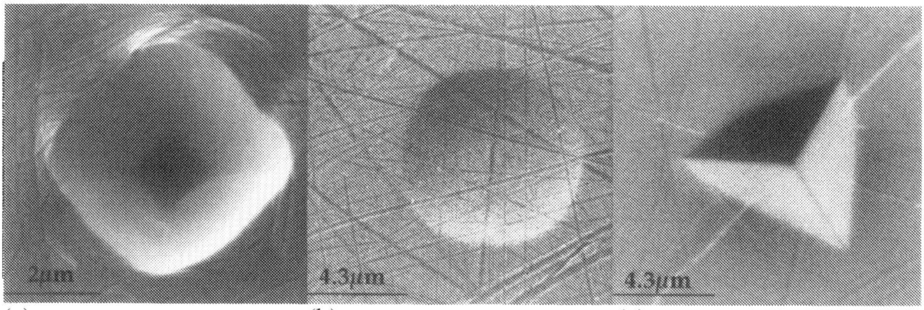

(a) (b) (c)

Fig. 2. SEM images of indents from a 2 μm(a), a 20μm(b), and a Berkovich(c) indenter in titanium alloy.

Further inspection of the indentation sites with the Scanning Force Microprobe revealed that only the Berkovich indenter conformed to the expected shape, see Fig. 3. The 2 μm radius spherical indenter was very irregular. The four sides of the pyramidal shape could be traced to within about 300 nm of the end of the tip. The 20 μm spherical indenter was actually tri-lobal as judged by the material piled up which could easily be determined with the amplification in the z-direction of the SFM. Furthermore, at the very tip there seemed to be a very small residue of a three-sided pyramid or facet. In addition, the indenter tip showed many other irregularities.

By measuring software generated cross sections of the SFM images from the impressions in the titanium alloy, estimations of the actual radii of the two spherical indentations were made. This was a time consuming and difficult task, especially for the smaller indenter for which the impressions were definitely not spherical. Measurements of several indentations were made, and the radius of the nominally 20 μm indenter was found to be 41 μm and the radius of the nominally 2 μm indenter was found to be 2.7 μm. Using these values in the equations for calculating the elastic moduli resulted in the values shown in Table III. To keep it simple, elastic recovery, which was proportional to the area of surface contact, was disregarded in this study. The elastic recovery portion of the indentation contributed to the reported larger diameter measurements in Table III.

Table III. E modulus calculated using the measured sphere radii.

Material	E-Modulus* 41 μm Indenter [GPa]	E-Modulus+ 41 μm Indenter [GPa]	E-Modulus* 2.7μm Indenter [GPa]	E-Modulus+ 2.7μm Indenter [GPa]
Silicon Nitride	250.9	204.7		
Sapphire	305.9	240	236.9	195.4
Alumina	308.2	241.4	217.5	181.9
Glass	62.5	59.1	36	34.9
Fused Silica	58.7	55.7	42.9	41.3
Silicon	128.2	115	90.1	83.3
Sintered SiC	264.8	213.9	236.7	195.2

*Hertz Theory
+ According to Yoffe,[12] assuming infinitely stiff indenter.

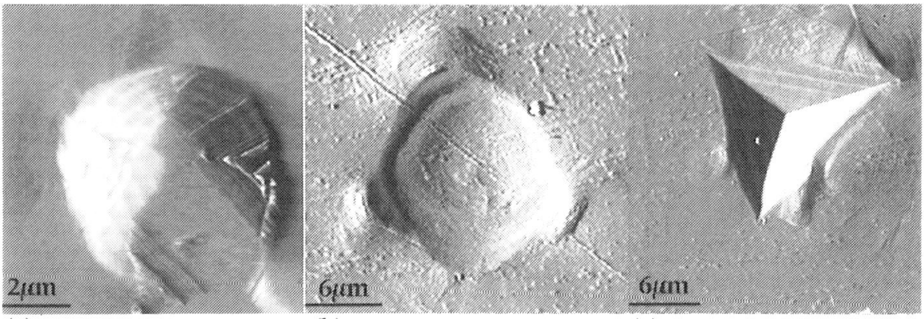

(a) (b) (c)
Fig. 3. SFM images of indents from a 2μm(a), a 20μm(b) and a Berkovich(c) indenter in titanium alloy.

SUMMARY

It is entirely possible to measure the elastic modulus of any material using a spherical or just about any shape indenter in the Nanoindenter. The experimental procedures need some refinement; for example, more data points at very low loads will help avoid errors in calculating the slope of the loading curve. There also was a transition observed in the load/displacement curves on the soft titanium alloy when the indenter "switched" shapes during penetration. Because of the irregular shape even at the very tip, it is difficult to assess how much surface area of the indenter is actually in contact with the substrate in the elastic region of the indentation process. Despite these difficulties, if indenters are carefully characterized, and if constants are calculated based on standard materials, blunt or spherical indenters are useful tools in obtaining elastic moduli of extremely thin films where large surface area contact is needed rather than penetration.

ACKNOWLEDGEMENT

Research sponsored by the U. S. Department of Energy, Assistant Secretary for Energy Efficiency and Renewable Energy, Office of Transportation Technologies, as part of the High Temperature Materials Laboratory User Program, under contract DE-AC05-84OR21400 with Lockheed Martin Energy Research Corp.

REFERENCES

[1]W. C. Oliver and G. M. Pharr, "An improved technique for determining hardness and elastic modulus using load and displacement sensing indentation experiments," J. Mater. Res. **7** [6] 1564-1583, 1992.

[2]R. Hill, B. Storåkers and A. B. Zdunek, "A theoretical study of the Brinell hardness test," Proc. Roy. Soc. A **423** 301-330, 1989.

[3]J. S. Field and M. V. Swain, "A simple predictive model for spherical indentation," J. Mater. Res. **8** [2] 297-306, 1993.

[4]M. V. Swain and J. Mencik, "Mechanical property characterization of thin films using spherical tipped indenters," Thin Solid Films **253**, 204-211, 1994.

[5]H. Hertz, "Hertz's Miscellaneous Papers," Chapters 5 and 6. Macmillan, London 1896.

[6]M. T. Huber, "Zur Theorie der Berührung fester elastischer Körper," Ann. Physik **14** 153-163, 1904.

[7]A. Kelly and N. H. Macmillan, **Strong Solids**, 3rd edition, Clarendon Press, Oxford, 1986.

[8]B. R. Lawn, "Hertzian Fracture in Single Crystals with the Diamond Structure" J. Appl. Phys, **39** 4828, 1968.

[9]R. Warren, "Measurements of the fracture properties of brittle solids by Hertzian indentation," Acta Metall. **26** 1759-69, 1978.

[10]K. Zeng, K. Breder, and D. J. Rowcliffe, The Hertzian stress field and formation of cone cracks - I. Theoretical approach," Acta Metall. Mater. **40** [10] 2595-2600, 1992.

[11]K. Zeng, K. Breder, and D. J. Rowcliffe, The Hertzian stress field and formation of cone cracks - II. Determination of fracture toughness," Acta Metall. Mater. **40** [10] 2601-2605, 1992.

[12]E. H. Yoffe, "Modified Hertz theory for spherical indentation," Phil. Mag. A, **50** [6] 813-828, 1984.

[13]K. Zeng, K. Breder, D. J. Rowcliffe, and C. Herrström, "Elastic modulus determined by Hertzian indentation," J. Mater. Sci. **27** 3789-3792, 1992.

KEYWORD AND AUTHOR INDEX